ENGINEERING COMMUNICATION

Charles W. Knisely
BUCKNELL UNIVERSITY

Karin I. Knisely
BUCKNELL UNIVERSITY

Australia • Brazil • Japan • Korea • Mexico • Singapore • Spain • United Kingdom • United States

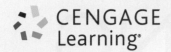

Engineering Communication
International Edition
Charles W. Knisely
Karin I. Knisely

Publisher: Timothy L. Anderson

Senior Developmental Editor:
Hilda Gowans

Senior Editorial Assistant:
Tanya Altieri

Content Project Manager:
D. Jean Buttrom

Production Director:
Sharon Smith

Media Assistant:
Ashley Kaupert

Rights Acquisition Director:
Audrey Pettengill

Rights Acquisition Specialist, Text and
Image: Amber Hosea

Text and Image Researcher:
Kristiina Paul

Senior Manufacturing Planner:
Doug Wilke

Copyeditor: Patricia Daly

Proofreader: Emily Jones

Indexer: Shelly Gerger-Knechtl

Compositor: MPS Limited

Senior Art Director: Michelle Kunkler

Internal Designer: Stratton Design

Cover Designer: Patti Hudepohl

Cover Image: © maymak/Shutterstock

Library of Congress Control Number: 2013954998

International Edition:

ISBN-13: 978-1-285-43604-3

ISBN-10: 1-285-43604-0

Cengage Learning International Offices

Asia
www.cengageasia.com
tel: (65) 6410 1200

Australia/New Zealand
www.cengage.com.au
tel: (61) 3 9685 4111

Brazil
www.cengage.com.br
tel: (55) 11 3665 9900

India
www.cengage.co.in
tel: (91) 11 4364 1111

Latin America
www.cengage.com.mx
tel: (52) 55 1500 6000

UK/Europe/Middle East/Africa
www.cengage.co.uk
tel: (44) 0 1264 332 424

Represented in Canada by
Nelson Education, Ltd.
tel: (416) 752 9100 / (800) 668 0671
www.nelson.com

Cengage Learning is a leading provider of customized learning solutions with office locations around the globe, including Singapore, the United Kingdom, Australia, Mexico, Brazil, and Japan. Locate your local office at: **www.cengage.com/global**

For product information: **www.cengage.com/international**
Visit your local office: **www.cengage.com/global**
Visit our corporate website: **www. cengage.com**

Unless otherwise noted, all items © Cengage Learning

Chapters in this book have been adapted from *A Student Handbook for Writing in Biology*. Used with permission from the publisher, Sinauer Associates, Inc. and W. H. Freeman & Co.

Printed in Canada
1 2 3 4 5 6 7 19 18 17 16 14 13

Contents

Preface

The purpose of this book is to provide engineering students and entry-level engineers with a practical, readable resource for communicating their knowledge according to the conventions of their profession. While engineering is focused on problem-solving, it is not *just* about problem-solving. Engineers must communicate both their problem-solving methods and their solutions. Most engineers in the workforce spend more than half of their time communicating technical information; the amount of time spent on reading, writing, preparing, listening to, and presenting communications increases with seniority. Yet according to a recent survey published by ASME Vision2030 (2012), only 10% of supervisors in industry thought that college graduates were well prepared to communicate in the workforce. One of our goals in writing this book is, therefore, to bridge the gap between academic and industrial expectations. Specifically, we attempt to convince engineering students and entry-level engineers that writing well will help them advance in their career.

Throughout the book, we refer to the importance of developing a professional voice. By professional voice, we mean the tone, the style, and the vocabulary engineers use when they write or talk about their engineering projects. Developing a professional voice is a lifelong endeavor, which requires reading the kinds of communications that the job requires, practicing writing, receiving timely feedback from an experienced mentor, and making a conscious effort not to repeat previous mistakes. Similarly, we provide instructors with the tools they need to teach technical writing as a stand-alone class or to incorporate writing into their engineering courses. These tools include examples and exercises on a variety of engineering topics, along with rubrics that help instructors give students precise and timely feedback on their written and oral assignments. For both students and instructors, we go beyond describing *what* constitutes an effective technical communication. We explain *how* to prepare and present engineering reports, proposals, and other types of engineering communications.

One of the difficulties in writing a book about technical communication is that there is no single uniform standard embraced by all branches of engineering. Technical organizations, journal editors, and professional societies all offer slightly different publication guidelines, yet there are many common elements. Thus, another goal of this book is to heighten awareness of the organizations that establish formatting standards in their discipline, highlighting both the common elements and the differences among them.

This book is divided into four parts:

- Part 1 Introduction to finding, reading, and citing technical resources
- Part 2 Preparing technical reports
- Part 3 Other types of professional writing
- Part 4 Oral presentations and poster preparation

The first three chapters comprise Part 1. In Chapter 1, we pose the question, "Why do engineers need to communicate?" and then provide evidence that communicating well is essential for a satisfying and rewarding career. In Chapter 2, we discuss the difference between primary and secondary sources and explain how to find technical information efficiently. We introduce engineering databases as a vehicle for finding reliable information on a variety of engineering topics, describe citation and reference formats, and explain that citing sources not only gives credit to those who generated the knowledge but also adds credibility to your own work. We suggest strategies for avoiding plagiarism. Chapter 3 introduces the structure of technical reports, one of the most common types of communications in engineering. The subdivision of reports into sections reflects the systematic approach used by engineers to solve problems and allows engineers to find information easily. Strategies for reading technical reports as well as engineering textbooks are also presented.

The subsequent three chapters, Chapters 4 to 6, comprise Part 2. In Chapter 4, we give step-by-step instructions for preparing each section of a lab report, a technical document whose structure is similar to that of a technical report. We provide examples of faulty writing along with revisions. In the annotated student lab report at the end of the chapter, we alert readers to specific elements of style and format. The Engineering Lab Report Checklist lists the most important components of the report and can be used at all stages of the writing process. Chapter 5 presents a systematic approach to revision, in other words, reading a draft critically for technical content and organization and correcting errors in grammar, punctuation, spelling, and format. Sidebars provide a "just in time" review of grammatical structures and rules for both multilingual writers and native speakers. At the end of the chapter, we include a Laboratory Report Mistakes checklist that can be used to self-evaluate your writing or peer review someone else's work. Prompts on the self-assessment and peer review forms are intended to start a constructive conversation between authors and reviewers. We provide specific suggestions for making the peer review process productive. Instructors will also find these materials useful for providing feedback on students' writing.

Chapter 6 is entitled "Engineering Toolbox and Visual Communications." In this chapter we discuss standard engineering symbols, units, and dimensions; rules for significant figures and rounding; different types of visual elements and when to use them; guidelines for formatting figures; graphs of standard functions; graphical analysis; and software and tools for preparing illustrations and editing images. Visual elements that are prepared according to accepted standards are easier for your audience to understand and enhance your credibility as an engineer.

After laying the foundation for reading and writing technical reports in Parts 1 and 2, we build upon it in Part 3. In Chapters 7 through 10, we provide guidance on preparing memos, business email, letters, resumes, proposals, progress reports, design specifications, patent applications, overview reports, site visits, white papers, and trade journal articles. Because engineering projects in the workplace are almost always done collaboratively, we discuss topics that need to be agreed on and adhered to by members of the group to ensure successful collaboration. These strategies can also be applied to group projects in college, and their successful implementation will go far to prepare students for the workforce.

In Part 4, we discuss oral presentations (Chapter 11) and posters (Chapter 12), forms of communication that rely heavily on the presenter's delivery. More than just describing the differences between written and oral communications, we show you how to prepare visual elements that enable the audience to grasp important concepts quickly. Evaluation forms for oral presentations and posters are provided at the end of each chapter. Sample posters illustrate effective poster layouts and formats.

Today's students have used computers for most of their lives, but most lack the keyboarding skills required by workplace engineers. These skills include writing formulas that contain Greek letters, mathematical symbols, and sub- and superscripted characters, and producing graphs, tables, and drawings in a format expected by supervisors and clients. In the appendices, which make up 20-30% of the book, we explain how to carry out these functions in Microsoft Word, Excel, and PowerPoint. These Microsoft programs are ubiquitous, and most engineering students are familiar with at least some of the menus. We recognize that the Microsoft Office suite is not the only option; engineers will become proficient with competitors' programs as well as with other, more specialized programs as needed for their job. By learning the commands in the MS Office suite, young engineers will be better prepared to use similar features in other software.

PEDAGOGICAL AIDS AND FEATURES

Teaching technical writing at the college level is challenging for several reasons. First, there is hardly enough time during the semester to teach content, let alone writing, in engineering courses. Second, most engineering students are not excited about writing, and they do not realize how important it is for their career to practice discipline-specific oral and written communication while still in college. Third, it is challenging for instructors to come up with interesting topics for a mixed class of mechanical, civil, chemical, electrical, and other engineering students. Fourth, many instructors, especially those whose mother tongue is not English, find it challenging to teach grammatical concepts and writing style to student populations with diverse cultural backgrounds and aptitudes for the English language. This book provides a number of tools to help instructors teach technical writing as a stand-alone class or to incorporate writing into their engineering courses.

When writing instruction is incorporated into a course with technical content, the students have a common foundation for their writing assignments. Instructors can assign lab reports, oral presentations, and other forms of communication appropriate to the technical content of the course. Rather than using precious in-class time to explain the structure and style of different communications, the instructor would refer students to specific chapters in this book. This book is also intended to be used as a look-up reference to help students find information on citation systems, formatting graphs, and all the other details involved in preparing technical communications.

When technical writing is a stand-alone class, the instructor faces two important challenges. First, students need something relevant to write about. Second, students enrolled in these kinds of classes typically have different technical backgrounds (majors) or English language ability. To address these challenges, we have designed a variety of exercises at the end of each chapter. Some of the exercises are generic, such as preparing detailed instructions for carrying out a specific procedure, explaining how specific devices and machines function, convincing a consumer of the advantages of one product from one vendor over that from a competitor, writing a summary of an emerging area of technology, or discussing a contemporary issue and its significance to each student's major. Other exercises are discipline-specific, such as finding information about heat removal technologies for cooling microelectronics, planning a site for a commercial wind farm, and using genetic engineering to design plants that glow in the dark. These exercises have been tested by and used successfully in several different classes of undergraduate engineering students.

Laboratory reports and experiments from other courses can also be used as a springboard for writing assignments in stand-alone writing classes. A lab report could be restructured as a detailed technical report, an oral presentation, or a poster. The data collected from a previous experiment can be analyzed with the goal of choosing an appropriate visual element that displays the most important results. Drafts of writing assignments provide ample material for peer review, which is beneficial because it is often easier to find mistakes in other people's writing than your own. Through repetition and revision, students learn strategies for writing clearly, concisely, and in the appropriate style and format.

Other features of this book that instructors will find beneficial include:

- Straightforward language and practical instructions
- Numerous examples that illustrate how and how not to write
- Checklists for students to evaluate drafts of their lab reports, oral presentations, and posters. These same checklists can be used by instructors to provide precise, detailed, and timely feedback on assignments.
- Chapter summaries at the end of each chapter to reinforce the take-away concepts
- A review of English grammar in Chapter 5 with tips for multilingual learners
- Detailed instructions for preparing visual elements for technical reports, which is consistent with the majority of technical society guidelines, in Chapter 6
- In Part 3, an introduction to advanced writing formats such as proposals, progress reports, and design specifications. These communications may be used in upper level engineering courses. Other topics such as patent applications, overview reports, site visits, white papers, and trade journal articles may or may not be covered in undergraduate courses. Some of these formats may be uncommon in some branches of engineering. However, we include these forms of technical communication to build awareness that they exist. Graduate students and engineers entering the workforce will probably encounter them at some point in their careers. Those who want to know more about these kinds of documents will find useful introductory information in this book.
- Appendices with detailed instructions for using MS Word, PowerPoint, and Excel commands to produce the formulas, symbols, graphs, tables, and drawings that are characteristic of technical reports. These instructions are the key to empowering students and their instructors with the "how" to achieve the "what."

ANCILLARY MATERIALS

This book is supplemented with ancillary materials on the Cengage website.
- For students
  - » More samples of engineering reports, specifications, and patents
  - » Links to websites for exercises, more sample documents, and templates
- For instructors
  - » Additional exercises
  - » Links to relevant sources for in-text exercises
  - » Grading rubrics
  - » Samples of edited student writing
  - » Access to a secure site with samples of possible responses to in-text exercises

In addition to the print version, this textbook will also be available online through MindTap, a personalized learning program. Students who purchase the MindTap version will have access to the book's MindTap Reader and will be able to complete homework and assessment material online, through their desktop, laptop, or iPad. If your class is using a Learning Management System (such as Blackboard, Moodle, or Angel) for tracking course content, assignments, and grading, you can seamlessly access the MindTap suite of content and assessments for this course.

In MindTap, instructors can:

- Personalize the Learning Path to match the course syllabus by rearranging content, hiding sections, or appending original material to the textbook content
- Connect a Learning Management System portal to the online course and Reader
- Customize online assessments and assignments
- Track student progress and comprehension with the Progress app
- Promote student engagement through interactivity and exercises

Additionally, students can listen to the text through ReadSpeaker, take notes and highlight content for easy reference, and check their understanding of the material.

ACKNOWLEDGMENTS

Many people have helped us develop our individual communication skills. For Charles, these people include Mrs. Donna Dudley Rowe, Mrs. Edna Claunch, Prof. Donald Rockwell at Lehigh University, many generations of students at universities in Karlsruhe, Germany, in Kyoto and Osaka, Japan, and, in far greater numbers, those at Bucknell University. For me, the greatest impact comes from my most stringent critic and exceptional editor, my wife and co-author, Karin Knisely. Over the years, we have engaged in many hours of productive discussions on technical communication. Even after writing this book together, we remain the best of friends!

Karin would like to thank her mentors at the University of New Hampshire, Professor John J. Sasner, Jr. and the late Miyoshi Ikawa, for introducing her to the genre of scientific writing and to Randy Wayne at Cornell University, who first encouraged her to publish the materials she had written for her biology students. I would especially like to thank Andy Sinauer and all of the professionals at Sinauer Associates and W. H. Freeman who helped with the production of *A Student Handbook for Writing in Biology*. This book provided the idea and structure for *Engineering Communication*. We sincerely appreciate the permission granted by Sinauer Associates and W. H. Freeman to adapt chapters from *Writing in Biology* for an audience of engineers.

Living and working in Europe and Japan for a number of years has given both authors a unique perspective on technical communication. We have gained valuable experience on how non-native speakers write papers in English. Conversely, we have learned how German and Japanese culture affects communication in those languages. Furthermore, working as a freelance German to English technical translator has had a profound impact on Karin's perspective on written communication. I appreciate the challenge of choosing just the right words to communicate information faithfully, in a clear and concise manner, and according to the conventions familiar to the reader. We are indebted to our colleagues and clients for insight on how to produce high-quality communications.

The prospectus for this book was written during Charles' sabbatical in Christchurch, New Zealand. We thank the Bucknell University administration for providing the release time and the support for this project. We also gratefully acknowledge the hospitality of the

Mechanical Engineering Department at the University of Canterbury, in particular, Prof. Keith Alexander, Ms. Rebecca Morgan, and Department Head Prof. Milo Kral; we thank Prof. Ian Wood of the Civil Engineering Department, who made introductions for us with the Mechanical Engineering Department.

We would also like to acknowledge colleagues at Bucknell University, who provided advice on various portions of this book: Professors Stu Thompson, Dave Kelley, Peter Jansson, Dan Hyde, Charles Kim, Emily Geist, and Christine Buffinton. We thank Dean Keith Buffinton and the Bucknell University administration for allowing us to use the survey data from mechanical engineering alumni contained in Chapter 1 and the excerpt from a construction specification in Chapter 9. We also thank former and current students who provided feedback, exercises, and samples of writing materials. In particular, Matt Mosquera volunteered to read Chapter 6 and provided useful suggestions for revision.

We sincerely appreciate the feedback and suggestions made by the following individuals, who read some or all of the first draft of the manuscript: Michael Berry, Montana State University; George Corliss, Marquette University; Stephen Kampe, Michigan Technological University; Robert D. Knecht, Colorado School of Mines; Jesa Kreiner, California State University, Fullerton; Stephanie Loveland, Iowa State University; Carole M. Mablekos, Drexel University; Mark Snyder, Illinois Institute of Technology; and Chad A.B. Wilson, University of Houston. In particular, the authors wish to thank Prof. George Corliss and Prof. Mark Snyder for their thorough commentary on each chapter of the book. Their attention to detail and suggestions were greatly appreciated and certainly improved the manuscript.

The authors also acknowledge the encouragement of Randall Adams, former acquisitions editor at Cengage Learning, who read our prospectus and recommended the publication of our book by Cengage. Hilda Gowans at Cengage Learning was a constant source of encouragement as the project progressed from prospectus to final manuscript. We could not imagine a better person to guide us through the steps involved in producing this book. Our production editor, Rose Kernan, was exceptionally vigilant in correcting our errors and inconsistencies.

Finally, we would like to thank our parents for instilling in us a love of reading and learning, an eagerness to acquire knowledge and share it with others, and a desire always to do our best. Our children, Katrina, Carleton, and Brian, have given us insight into what motivates today's college students and we thank them for keeping us current on the latest technological developments. In particular, we thank Brian for reading several of the chapters and making helpful suggestions.

CHARLES W. KNISELY

KARIN I. KNISELY

Chapters in this book have been adapted from *A Student Handbook for Writing in Biology.* Used with permission from the publisher, Sinauer Associates, Inc. and W. H. Freeman & Co."

This book is dedicated to those who taught us always to strive for greater knowledge and to those who embrace the lessons we strive to pass on to them.

About the Authors

Professor Charles Knisely has been a faculty member in the Mechanical Engineering Department at Bucknell University since 1990. Professor Knisely teaches courses in technical communication, mechanical engineering laboratory practice, thermodynamics, fluid mechanics, heat transfer, and renewable energy, as well as a number of other upper level courses. Many of the courses he teaches have writing components. He served as the department chair from 2010 to 2014.

Professor Knisely received his Ph.D. in Mechanical Engineering from Lehigh University in Bethlehem, Pennsylvania in 1980, and subsequently worked in industrial research in Switzerland and at a government research lab in Bethesda, Maryland. He held positions at the University of Karlsruhe (Germany) and at Kyoto University in Japan before coming to Bucknell University. He subsequently held visiting academic positions at the Osaka Electro-Communication University (Japan), the University of Siegen (Germany), and the University of Canterbury (New Zealand). Professor Knisely has served as an ABET Program Evaluator (PEV) since 1999.

His main areas of research are flow-induced vibrations, fluid dynamics, scroll compressors, and hydraulic gate instabilities. He has authored or co-authored over a hundred journal articles and conference papers, many of which were prepared with international collaborators. He holds six patents.

Professor Knisely's interests outside of his professional activities include playing squash, traveling, zymurgy, woodworking, gardening, and landscaping. In 1999, along with his wife and two other families, he founded the Central Susquehanna Lacrosse Club and served as president and head boys' coach. He spearheaded efforts to establish local high school lacrosse teams. Until becoming department chair in 2010, he coached, officiated, and trained officials for boys' high school lacrosse in north central Pennsylvania.

Karin Knisely has been Lab Director of Core Course Biology at Bucknell University since 1994. She teaches five lab sections a week each semester, alternating between Introduction to Molecules and Cells and Organismal Biology. In addition to her teaching responsibilities, she orders the supplies and prepares the chemicals and equipment for 160–220 students each semester and collaborates with faculty in the Biology Department to write and revise the lab exercises.

Since earning her M.S. in Zoology from the University of New Hampshire in 1981, Mrs. Knisely has held positions as a German Academic Exchange Service (DAAD) research fellow in Constance, Germany; worked in a translation agency in Stuttgart, Germany; worked as a research associate in a biotechnology company in Gaithersburg, Maryland; taught weekly classes in English conversation to engineers at a university in Osaka and at a manufacturing company in Kyoto, Japan; proofread and edited manuscripts for a biennial conference on chemical water and wastewater treatment for over 20 years; has translated technical documents from German into English for over 30 years; and, together with her husband, taught three short courses to Japanese engineers on how to write technical papers in English.

In her capacity as Lab Director, Mrs. Knisely has graded thousands of lab reports and is quite familiar with the kinds of mistakes students make when writing biology lab reports for the first time. To give students guidance on how to write lab reports according to the style and format of journal articles, she prepared handouts that ultimately resulted in the publication of *A Student Handbook for Writing in Biology*, co-published by Sinauer Associates, Inc. and W.H. Freeman and Co., now in its Fourth Edition.

Outside of academic pursuits, Mrs. Knisely enjoys traveling and spending time with her family. She enjoys outdoor activities such as biking, walking, and skiing, and has coached, officiated, and trained officials for girls' lacrosse in the Susquehanna Valley since 1999.

Professor and Mrs. Knisely have three children and live in Lewisburg, PA.

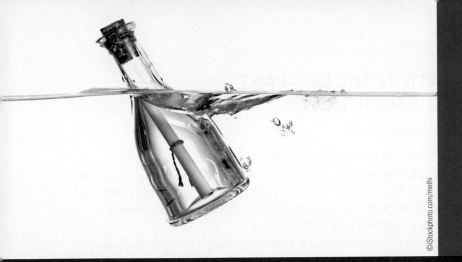

Chapter 1

WHY DO ENGINEERS NEED TO COMMUNICATE?

Most engineers want to contribute to improving the quality of life for the human race. Regardless of the quality of your ideas, if you cannot effectively explain your ideas to other people, no one will ever benefit from your ingenuity and creative abilities. Engineers also want to be rewarded for their contributions. Without well-written, compelling arguments, your proposals for new devices, systems, and processes will never be implemented. Without technical communication skills, you may well be destined to implement other people's ideas for your entire career. You will be an unlikely candidate for promotion. Technical communication is essential for a satisfying and rewarding career in engineering.

COMMUNICATION SKILLS—PART OF THE JOB DESCRIPTION

Are you aware of the following facts?

- Engineers spend 50 to 90% of their time on communications.
- Engineers must be able to communicate effectively to advance within their organization.
- The American Society of Mechanical Engineers (ASME Vision2030, 2012) found employer dissatisfaction with the communication skills of entry-level engineers.
- Frequent careful reading improves your writing ability.
- Writers anticipate the needs of their audience, structuring their communications for ease of comprehension.

If you are an engineering student, you will receive writing and communications instruction at some point in your studies. You have options: You can do the minimum to get through the assignments, or, with a forward-looking perspective, you can embrace the need for engineers to communicate and you can use communication assignments to hone your skills in preparation for a career.

> The engineering curriculum was strenuous and challenging. At the time, I didn't see the benefits. I was just trying to pass my classes (and, of course, enjoy the university experience). My biggest weakness after graduating was my lack of proficient presentation and communication skills. (*Anonymous response from an engineering graduate in a recent survey*)

If you are a working professional who has not acquired sufficient communication skills before beginning your employment, you can still acquire these skills if you recognize the need for them and make a commitment to acquire them. Carefully study reports, proposals, and oral presentations prepared by your older, more experienced colleagues. Identify a potential mentor from among these colleagues, someone with excellent communication skills and with whom you have a good working relationship. Ask your colleague to critique your writing before you submit your reports.

> The most intelligent engineer is of no value to others without the ability to communicate, both orally and in writing. (*after Davis, 1977a*)

Study technical reports, read online work-related materials, and use an instructional engineering communication guide such as this book to become familiar with various communication genres. Read classic novels, science fiction books, and any nonfiction books you enjoy in your nonworking hours. Your efforts will pay off. In all of your reading, pay attention to word choice, grammatical structures, and the logic flow. As you become more familiar with communication skills, the preparation of your reports, proposals, and presentations will become routine.

In this chapter, we will explore various types of engineering communications. We will show you that the two skills most desired in newly hired engineers are communication and problem solving. Finally, we will try to convince you that your career success may depend as much on your communication skills as on your analytical and creative abilities.

A DAY IN THE LIFE OF AN ENGINEER

Imagine yourself as a working professional, having landed the job of your dreams right out of college. What do you do in the course of the day, and why do you do it?

A typical engineering day might consist of the following:

- Reading and responding to your business email several times a day
- Responding to queries and requests from your supervisor
- Working on a project, or multiple projects, which may include
  - » Hands-on engineering—designing, testing, and measuring
  - » Modeling or simulating processes on a computer
  - » Interacting with subordinates and superiors
  - » Writing notes on your activities
  - » Preparing formal project reports
- Meeting with administrators/clients/prospective clients and
  - » Presenting a proposal
  - » Listening to a statement of the problem or an explanation of the needs
  - » Discussing issues related to a project
  - » Providing input on potential solutions
  - » Discussing related issues and technical literature
- Reading your printed mail
- Making and responding to telephone calls

While this bulleted list may not be all-inclusive, engineers might summarize their daily activities as follows: "I will be using my intellectual abilities to improve human existence and at the same time to help my company make a profit." Now ask yourself, *how will other people know* what you do? Reflecting on the answer to this question will help you understand the importance of communication in realizing your ideas and earning respect from your colleagues.

Many studies over several decades have examined the importance, frequency, and type of communications used by professional engineers (see Davis, 1977a, 1977b, 1978; Schiff, 1980; Spretnak, 1983; Pinelli et al., 1995a, 1995b; Kreth, 2000; Woody, 2010; and ASME Vision2030, 2012). You might be surprised by some of the results, especially if you are or were a typical engineering student who decided to major in engineering because "I like math and science, but I do not like all the reading and writing in English." A recent poll of mechanical engineering alumni from the classes of 2000 to 2012 indicates that 79% of the respondents engage in multiple written and oral professional communication activities every day (see Figure 1.1), and 92% engage in multiple communication activities over the course of a week.

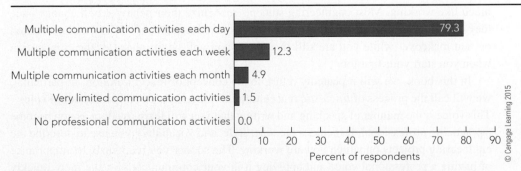

FIGURE 1.1 | Response to the request "Please characterize the frequency of your professional communication activities since graduation" from a survey of 203 mechanical engineering alumni in the classes of 2000 to 2012

TABLE 1.1 | Estimated time spent on communication

Source	Kreth (2000) "Co-op"	Spretnak (1983)	Pinelli et al. (1995a)
Years of experience	3 to 4	2 to 32	23.5 (avg)
Percentage of time spent preparing own communications	10 to 60 35 (avg)	32 (avg)	49 to 58
Percentage of time spent on someone else's communications	Not assessed	11 to 29	38 to 49
Average percentage of time spent on all communications	35	Over 50	90

As shown in Table 1.1, there is evidence that the average time spent on professional communication has increased slightly over the past 35 years. Even more compelling is the trend that the time spent on communication increases with years of professional experience.

The message is very clear:

Engineers must communicate effectively.

While the tone, style, and format of engineering communications differ from those in the social sciences and humanities, proper grammar and punctuation are essential. Without a command of the structure of the language, your writing will lack clarity. From the perspective of a practicing engineer, time is money. Communications that are direct, concise, and unambiguous save time and are invaluable in engineering practice.

Developing effective communication skills while you are still in college may be the first step in *reacculturation* to the workplace. **Reacculturation** means adopting new habits, practices, and processes to fit into an established culture. For example, if your new company has a dress code (either formally written or informally recognized), you will not show up for work in your favorite T-shirt. Similarly, you will need to modify other habits, such as sleeping in during the week and simply taking an afternoon off because you did not feel much like working. Most engineering students recognize their behavior will change once they enter the workforce. Your language will also need a makeover. If you can get started on that makeover while you are still in college, you will be far ahead of the competition when you start your first job.

In this book, we will repeatedly return to the concept of reacculturation. In particular, we will call the process of *linguistic* reacculturation "developing your **professional voice**." This voice is the manner of speaking and writing that you will learn to adopt as you become a practicing professional. It includes the tone, style, and vocabulary you use to describe the engineering projects on which you are working. The sooner you recognize the importance of having a professional voice and applying it in your communications, the more quickly you will advance in your career.

TYPES OF TECHNICAL COMMUNICATIONS

Technical *writing* takes many forms, as shown in Figure 1.2. As an undergraduate engineering student, you will probably need to write laboratory reports, memos, feasibility studies, project progress reports, and design reports. You might also need to summarize journal articles and prepare literature surveys on topics of interest. Upper-class students may write research proposals for honors work and then complete their projects by submitting honors theses.

Graduate students typically write master's theses and doctoral dissertations, defending their written work with oral presentations. Professors write lectures, letters of recommendation for students, journal articles, grant proposals, reviews of journal articles and research proposals submitted by their colleagues, and collaboratively write assessments of colleagues for promotion and tenure reviews.

In business and industry, technical writing also includes, but is not limited to, product descriptions, operating manuals, operating procedures, white papers, site visit reports, progress reports, project proposals, feasibility studies, engineering specifications, and sales and marketing material. At various stages in your career, you will prepare resumes and cover letters for employment. Almost all writing projects will require some form of research, in which you will search engineering databases and other printed as well as online resources for the most current information.

Technical communications may also take the form of *oral presentations*, ranging from a one-on-one telephone conversation to a formal speech with slide show. Regardless of the medium, engineers need to *tailor communications to their audience*, which may be another engineer, a company executive, or a client; an entire engineering department, technicians who are going to build the product, or a custodian who is going to perform maintenance; a class of elementary school children, college engineering students attending a seminar, or

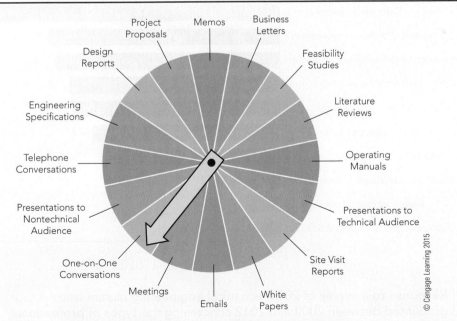

FIGURE 1.2 | Wheel of fortune for engineering communication. Do you know with certainty where your career will take you?

graduate students on a recruiting visit. In short, engineers should be prepared to tackle a wide variety of written and oral communications, as shown in Figure 1.2.

You might think that you do not need to master certain forms of communications because you are certain that you will not seek employment that requires those skills. Beware: The career path of many engineers is frequently like a wheel of fortune (and hence the theme of Figure 1.2). Opportunities arise, and engineers end up in careers they did not envision while they were in school.

What are the most common types of communications undertaken by engineers? The results of a survey of recent mechanical engineering graduates are given in Figure 1.3. Email is clearly pervasive, but two-thirds of the engineers in this survey also spend time on the telephone and on the preparation of project reports, proposals, and presentations of project outcomes. Only mechanical engineering alumni were included in the survey, so feasibility studies, which are commonly undertaken in civil engineering, computer engineering, and project management, but not typically in mechanical engineering, are absent. The data in Figure 1.3 represent just a snapshot of the work environment of one branch of the engineering profession; they are, however, consistent with information from previously published reports (see Schiff, 1980; Kreth, 2000; and Ruff and Carter, 2009; among others).

Email has emerged as the predominant form of professional communication in academia and industry. The primary advantage of communicating by email rather than by phone or in a face-to-face meeting is convenience. Emails make it possible to

- Send and read messages when it suits your schedule
- Attach electronic documents that can be read immediately by the recipient

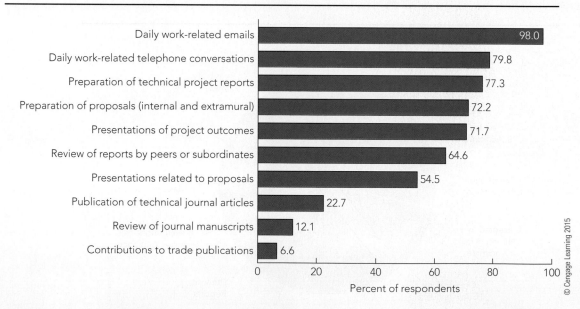

© Cengage Learning 2015

FIGURE 1.3 | Response to a survey of 203 mechanical engineering alumni who graduated between 2000 and 2012 concerning the types of professional communications they have used in their career. Selected comments from this survey are reproduced in the accompanying **sidebar** on page 8.

- Have a written record to refer back to, rather than relying on memory or taking accurate notes
- Convey information without the intrusiveness of a phone call.

While email messages are typically much less formal than other forms of technical communication, they should still be informative (state clearly what is needed and when), concise, grammatically correct, and without spelling mistakes. Emails with these characteristics show consideration for your colleagues, as illustrated by comment 12 by one recent graduate in the sidebar "What Early Career Engineers Say about Workplace Communication." Furthermore, you never know when your emails may be passed up the management chain, so you want to put the best possible polish on your image. When you are employed in an organization, it is good practice to keep a separate private email account for your personal business. Do not use your work email to send personal communications.

Email certainly has many advantages, but phone calls and face-to-face meetings are perceived as more personal and may be preferred in certain situations. For example, hearing someone's voice or watching their facial expressions and gestures may help you gauge their true position on a particular topic much more readily than reading their email response alone. In addition, when the communication consists of multiple rounds of questions and answers, phone calls and meetings ultimately may be more efficient than emails. Because the more traditional forms of communication are still widely used and have not diminished in importance, engineers need to practice carrying on meaningful oral interactions with fellow students, professors, colleagues, and clients. Comments 3, 6, 8, and 10 in the **sidebar** on the next page illustrate this point.

Another significant aspect of technical communication is that it is becoming increasingly collaborative, both in academia and in industry (see Table 1.2). In higher education, this trend resulted in part from teamwork requirements for accreditation, which were driven by industrial expectations for engineering graduates. In industry, the complexity and size of most modern engineering systems make it impossible for one person to undertake all analyses and reporting. Software developments during the past 20 years have made collaboration even more prevalent. One might speculate that almost all design and manufacturing communications are prepared collaboratively. Thus, entry-level engineers need to be able to collaborate when preparing engineering communications.

TABLE 1.2 | Reported percentage of communications prepared collaboratively by practicing engineers

Number of Collaborators	Design/Development Engineers	Manufacturing/ Production Engineers
With at least 1 other person	45%	83%
With at least 3 to 5 people	45%	66%
With more than 5 people	39%	29%

Based on Pinelli et. al., 1995a

What Early Career Engineers Say about Workplace Communication
(From a survey of mechanical engineers who graduated 2000–2012)

1. The majority of my job involves technical communication between the 200+ engineers working on our satellite. I ensure requirements are met, authorize troubleshooting/repair work, ensure documentation and test reports are correct, perform trade studies, and serve as the primary contact for technical issues between the customer and the engineering team. **Professional communication is the most important part of my job, and my expertise in this area has resulted in a promotion and recognition** from numerous managers from various organizations within the company.

2. [I] contribute to many court filings, including briefs, motions, and pleadings.

3. [My activities include] leading meetings and participating **in community presentations** geared toward explaining the rollout of the Smart Grid initiative.

4. In my experience, the ability to communicate effectively is the only skill that has outweighed problem solving.

5. Among the more technical pieces of communication I write are design and clinical risk management documents, functional requirements documents, verification and validation test protocols, test reports, and engineering rationales.

6. Our technical communications are accomplished via design reviews. These are often technical presentations with pictures, drawings, analysis results, schedule, budget, etc.

7. [I] write standard language for ASTM F24 [and] RFPs for contracts. I have written over 10 patents and have been awarded 5 to date.

8. [I] brainstorm in face-to-face meetings [and discuss] casual design concepts face to face.

9. [I develop] various best practice/standard work/helpful tip guides **to lower the learning curve for colleagues**.

10. [In] each of my projects [I create] a report, which is then presented to the customers, as well as [conduct] interviews for project selection. My projects also [involve] workshops with both internal and external customers, which **require a high level of communication and communication with different levels of audience. . . ranging from maintenance employees and custodians to CFOs** [requiring] presentations appropriate to the audience.

11. Preparing drawings ("blueprints") and specifications.

12. The biggest waste of time I see in my profession [results from] unclear or vague emails. It is important to stress clear communication — who you are addressing, what you need from them, and when you need it by.

13. [The] improvement I see needed in most students is in their ability to communicate clearly and properly both internally and externally with clients in a clear, professional, well-written, and timely manner.

14. [Students need] more frequent [projects] requiring independent research. It was a shock at first in the professional world to be [assigned] a task without guidance on how to complete it.

15. Scientific [and] technical communication with non engineers and with engineers and scientists in other disciplines is essential.

(Bold added for emphasis)

WANTED BY INDUSTRY: ENGINEERING STUDENTS WITH COMMUNICATION SKILLS

As discussed in the previous section, engineering students and professional engineers produce many different kinds of communications. How well do academic institutions prepare engineering students for writing in the workforce? A recent survey published by ASME Vision2030 (2012) shows that around 40% of engineering educators think that their graduates are well prepared to communicate in the workforce, yet only 10% of supervisors in industry think so (Figure 1.4). Similarly, a much higher percentage of educators than supervisors think that early career engineers have strong interpersonal skills and work well in teams. Only in the information processing category were supervisors more impressed than educators with the abilities of the new hires. The self-assessment of the early career engineers was similar to that of the educators.

Why do engineering graduates seem to be ill-prepared for some of the work they do in industry? Pinelli et al. (1995a) suggest that graduates lack communication skills either because they do not receive sufficient instruction or because the technical communication instruction they do receive "is inappropriate for the workplace; that is, there is a 'disconnect' between academic perceptions of workplace communications and the realities of workplace communications." According to McKee (2012), this "disconnect" between academic and industrial perceptions of engineering students' writing ability has persisted for the past 40 years.

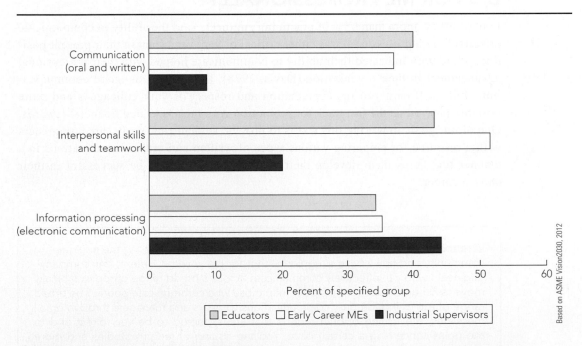

FIGURE 1.4 | Assessment of strengths of entry-level engineers by their industrial supervisors, the early career engineers themselves, and engineering educators

Do engineering students receive sufficient instruction? Certainly, all of the engineering programs we have examined require some basic writing courses. One problem is lack of time to give students enough practice writing all of the different types of communications used in industry. Another problem is trying to achieve consistent and uniform standards for good communication in all engineering courses. A third problem is retention, the ability of students to remember what they learned from semester to semester. Instruction will only be "sufficient" when students have frequent writing assignments every semester, receive consistent and detailed feedback, and are motivated to improve.

Based on the recent ASME survey and McKee's dissertation, clearly there is a "disconnect" between academic perceptions and the realities of the workplace. The disconnect may in part be generational. Young engineers have grown up in a culture in which immediacy quite often trumps quality; experienced supervisors in industry have other priorities. A successful synthesis of multiple perspectives within the engineering group is invaluable in turning a great idea into a great product. The collective ability of the engineering group to communicate the novelty of a product or idea to company executives helps them craft their marketing strategy. The fortunes of the company ride on its ability to convince society at large that the new product or idea is better than what is currently available. Thus communication is crucial at all levels of this process.

WHAT CAN GOOD COMMUNICATION SKILLS DO FOR ME PROFESSIONALLY?

In one survey, more than 95% of practicing engineers said the ability to communicate effectively as a professional was either critical or very important in their present position, while 96% indicated their ability to communicate helped with their professional advancement in their organization (Davis, 1978). In other words, good communication skills will earn you the appreciation and respect of your colleagues and quite possibly put you on the fast track for promotion with the associated financial rewards. One of our goals in writing this text is to provide students and early career engineers with a resource for preparing various types of written and oral communications in a manner that helps them develop their professional voice and be successful in their chosen career.

In retrospect, I realize the lack of writing ability I had when entering the engineering community and how I have struggled with it for many years. Here at [*my company*], engineers who can effectively communicate across a broad range of topics typically move faster up the corporate ladder than those who communicate poorly. The broad range of topics include daily communications to peers and managers through Email messages, the issuing of brief reports that are expected to be very direct and to the point concerning a certain issue, and the issuing of lengthy studies and more global reports covering a broad range of topics. A "rambling report" or one filled with emotion is often quickly discarded and the engineer's credibility is diminished.

– Engineer with 4 years of experience (quoted in Kreth, 2000)

HALLMARKS OF TECHNICAL WRITING*

What distinguishes technical writing from other kinds of writing? One difference is the motive. Technical writing aims *to inform* rather than to entertain the reader. The reader is typically a technically trained person who intends to use this information, for example, to learn more about a process, to improve a product, or to operate a machine or instrument. However, engineers must also write for nontechnical audiences, as shown by the comments from the alumni survey in the sidebar. Engineers must communicate at a level that makes the information understandable for the intended audience.

A second difference between technical and nontechnical writing is the style. *Brevity, precision, a standard format, and proper use of grammar* and *punctuation* are the hallmarks of well-written technical papers. The authors have something important to communicate, and they want to make sure that others understand the significance of their work. Flowery language and "stream of consciousness" writing are not appropriate in technical writing, because they can obscure the writer's intended meaning.

A third difference is the use of the passive voice to emphasize *actions, measurements, processes, or devices,* not the individual who undertook the work. Procedures that can be performed by any trained individual are described in the passive voice, because *who* did the procedure is not as important as *what* was done.

A fourth difference between technical and other types of writing is the *tone*. Technical writing is factual and objective. The writer presents information without emotion and without editorializing, letting the facts speak for themselves.

Table 1.3 summarizes characteristics of good technical writing. Use this table as a checklist to assess your own writing.

TABLE 1.3 | Characteristics of good technical communication

Good Technical Writing Is	Good Technical Writing
• Clear and precise	• Addresses the needs of the audience
• Concise	• Adheres to standards of the profession
• Well organized	• Uses correct and appropriate units
• Well designed and laid out	• Contains effective graphics
• Grammatically correct	• Avoids slang, clichés, and verbosity
• Factual and objective	• Makes appropriate use of passive voice

© Cengage Learning 2015

WHAT CAN I DO TO BECOME A BETTER WRITER?

One obvious answer to the question of how to become a better writer is to *practice writing*. As you practice your writing, recall one of the axioms of sports:

This section has been adapted from *A Student Handbook for Writing in Biology*, Chapter 3. Used with permission from the publishers, Sinauer Associates, Inc. and W. H. Freeman & Co.

Practice without improvement is the same as no practice at all.

Not only do you need to seek feedback and constructive criticism of your writing, you need to remember your mistakes from the previous report as you begin your next writing effort. Keeping a journal of "errors I made" serves as a reference for what to avoid next time you write.

A less obvious and perhaps surprising way to become a better writer is to become a better and more frequent *reader*. Spretnak (1983) describes an experiment involving two groups of high school students. Each student in the first group, called "the writers," wrote the equivalent of one theme per week, which was carefully corrected by the instructor, returned to the student, and then rewritten by the student. Each student in the second group, called "the readers," wrote the equivalent of one theme every three weeks (i.e., one-third as much writing), received similar feedback on the writing, and rewrote their themes; however, they spent class time in the intervening weeks reading books they had selected. An assessment of the writing skills of both groups of students was administered at the beginning and end of the academic year using a standardized test for spelling, style, diction, and the mechanics of writing (the STEP Writing Test, Form 2A and then 2B, designed by the Educational Testing Service), as well as essays graded by three experienced standardized test graders. Table 1.4 shows that the overall improvement of the "readers" was almost double that of the "writers."

Spretnak speculates that "the brain somehow assimilates examples of economical prose, extensive vocabulary, and effective ordering which the reading of good writing provides. Later, the student, when writing, seems to draw creatively from his or her 'data bank' of rhetorical possibilities." Spretnak supports her hypothesis with comments made by engineering alumni on a survey she conducted. Comments such as "If one does not read, it is difficult to write well," and "Reading technical papers helps engineers learn to communicate" were common.

If you carefully read your engineering textbooks, looking at the way the material is presented, thinking about the word choices the authors make, and seeing the big picture of their writing styles, your writing will benefit from your reading. If, instead, you skim through the textbook and try to find example problems to guide you in your homework solutions, your writing will benefit substantially less. Even more interesting is Spretnak's finding that there is a positive correlation between writing skills and amount of time spent in *leisure reading*. It does not matter what you read, as long as you read carefully and thoughtfully.

TABLE 1.4 | Improvement in standardized writing test scores over one academic year

Area of Assessment	Readers	Writers
Content and organization	+0.7	+0.45
Mechanics	+0.38	+0.11
Diction and rhetoric	+0.7	+0.07
Overall improvement—includes essay grading	+6.5	+3.5

Based on Spretnak, 1983

Spretnak suggests that instructors should discuss the writing formats and various models of writing (for example, technical reports, design and feasibility studies, and journal articles) as they are used in class. She writes, "when students see that word choice is both efficient and strategic, they begin to develop an interest in the precision of language. Nurturing this attitude is important since engineering students often enter the course with a low regard for prose as being more arbitrary, vague, and imprecise than mathematics."

Reading that Informs Writing

Reading can assist your writing if you pay attention to:

- Word choice
- Logic flow
- Patterns of organization
- Format

Honing your critical reading skills will also help with your career advancement. Spretnak (1983) cites the need for engineering managers, supervisors, project leaders, department heads, and division directors to be able to read and critique the writing of the engineers in their charge. Such critical reading can constitute up to 20% of their working hours. "Critical reading skills, then, may be seen as a requisite for such advancement," she concludes.

Obviously, reading without doing any writing will not necessarily improve your writing. In engineering terms, we might characterize reading as a catalyst. Reading by itself does not produce the reaction (your writing), but careful and thorough reading (adding the catalyst) along with writing practice serves to accelerate the development of your writing.

BEFORE YOU START WRITING

Most of us recognize good writing when we see it, without making a conscious effort to understand what makes the writing good. When you are in the position of the writer, however, you need to have some practical guidelines to get you off to a good start. Keep in mind that the purpose of technical writing is to convey the knowledge and expertise you acquired on a particular subject to another person.

As illustrated in Figure 1.5, in order to convey technical knowledge clearly, accurately, and concisely you need to

- Think carefully about the needs of your audience
- Use the established format and standards that are familiar to your audience
- Understand the subject matter
- Adhere to accepted grammatical rules.

For whom are you preparing this technical communication—your professor, other engineers with a background similar to your own, business people, or a general audience with little technical expertise? What does your audience hope to gain from reading your work or hearing your presentation? Answers to these questions will determine the focus of your writing project, the level of detail, and the language you choose.

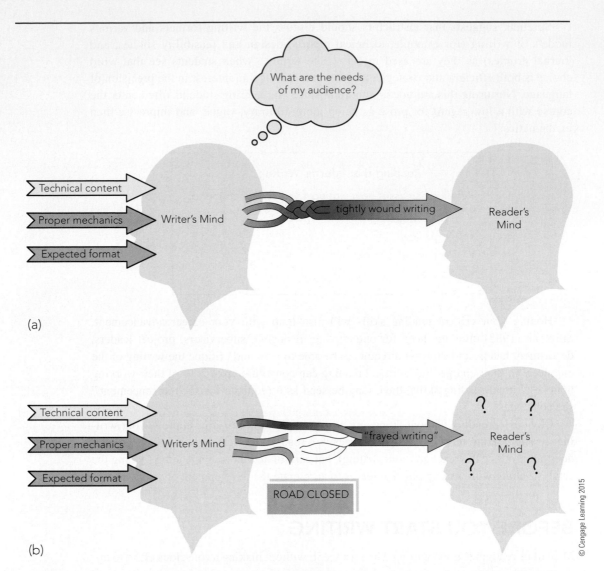

FIGURE 1.5 | **(a)** A good writer anticipates the needs of the target audience and weaves the threads of technical content, proper mechanics, and the expected format into a tightly braided structure that facilitates the flow of information from the writer's mind to the reader's mind. **(b)** Ignoring the needs of the audience or failure of one or more of the threads creates roadblocks, leading to congestion and inefficient information transfer.

Your audience will also dictate the format and style of your writing. For example, if you are an undergraduate engineering student, your professor may provide a lab report rubric that prescribes the format and content of the assignment. If you are a graduate student who would like to publish a paper in a technical journal, the "Instructions to Authors" section provides details on length, sections, tables and figures, references, and so on. If you are a professional engineer, your employer may have specific templates for formatting internal memos, oral presentations, and other types of communications. The purpose of using a standard format is to allow members of your audience to obtain knowledge quickly and efficiently.

The necessity of understanding the subject you are writing about should be obvious: The information you communicate to others must be accurate. By putting your knowledge into words, and particularly by receiving feedback on your written work, you become aware of what you understand and do not understand about the subject. Therefore, while the writing process improves your understanding of the subject, you must have sufficient background knowledge before you begin.

Good technical writing also requires an understanding of the mechanics of writing (grammar, punctuation, verb tense and voice, word choice, parallelism, and so on). Creative use of grammar, punctuation, and capitalization is not productive because it obscures the meaning of the content. Grammatically correct communications are authoritative and enhance the credibility of their authors.

In summary, good writing is the product of attention to the needs of your audience, adherence to the accepted format and standards, technical understanding of the material, and proper use of writing mechanics. The writer weaves these threads into a tightly wound braid that the reader perceives as a single strand, as shown in Figure 1.5. When any one of the three threads breaks, the remaining ones must carry an overload. As on a highway, when one lane closes, congestion results and it becomes harder to get to your destination. Similarly, a breakdown in any one of the three threads of technical writing slows the flow of information to your reader's mind, with the unfortunate consequence that some readers may decide that it is simply not worth the effort to put up with the congestion.

In my long experience and association with scientists and engineers, I can't remember a single instance of anyone advancing to a position of significance who could not [communicate] effectively.

(Davis, 1978)

PREVIEW OF UPCOMING CHAPTERS

Now that we have introduced you to engineering communication and how good writing will benefit your career, it is time to turn our attention to the process of acquiring information and citing sources, assimilating that information, and preparing written and oral communications that target your audience. Chapter 2 presents strategies for finding information, both online and print material, and appropriately acknowledging these sources. Reading technical articles is not only a way to acquire knowledge; the reading process is a catalyst for learning to write well. Chapter 3 provides guidelines for reading certain types of technical writing in preparation for writing technical reports. Chapter 4 provides instructions for writing laboratory reports, and an annotated model lab report offers tips on format and style. Chapter 5 provides instruction and strategies for revising your work. Chapter 6 is an engineering student's toolbox, which contains information on nomenclature, format, and analysis of graphs, and other tools of the trade. Chapters 7 to 10 cover many different forms of writing engineers might encounter in their careers–the writing of letters, memos, email, resumes, proposals, design specifications, literature reviews, progress reports, site visit reports, and white papers. Chapter 11 provides a guide for preparing and presenting oral presentations using PowerPoint slides, while Chapter 12 does the same for poster preparation and presentation.

The three appendices cover the use of Microsoft Word, Excel, and PowerPoint. They will be useful references for you as you prepare your papers, presentations, and posters.

SUMMARY

As you work through this text, you will receive instruction on many aspects of engineering communication. Making mistakes is one of the best ways to learn, provided you do not simply keep repeating your mistakes. Remember the sports axiom: "Practice without improvement is the same as no practice at all." Practice writing like your job depended on the way you write, because soon it might. No matter what type of communication you practice, strive to develop and use your professional voice.

EXERCISES

1. Explain as best you can how the artwork on the cover of this book relates to the material described in this chapter. What is the relevance of the artwork? Does it have any obvious or subtle relationship to technical writing? Can you use the image as a mnemonic aid in remembering any point discussed in this chapter?

2. Watch the YouTube video of Stephen King's advice on writing found at <http://www .youtube.com/watch?v=hqp7A0B7abc>. Write a concise, well-organized paragraph summarizing the writing advice given in the short video clip.

3. Through the alumni office, your department office, the dean's office, personal connections, or a more experienced colleague, find an employed professional in your discipline. Conduct a face-to-face interview, a telephone interview, or an interview via electronic media. Determine the following: (a) the types of communications this person writes, (b) the frequency of the communications, and (c) the strategies used by this

individual to become a better writer. Ask for suggestions on what you might do to develop your professional voice. Write a short report to document the results of the interview.

4. Search the classified ads for engineering positions in your field in a major newspaper, a professional society news periodical, or an online job advertisement. Examine at least 10 position descriptions and tabulate how many specifically list communication skills as a required or desired prerequisite for the position. Summarize your findings in a table and describe them in a concise, well-organized paragraph.

REFERENCES

ASME Vision2030. 2012. ASME Vision 2030 Project: Drivers for change; data, actions & advocacy. Accessed 7 January 2013 at <http://files.asme.org/asmeorg/Education/College/Faculty/ME/33615.pdf> .

Davis, R.M. 1977a. Technical writing—who needs it? *Engineering Education* 68 (2): 209–211.

Davis, R.M. 1977b. Technical writing in industry and government. *Journal of Technical Writing and Communication* 7(3): 235–242.

Davis, R.M. 1978. How important is technical writing? A survey of opinions of successful engineers. *Journal of Technical Writing and Communication* 8(3): 207–216.

Kreth, M.L. 2000. A survey of the co-op writing experiences of recent engineering graduates. *IEEE Transactions on Professional Communication* 43(2): 137–152.

McKee, C.D. 2012. Information design: A new approach to teaching technical writing service courses, Ph.D. Dissertation, Oklahoma State University, Stillwater, OK 74078.

Pinelli, T.E., Barclay, R.O., Keene, M.L., Kennedy, J.M., and Hecht, L.F. 1995a. From student to entry-level professional: Examining the role of language and written communications in the reacculturation of aerospace engineering students. *Technical Communication* 42(3): 492–592.

Pinelli, T.E., Kennedy, J.M., and Barclay, R.O. 1995b. Workplace communications skills and the value of communications and information use skills instruction—engineering students' perspectives. 1995 IEEE International Professional Communication Conference, *IPCC 95 Proceedings*, Savannah, GA: 161–165.

Ruff, S. and Carter, M. 2009. Communication learning outcomes from software engineering professionals: A basis for teaching communication in the engineering curriculum. 38th ASEE/IEEE Frontiers in Education Conference, 18–21 October, San Antonio, TX, pp. W1E1–W1E6.

Schiff, P.M. 1980. Speech: Another facet of technical communication. *Engineering Education* 71: 180–181.

Spretnak, C.M. 1983. Reading and writing for engineering students. *Journal of Advanced Composition* [Internet], 4(1): 133–137 [c2006]. Accessed 21 September 2013 at. <http://www.jaconlinejournal.com/archives/vol4/spretnak-reading.pdf>.

Woody, D. 2010. Improving engineering work flow in technical communications: A white paper from ZAETRIC® Business Solutions, LLC. [Internet, c2010]. Accessed 15 September at <http://goo.gl/CshR0>.

Part 1

INTRODUCTION TO FINDING, READING, AND CITING TECHNICAL SOURCES

©iStockphoto.com/melhi

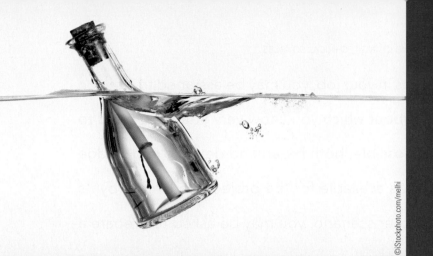
©iStockphoto.com/melhi

Chapter
2

FINDING INFORMATION AND CITING SOURCES

The development of research skills, knowing where to find reliable print and online information sources, and learning how to document these sources appropriately is invaluable at all stages of an engineer's career. As a student, you may be assigned a project to complement the classroom activities in a course. If you are a self-driven person, you may want to undertake independent study in an area of special interest to you. You will need to find the most current information. As an early career engineer, you may find that your laboratory measurements do not compare well with theory and that you cannot find any obvious reason for the discrepancy. A starting point in resolving the discrepancy would be to determine whether others have found similar results and how they resolved the issues.

The format of parts of this chapter has been adapted from *A Student Handbook for Writing in Biology*, Chapter 2. Used with permission from the publishers, Sinauer Associates, Inc. and W. H. Freeman & Co.

72 hits in WoK, most of them related to the physiological effects on the pitchers' arms. Many references addressed the colloquial use "throwing a curveball," meaning to misguide or deceive.

Use the Same Keywords in a Different Database or Search Engine

If you are not having any success with different keyword combinations in one search engine or database, try a different one. Google Scholar and Scirus search the entire Web and may find links to published journal articles on scientists' homepages or course websites. These media are not included in Engineering Village or Web of Knowledge database searches. Take advantage of the resources at your disposal. Remember that once you find that one good journal article, it will be much easier to find more (see the section "Finding Related Articles").

EVALUATING SEARCH RESULTS

After you type a keyword string into the search box, the search engine goes to work. The result is a page that lists the records by publication date (most recent first), relevance, or another criterion. Each journal article record contains the article title, the authors' names, the name of the journal, the volume and issue numbers, the pages, and the publication date. The title will determine whether you want to read the abstract. After having read the abstract, you will decide whether you want to read the entire paper. This iterative process is summarized in Figure 2.4.

The formats of results pages for WoK and EV differ slightly, but both contain the same basic information about the journal articles (Figure 2.5). You will need this information when you cite the article in your lab report or another written communication (see the section "Common Engineering Citation Formats" later in this chapter).

Skim through the titles of the first 20 records. If the titles seem to be unrelated to your topic, start a new search with different keywords using the strategies described previously (see the section "Choose Effective Keywords"). If a title seems promising, click it to open a page that contains the abstract (Figure 2.6). On the basis of the title and the abstract, decide whether you want to read the entire article.

Finding Related Articles

Once you have found a good article, Web of Knowledge makes it easy to find related articles. In the **Related Records** section on the right side of the screenshot in Figure 2.6, papers are listed, which cite references that were also cited in the article. Common references indicate that the authors were pursuing a similar research topic. In the **References** section, you can view the references listed in the article. Browsing the list allows you to find related papers with a slightly different focus. Clicking

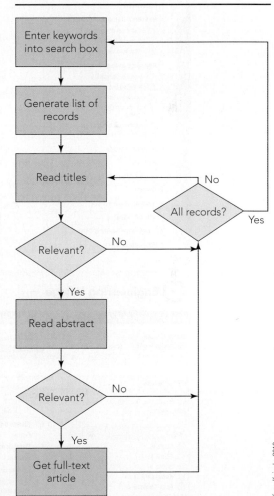

from Knisely, 2013

FIGURE 2.4 | Evaluating database or search engine results is an iterative process

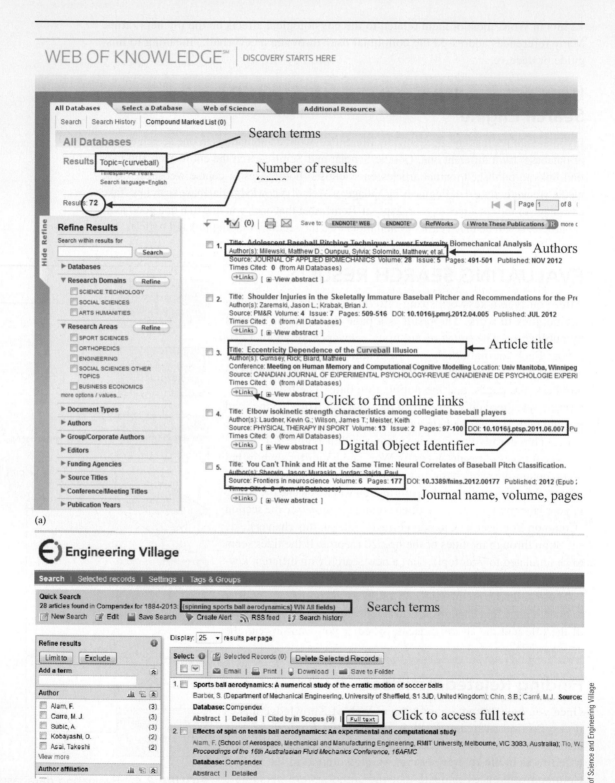

FIGURE 2.5 | Results page from (**a**) Web of Science for *curveball* and (**b**) Engineering Village for *spinning sports ball aerodynamics*

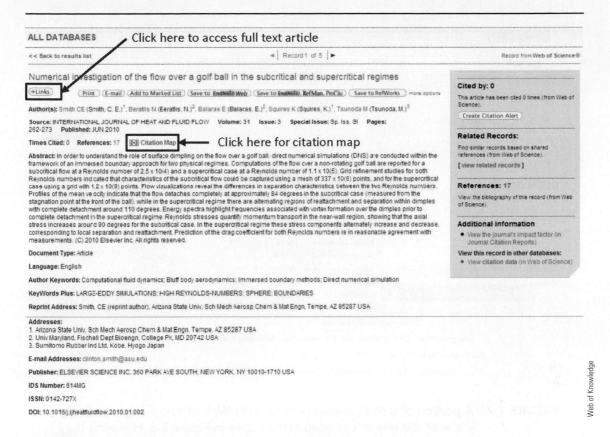

FIGURE 2.6 | Detailed record from the Web of Knowledge showing the abstract, a link to the full-text article, a link to the citation map, and links to related articles

the **Citation Map** button lets you generate forward and backward citation maps, as shown in Figure 2.7. Forward citation maps show more recent papers in which the present article was cited, and backward citation maps display links to articles cited in the present article. Hovering over an author's name on the citation map opens a drop-down box that displays the full citation for that paper. Clicking on an author's name displays the full citation for that referenced paper in the bottom right corner of the page. Web of Knowledge permits citation mapping for two forward and two backward generations.

Finding a **review article** is equivalent to hitting the jackpot. **Review articles** are secondary references that summarize the findings of all major journal articles on a specific topic since the last review. You can find background information, the state of current knowledge, and a list of the primary journal articles authored by researchers who are working on this topic. If you are unable to find a relevant review article in a database, go directly to the Annual Reviews website <http://www.annualreviews .org/> and search for your topic. If you find a promising review article on this website, you may be able to obtain a copy through your academic library.

Most of the article databases and search engines also have an **Advanced Search** option. Advanced search makes it possible for you to limit your search by specifying one or more authors, publication years, journals, and other criteria.

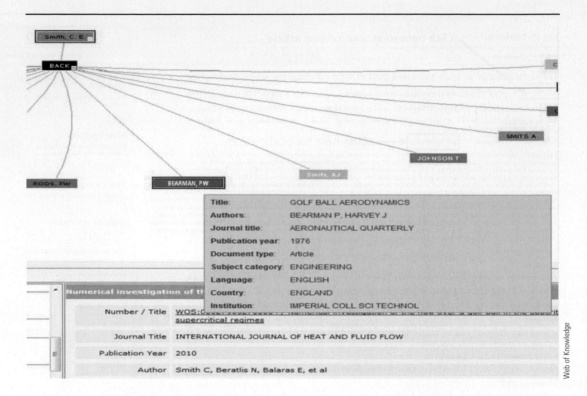

FIGURE 2.7 | A portion of a citation map generated in Web of Knowledge, showing some of the references cited in the paper in Figure 2.6. Hovering the cursor over one of the name blocks (outlined in red) produces a dropdown menu with the details of the referenced paper (see the box with GOLF BALL AERODYNAMICS in the first line).

Obtaining Full-text Articles

If the title and the abstract of an article sound promising, you will want to obtain the full-text article. The WoK and EV abstract pages have links to full-text articles, which, unless they are identified as preprints, are the same as those found in the print version of the journal. To access the full-text article in WoK, click the →**Links** button on the left side below the article title (see Figure 2.5a).

Both the EV and WoK databases lead you to the journal publisher's homepage, where you can download the full-text article as a PDF or HTML file. PDFs preserve formatting, while HTML files contain hyperlinks that make it easy to access other references. Save the full-text article to your computer or another storage medium to read later. Copy the URL and note the download date, using a program like Notepad or manually on paper because you may need this information when citing the source.

Although the abstract is usually free, some publishers charge a fee to access the full-text article. Fortunately, academic libraries and institutions often purchase subscriptions so that faculty, staff, and students have free access to many electronic journal articles. If your library does not have a subscription and you are not in a hurry to get the article, you may be able to use interlibrary loan. **Interlibrary loan** is a way for a library to borrow or obtain materials that it does not own from another library or organization. Ask your reference librarian about this service.

COMMON ENGINEERING CITATION FORMATS

In the engineering community, there are almost as many variations in citation style as there are journals. Many of these citation systems can be found through a search of the Web for "engineering citation formats" or at a site such as the Lehigh University (2012) website. Professional societies such as the American Institute of Chemical Engineers (AIChE), the American Society of Civil Engineers (ASCE), the American Society of Mechanical Engineers (ASME), the Association for Computing Machinery (ACM), and the Institution of Electrical and Electronics Engineers (IEEE) also have their own citation formats. Some societies have multiple formats depending on specific journals and conferences. The APA and the Chicago (name-year) formats are used in some technical publications, usually with the omission of page numbers in the in-text citation.

The life sciences most commonly and almost uniformly use the Council of Science Editors (CSE) formats. The humanities and social sciences use MLA or Chicago style (citation-sequence) reference formats. The citation system you use will depend on your instructor's preference or on the format specified by the particular organization or journal publishing your work. This section provides an overview of two of the most common systems: the citation-sequence system and the name-year system.

The Citation-Sequence System

The *in-text citation* consists of a number in square brackets or parentheses following the cited sentence (some journals require this number to be a superscripted endnote, but never a footnote). The first reference cited is number 1, the second reference cited is number 2, and so on. On the page of references that follows the Discussion or Conclusion section, the sources are listed in **numerical order**, and the full reference is given. The order of the elements in the *end reference* depends on which professional society's format you use, but in any given reference list, keep the order consistent.

Examples of Citation   Three possible examples of in-text citation are as follows:

IMPLICIT CITATION		EXPLICIT CITATION
…was confirmed in a previous study [1].	*or*	…was confirmed by Rice [1].
…was confirmed in a previous study (*1*).	*or*	…was confirmed by Rice (*1*).
…was confirmed in a previous study.[1]	*or*	…was confirmed by Rice.[1]

(When a superscript is used for the citation, as in the last example above, the number is placed *after* the period ending the sentence. The bracketed numbers, as in the first two examples, are inserted *before* the period.)

The corresponding examples of end references for each of these in-text citation formats are as follows:

[1] Rice W, 1965, "An analytical and experimental investigation of multiple disk turbines," *J. Eng. Power* 87, 29–36.   **[ASME citation format]**

1.   **Rice, W**., "An analytical and experimental investigation of multiple disk turbines," *J. Eng. Power*, **87**, pp. 29–36 (1965).   **[AIChE citation format]**

[1] W. Rice, "An analytical and experimental investigation of multiple disk turbines," *J. Eng. Power* vol. 87, pp. 29–36, 1965.   **[IEEE citation format]**

Note that in the "big picture view" all end references contain the same information. The differences are in the punctuation, line spacing, font face (bold, italic, or regular), and order in which the information is presented.

The Name-Year System

The *in-text citation* consists of author(s) and year. The citation may be given in parentheses following the cited sentence (parenthetical implicit citation), or the author(s) may be the subject of the sentence (direct explicit citation). In the *end references*, the sources are listed in **alphabetical order** according to the first author's last name.

Examples of Citation   If the paper has one or two authors and is cited parenthetically, list the last name(s) followed by a comma and the year of publication:

> …was presented in a previous study (Romanin and Carey, 2011).

> (When a parenthetical citation is used in the name-year system and occurs at the end of the sentence, as above, the parenthetical citation is inserted just *before* the period that ends the sentence.)

If the citation is explicit, cite the last name(s) followed by the year of publication in parentheses:

> Romanin and Carey (2011) suggest that disk turbines with efficiencies as high as 80% might be possible…

If the paper has three or more authors, cite only the last name of the first author, then write the Latin phrase et al. followed by a comma, and then give the year of publication. In science publications, the English "and others" is preferred to et al. (Council of Science Editors Manual, 2006).

> …was examined in a previous study (Kandlikar et al., 2005).

Note that some citation formats require the listing of all authors in the first in-text citation and then the use of the first author's name followed by et al. in subsequent citations of the same source.

If the same author or group of authors has published more than one paper in the same year, add a letter after the year:

> …was described by Anami and Ishii (1998a, 1998b).

These letters would also appear in the end references, with the 1998a reference first, followed by the 1998b reference, and so on.

The name-year system has the advantage that people working in the field will know the literature and, upon seeing the authors' names, will understand the reference without having to check the reference list. With the citation-sequence system, the reader must turn to the reference list at the end of the paper to gain the same information. The disadvantage of the name-year system is that a list of names and dates disrupts the flow of the text more than a shorter list of sequential numbers. Regardless of which system you choose, use it consistently throughout your paper.

When you use the citation-sequence system, it is prudent to cite sources using the name-year system in early drafts. Once you finish the revisions, you can execute a sequential search-and-replace for each new reference you encounter. In other words, the first reference cited would be designated *1*; using the "replace all" command, you would search for all instances of this name-year and replace it with the number *1*. For this approach to be successful, each source must be cited consistently.

An even easier way to manage citations and references is to use a program such as RefWorks, ProCite, EndNote, EndNote Web, or Reference Manager. Once your report is finished, you can have your citations and reference list generated automatically in whichever formatting style you choose.

Personal Communication

Unpublished information obtained during a discussion or a lecture should be acknowledged when you use it in your written work. The in-text citation includes the authority, the date, and the words "personal communication" or "unreferenced."

Example of Citation

…may be explained by considering parametric excitation of the arm (C. Peterson, personal communication, 25 February 2012).

It is **not necessary** to include personal communications in the References section.

END REFERENCE FORMAT

There are a variety of end reference formats; they differ mostly in punctuation and in the order of the elements. Tables 2.5 to 2.7 give examples of the end reference format for seven different types of publications in each of the following systems:

- ASME (2011) Citation-Sequence System
- ASCE (2012) Name-Year System
- IEEE (2006) Citation-Sequence System
- AIChE (2009) Citation-Sequence System
- APA (2012) Name-Year System
- Chicago Manual (2011) Name-Year System.

Information about referencing other types of publications can be found at each technical society's website.

Format for a Journal Article

Journal articles are a commonly cited form of publication in technical reports. In the following paragraphs, the ASME Citation-Sequence System and the ASCE Name-Year System will be used to illustrate the fundamental differences in the end reference format between these two systems.

Journal Article General Format

C-S   Number of the citation. All authors' names (LastName, F.M.) separated by commas with *and* before the last author, Year, "Article title," *Journal title*, volume number (issue number) pp. inclusive pages.   **[ASME format]**

N-Y   All authors' names (LastName, F.M.) separated by commas with *and* before the last author. (Year of publication). "Article title." *Journal title*, volume number (issue number), inclusive pages.   **[ASCE format]**

Some N-Y systems use a hanging indent to separate the individual references.

TABLE 2.5 | Sample end references using ASME (2011) and ASCE (2012) formats

Type of Publication	ASME (2011): Citation-Sequence System	ASCE (2012): Name-Year System
Journal Article	[1] Alaways, L. W., Mish, S. P., and Hubbard, M., 2001, "Identification of Release Conditions and Aerodynamic Forces in Pitched-Baseball Trajectories," *Journal of Applied Biomechanics*, 17(1) pp. 63–76.	Alaways, L. W., Mish, S. P., and Hubbard, M. (2001). "Identification of release conditions and aerodynamic forces in pitched-baseball trajectories." *Journal of Applied Biomechanics*, 17(1), 63–76.
Conference Proceedings	[2] Kensrud, J. R., and Smith, L. V., 2010, "In Situ Drag Measurements of Sports Balls," *8th Conference of the International Sports Engineering Association*, ISEA, July 12–16, 2010, Elsevier Ltd, Vienna, Austria, 2, pp. 2437–2442.	Kensrud, J. R., and Smith, L. V. (2010). "In situ drag measurements of sports balls." *8th Conference of the International Sports Engineering Association*, ISEA, July 12–16, 2010, Elsevier Ltd, Vienna, Austria, 2437–2442.
Book	[3] Knisely, K., 2013, *A Student Handbook for Writing in Biology*, 4th ed. Sinauer Assoc. Inc., Sunderland, MA.	Knisely, K. (2013). *A student handbook for writing in biology.* 4th ed. Sinauer Assoc. Inc., Sunderland, MA.
Edited Book	[4] Kolkman, P. A., 1979, "Development of Vibration-Free Gate Design: Learning from Experience and Theory," *IAHR/IUTAM Symposium on Practical Experiences with Flow-Induced Vibrations, Karlsruhe, West Germany*, Naudascher, E. and Rockwell, D., eds., Springer-Verlag, Berlin, pp. 351–385.	Kolkman, P. A. (1979). "Development of vibration-free gate design: Learning from experience and theory," *IAHR/IUTAM Symposium on Practical Experiences with Flow-Induced Vibrations, Karlsruhe, West Germany*, Naudascher, E. and Rockwell, D., eds., Springer-Verlag, Berlin, 351–385
Reports	[5] U.S. Bureau of Reclamation, 1996, "Forensic Report: Spillway Gate 3 Failure, Folsom Dam," Technical Service Center, Denver, CO.	U.S. Bureau of Reclamation. (1996). "Forensic report: spillway gate 3 failure, Folsom Dam," Technical Service Center, Denver, CO.
Theses and Dissertations	[6] Siegel, N. P., 2003, "Development and Validation of a Computational Model for a Proton Exchange Membrane Fuel Cell," Ph.D. Dissertation, Virginia Polytechnic Institute and State University, Blacksburg, VA.	Siegel, N. P. (2003). "Development and validation of a computational model for a proton exchange membrane fuel cell," Ph.D. Dissertation, Virginia Polytechnic Institute and State University, Blacksburg, VA.
Web Pages	[7] Roig, M. 2006, "Avoiding Plagiarism, Self-plagiarism, and Other Questionable Writing Practices: A Guide to Ethical Writing," <http://facpub.stjohns.edu/~roigm /plagiarism/Index.html> (accessed January 18, 2012).	Roig, M. (2006) "Avoiding plagiarism, self-plagiarism, and other questionable writing practices: A guide to ethical writing," <http://facpub.stjohns .edu/~roigm/plagiarism/Index.html> (Jan. 18, 2012).

TABLE 2.6 | Sample end references using IEEE (2006) and AIChE (2009) formats

Type of Publication	IEEE (2006): Citation-Sequence System	AIChE (2009): Citation-Sequence System
Journal Article	[1] L. W. Alaways, S. P. Mish and M. Hubbard, "Identification of release conditions and aerodynamic forces in pitched-baseball trajectories," *Journal of Applied Biomechanics*, vol. 17, pp. 63–76, 2001.	1. **Alaways, L.W.**, et al., "Identification of release conditions and aerodynamic forces in pitched-baseball trajectories," Journal of Applied Biomechanics, 17, pp. 63–76 (2001).
Conference Proceedings	[2] J. R. Kensrud and L. V. Smith, "In situ drag measurements of sports balls," *in 8th Conference of the International Sports Engineering Association,* ISEA, July 12, 2010–July 16, 2010. Vienna, Austria: Elsevier, 2010, pp. 2437–2442.	2. **Kensrud J.R., and L.V. Smith**, "In situ drag measurements of sports balls," presented at the 8th Conference of the International Sports Engineering Association, ISEA, Elsevier, Vienna, Austria, pp. 2437–2442 (July 2010).
Book	[3] K. Knisely, *A Student Handbook for Writing in Biology.* Sunderland, MA: Sinauer Assoc., 2013.	3. **Knisely, K.** "A student handbook for writing in biology," 4th ed., Sinauer Assoc. Inc., Sunderland, MA; (2013).
Edited Book	[4] P. A. Kolkman, "Development of vibration-free gate design: Learning from experience and theory," in *IAHR/IUTAM Symposium on Practical Experiences with Flow-Induced Vibrations, Karlsruhe, West Germany,* (E. Naudascher and D. Rockwell, Eds.). Berlin: Springer-Verlag, 1979, pp. 351–385.	4. **Kolkman, P.A.**, "Development of vibration-free gate design: Learning from experience and theory," in IAHR/IUTAM Symposium on Practical Experiences with Flow-Induced Vibrations, Karlsruhe, West Germany," E. Naudascher and D. Rockwell Eds., Springer-Verlag, Berlin, pp. 351–385 (1979).
Reports	[5] *Forensic Report: Spillway Gate 3 Failure, Folsom Dam.* U.S. Bureau of Reclamation, Technical Service Center, Denver, CO, 1996.	5. *U.S. Bureau of Reclamation*, "Forensic report: spillway gate 3 failure, Folsom Dam," USBR, Technical Service Center, Denver, CO (1996).
Theses and Dissertations	[6] N. P. Siegel, "Development and validation of a computational model for a proton exchange membrane fuel cell." Ph.D. Dissertation, Dept. Mech. Eng. Virginia Polytechnic and State University, Blacksburg, 2003.	6. **Siegel, N.P.** "Development and validation of a computational model for a proton exchange membrane fuel cell," thesis, presented to Virginia Polytechnic and State University in partial fulfillment of the requirements for the degree Doctor of Philosophy (2003).
Web Pages	[7] M. Roig. "Avoiding plagiarism, self-plagiarism, and other questionable writing practices: A guide to ethical writing." Internet: <http://facpub.stjohns.edu/~roigm/plagiarism/Index.html>, 2006 [January18, 2012].	7. **Roig, M.**, "Avoiding plagiarism, self-plagiarism, and other questionable writing practices: A guide to ethical writing," available via <http://facpub.stjohns.edu/~roigm/plagiarism/Index.html>. (2006) [Accessed 1/18/2012.]

TABLE 2.7 | Sample end references using APA (2012) and Chicago N-Y (2011) formats

Type of Publication	APA (2012): Name-Year System	Chicago (2011): Name-Year System
Journal Article	Alaways, L. W., Mish, S. P., & Hubbard, M. (2001). Identification of release conditions and aerodynamic forces in pitched-baseball trajectories. *Journal of Applied Biomechanics, 17*(1), 63–76.	Alaways, L. W., S. P. Mish, and M. Hubbard. 2001. Identification of release conditions and aerodynamic forces in pitched-baseball trajectories. *Journal of Applied Biomechanics* 17 (1): 63–76
Conference Proceedings	Kensrud, J. R., & Smith, L. V. (2010). In situ drag measurements of sports balls. *8th Conference of the International Sports Engineering Association, ISEA, July 12–July 16, 2010, 2,* 2437–2442. Vienna, Austria: Elsevier Ltd.	Kensrud, Jeffrey R., and Lloyd V. Smith. 2010. In situ drag measurements of sports balls. Paper presented at 8th Conference of the International Sports Engineering Association, ISEA, July 12–July 16, 2010. Vienna, Austria: Elsevier.
Book	Knisely, K. (2013). A student handbook for writing in biology (4th ed.). Sunderland, MA: Sinauer Assoc. Inc.	Knisely, Karin. 2013. *A Student Handbook for Writing in Biology*. 4th ed. Sunderland, MA: Sinauer Assoc. Inc.
Edited Book	Kolkman, P. A. (1979). Development of vibration-free gate design: Learning from experience and theory. In E. Naudascher, & D. Rockwell (Eds.), IAHR/IUTAM Symposium on Practical Experiences with Flow-induced Vibrations, Karlsruhe, West Germany (pp. 351–385). Berlin: Springer-Verlag.	Kolkman, P. A. 1979. Development of vibration-free gate design: Learning from experience and theory. In *IAHR/IUTAM Symposium on Practical Experiences with Flow-induced Vibrations, Karlsruhe, West Germany*, eds. Eduard Naudascher, Donald Rockwell, 351–385. Berlin: Springer-Verlag.
Reports	Forensic report: Spillway gate 3 failure, Folsom Dam. (1996). Technical Service Center, Denver, CO: U.S. Bureau of Reclamation.	Forensic report: Spillway gate 3 failure, Folsom Dam. 1996. Technical Service Center, Denver, CO: U.S. Bureau of Reclamation.
Theses and Dissertations	Siegel, N. P. (2003). *Development and validation of a computational model for a proton exchange membrane fuel cell.* (Unpublished doctoral dissertation). Virginia Polytechnic Institute and State University, Blacksburg, VA.	Siegel, N. P. 2003. Development and validation of a computational model for a proton exchange membrane fuel cell. Ph.D., Virginia Polytechnic Institute and State University, Blacksburg, VA.
Web Pages	Roig, M. (2006). Avoiding plagiarism, self-plagiarism, and other questionable writing practices: A guide to ethical writing. Retrieved January 18, 2012 from <http://facpub.stjohns.edu/~roigm/plagiarism/Index.html>.	Roig, M. Avoiding plagiarism, self-plagiarism, and other questionable writing practices: A guide to ethical writing. 2006 [cited 01/18/2012]. Available from <http://facpub.stjohns.edu/~roigm/plagiarism/Index.html>.

Journal Article with Two Authors

C-S [9] Neal, R. J., and Hubinger, L. M., 1990, "The effect of groove shape on the post impact kinematics of golf balls," *Australian Journal of Science and Medicine in Sport (AJSMS)*, 22(2), 39–43. **[ASME format]**

N-Y Neal, R. J., and Hubinger, L. M. (1990). "Effect of groove shape on the post impact kinematics of golf balls," *Australian Journal of Science and Medicine in Sport (AJSMS)*, 22(2), 39–43. **[ASCE format]**

Journal Article with Three or More Authors

In most name-year systems, all authors are listed in the end reference. In the AIChE format, when there are two authors, both authors are listed; however, when there are three or more authors, just the first author's name followed by et al. is used (see Table 2.6). In the APA format, the first six authors are listed, followed by et al. if there are more than six.

C-S [5] Always, L. W., Mish, S. P., and Hubbard, M., 2001, "Identification of Release Conditions and Aerodynamic Forces in Pitched-Baseball Trajectories," *Journal of Applied Biomechanics*, 17(1) pp. 63–76. **[ASME format]**

N-Y Always, L. W., Mish, S. P., and Hubbard, M. (2001). "Identification of release conditions and aerodynamic forces in pitched-baseball trajectories." *Journal of Applied Biomechanics*, 17(1), 63–76. **[ASCE format]**

Unpublished Laboratory Exercise

Unpublished material is usually not included in the References section. If your instructor asks you to cite the laboratory exercise instructions in your laboratory report, however, the format of the end reference could look like this:

C-S Number of the citation Author (Omit if unknown). Year, "Title of lab exercise," Course number, Department, University. **[ASME format]**

N-Y Author (Omit if unknown). (Year). "Title of lab exercise." Course number, Department, University. **[ASCE format]**

Example:

C-S 1. Knisely, C.W., 2009, "Transient Heat Conduction," MECH 312, Mechanical Engineering Department, Bucknell University. **[ASME format]**

N-Y Knisely, C.W. (2009). "Transient Heat Conduction." MECH 312, Mechanical Engineering Department, Bucknell University. **[ASCE format]**

When using a name-year system and no author's name is provided, you may list the reference as Anonymous (Date), or list the publishing organization, if known, as in ASME (2013), or use a shortened form of the title, as in Future of Geothermal Energy (2006).

Sources from the Internet and the World Wide Web

In the previous discussion of print sources, you learned that the in-text citation and end reference format differs for journal articles and other types of publications. These format differences also apply to online publications. For an online journal article, you should be able to locate the names of the authors, a title, the journal name, a date of publication, the volume and issue number, the extent (number of pages or similar), and possibly a digital object identifier (DOI). Besides this basic information, the CSE Manual (2006) recommends that you provide two additional items when your reference comes from the Internet: the URL (uniform resource locator) and the date accessed.

When URLs are used in text, they are enclosed in angle brackets (< >) to distinguish them from the rest of the text. (URLs can also be printed in color, in which case the angle brackets are not required.) Every character in a URL is significant, as are spaces and capitalization. Very long URLs can be broken before a punctuation mark (tilde ~, hyphen -, underscore _, period ., forward slash /, backslash \, or pipe |). The punctuation mark is then moved to the next line, as in the following example:

<http://mitei.mit.edu/publications/reports-studies
/future-natural-gas>

If you cut and paste such a URL from an electronic source, be aware that the line feed may be entered as an extra space in the URL, and the browser will not find the correct web page. Remove the space before the punctuation that begins the continued line and your browser will then find the correct web page.

A description of how to cite all possible Internet sources is beyond the scope of this book. Instead, guidelines and examples will be given for frequently encountered formats, such as journal articles and homepages. See Patrias (2007) for a comprehensive discussion of Internet citation formats along with many examples.

Journal Articles

The *in-text citation* for an online journal article is exactly the same as that for a printed journal article (see "Common Engineering Citation Formats"). A good approach for writing the *end reference* of an online journal article is to first locate the information you would need for a printed journal article and then add the Internet-specific items (CSE Manual, 2006). Choose one of the two systems described above—name-year or citation-sequence—and position the elements accordingly.

The **general format** for a *printed* end reference in the ASCE Name-Year System, including punctuation, is as follows:

> Author(s). (Date of publication). "Title of article." *Title of journal*, volume(issue), Inclusive page numbers

We were not able to determine if ASCE has a specific end reference format for online journal articles. Including the URL and the date accessed in the reference would indicate that the article came from the Internet, as shown below in boldface.

> Author(s). (Date of publication). "Title of article." *Title of journal*, volume(issue), Inclusive page numbers. **Available from URL (Date accessed)**

A screenshot of an online journal article webpage is shown in Figure 2.8, and the elements required for citation are labeled. The corresponding end reference with Internet-specific information in both C-S and N-Y formats is as follows:

C-S   1. Alexander, K.V., and Giddens E.P., 2008, "Microhydro: Cost-effective, modular systems for low heads, " *Renewable Energy* 33 (2008), pp. 1379–1391. Available from <http://www.sciencedirect.com/science/article/pii/S0960148107002364> (January 21, 2012).

N-Y   Alexander, K.V., and Giddens E.P. (2008). "Microhydro: Cost-effective, modular systems for low heads," Renewable Energy 33, 1379–1391. Available from <http://www.sciencedirect.com/science/article/pii/S0960148107002364> (January 21, 2012).

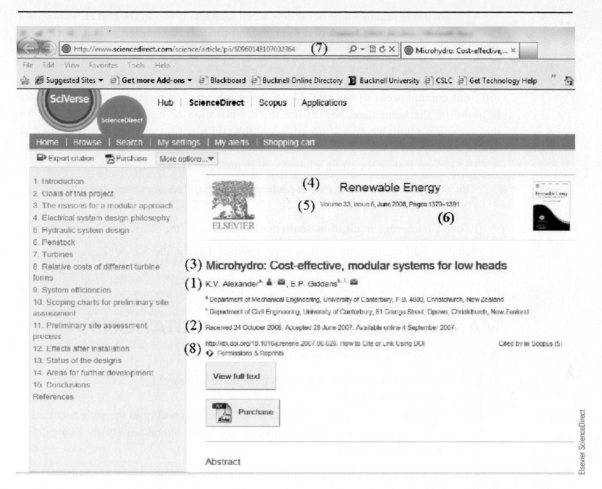

FIGURE 2.8 | Webpage citation for an online journal article. The basic information needed to cite a journal article includes (1) authors, (2) date of publication, (3) article title, (4) journal title, (5) volume and issue number (if given), and (6) inclusive pages. In addition, for an online journal article, include (7) the URL and the date retrieved in the end reference. Some publishers allow the digital object identifier (8) to be used in place of (7).

References Including Digital Object Identifier (DOI)

Since 2007, the Digital Object Identifier (DOI) has become the preferred means of referencing online journal articles, reports, books, and other media (Purdue OWL, 2012). The DOI provides a permanent link to an online source even when the URL changes.

An example of an end reference in APA format for an article from an online periodical with an assigned DOI is as follows:

Author, A. A., & Author, B. B. (Date of publication). Title of article. Title of Journal, volume number, page range. doi:0000000/000000000000

Anami, K., Ishii, N., & Knisely, C.W. (2012). Pressure induced by vertical planar and inclined curved weir-plates undergoing streamwise rotational vibration. Journal of Fluids and Structures, 29, 35-49. doi:10.1016/j.jfluidstructs.2011.11.007

APA no longer requires a retrieval date when a DOI is given in the end reference, and the date needs to be included only if the cited information is expected to change (Purdue OWL, 2012).

Homepages
A **homepage** is the main page of a website, which provides links to different content areas of the site. Most of the information required to cite a website is found on the homepage. Make sure the organization or individual responsible for the website is reputable and, if possible, confirm information on the site using another source.

The general format for citing a homepage in **name-year format** is

> Title of Homepage [Internet]. Date of publication. Edition. Place of publication: publisher. Available from URL (date cited).

To cite a homepage in **citation-sequence format**, move the date after the publisher:

> Number of the citation. Title of Homepage [Internet]. Edition. Place of publication: publisher; date of publication. Available from URL [date cited].

MANAGING REFERENCES AND CITATIONS

Reference management software makes it possible to

- Build your own collection of references from database searches.
- Insert citations into a paper.
- Format both the in-text citation and the end reference according to the style required by your professor, company, or the journal in which you intend to publish. Many different formats are available.

Some of reference management software programs are commercially available (e.g., ProCite, EndNote, and Reference Manager) and others have free access as long as you are affiliated with a subscribing institution (e.g., RefWorks and EndNote Web).

Many engineers, scientists, and other scholars rely heavily on reference management software to organize all of their references. Students will appreciate the convenience and ease of use of these programs as well.

RefWorks

RefWorks is used here as an example of reference management software. The following instructions for RefWorks are presented only to raise your awareness of the possibilities. If you like what you see, ask your librarian or IT department if you have access to reference management software at your institution.

Create an Account
Go to the RefWorks login page found at <https://www.refworks.com/refworks2/?r=authentication::init>. Enter your institution's Group Code to log in. Then create an individual account by clicking **Sign up for an Individual Account** and entering your personal information. Now return to the RefWorks Login Center page and log in.

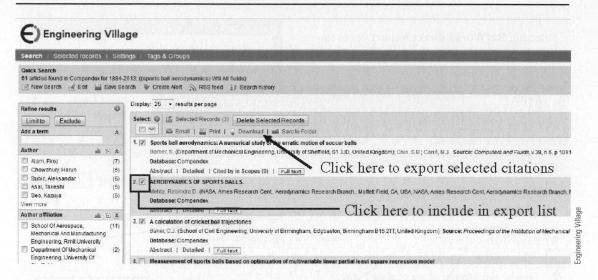

FIGURE 2.9 | Engineering Village citation list ready to download. Click the box in front of the desired citations and then click **Download**. Select **RefWorks direct import** as the output destination from the pop-up menu.

Download Citation from a Database   For this example, we will generate a reference list for *sports ball aerodynamics* in Engineering Village (EV). After entering EV and searching for *sports ball aerodynamics*, the results page should look similar to that in Figure 2.9.

1. Select the citations you wish to keep by clicking the box next to the citation number.
2. Click the **Download** link near the top center of the page.
3. Select **RefWorks direct import** from the dropdown menu on the **Download Selected Records** page and then click the **Download** button.
4. If the export manager cannot find your RefWorks account, you may be asked to log in again. Once logged into RefWorks you will see a message **Importing references, please wait . .** and then **Import completed - XX records imported**, where XX should be the number of records you selected.
5. In RefWorks, click **View Last Imported Folder** (bottom right corner of the **Import References** page, Figure 2.10).
6. Click **Create New Folder** button near the top left corner of the page and enter a name for the new folder. Click **Create**.
7. On the **Last Imported Folder** page, click the checkbox next to **Ref ID** or click the **All in list** button to select all records. Click the dropdown menu to display available folders. Select the new folder you just created. Your selected references will now be in a separate folder. With an increasing number of references, folders keep related references together.

Create an In-text Citation and End Reference List   When you write lab reports and other scholarly communications, you will cite the work of others and then, at

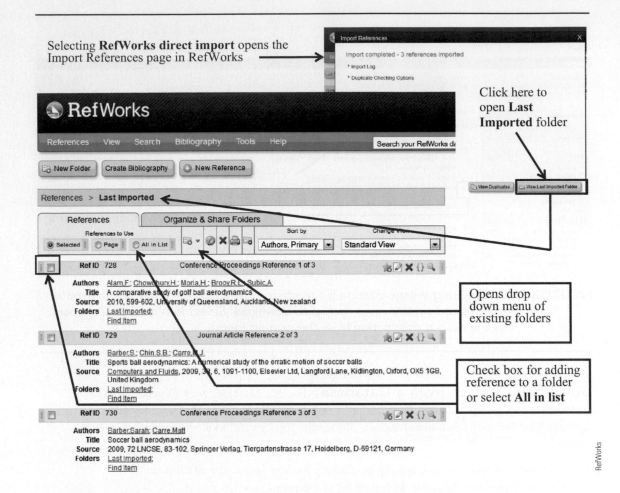

Selecting **RefWorks direct import** opens the Import References page in RefWorks

Click here to open **Last Imported** folder

Opens drop down menu of existing folders

Check box for adding reference to a folder or select **All in list**

RefWorks

FIGURE 2.10 | Downloaded references can be moved to folders to facilitate reference management

the end of your paper, list the full references. As described in the "Common Engineering Citation Formats" section, there are two commonly used citation systems:

- Citation-Sequence (C-S) and
- Name-Year (N-Y)

Different technical societies (e.g., ACM, AIChE, ASCE, ASME, and IEEE), as well as APA, Chicago (name-year) and CSE (Council of Science Editors), have slightly different styles within these systems. In the following examples, we will show you how to generate in-text citations and end references using the ASME and ASCE styles.

To cite sources saved in RefWorks, you first have to download a utility called Write-N-Cite. To do so, after opening RefWorks, click the **Tools** pulldown menu in RefWorks, select **Write-N-Cite**, and then download the version that is compatible with your operating system. Write-N-Cite only needs to be installed once, but Word must be closed when you run the executable installation program.

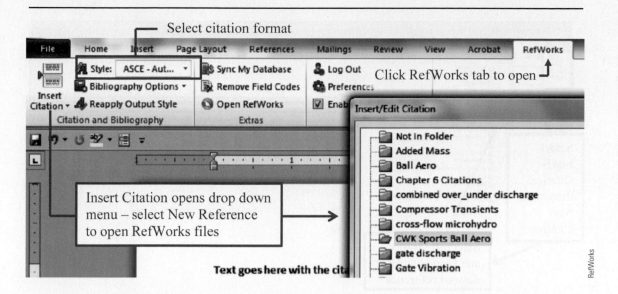

FIGURE 2.11 | Opening the folder in RefWorks that contains the records you wish to cite

Citation–Sequence (C-S)   In the citation-sequence system, in-text citations are numbered sequentially and the corresponding full reference is given in a numbered list at the end of the paper.

1. Open a Word document and begin typing your paper. Save the document after you come to a sentence in which you want to cite a reference.
2. Position the cursor either after the author's name (for an explicit direct citation), or after the sentence period to insert a citation in C-S format (for an implicit citation).
3. Launch **RefWorks** from the pulldown menu (Word 2003) or from the **RefWorks** add-in tab (Word 2007) at the top of the page (Figure 2.11).
4. Select a citation format from the dropdown menu in the **Style** box in the **Citation and Bibliography** group (see the example in the top left in Figure 2.11). Click **Insert Citation**, then click **Insert New**, and navigate to the folder that contains the reference you want to cite. Select the desired reference and then click **OK** (Figure 2.12).
5. In the MS Word document, position the cursor at the location of each subsequent citation and click **Insert Citation**. Select the reference to be cited and click **OK**.
6. While in earlier versions of Write-N-Cite it was necessary to save the Word document before generating the in-text citation and the end references list, in Write-N-Cite 4 this step is recommended but not required.
7. After all references have been cited in the text, position the cursor at the location desired for the end references and select **RefWorks** (→Citation and Bibliography) **| Bibliography Options | Create Bibliography**. If the group is not open, click the **RefWorks** tab again to reopen this group.
8. If you need to change the formatting style, select a different style in the **Style** box.
9. If you edit your document and need to add or remove citations, select **RefWorks** (→Citation and Bibliography) **| Bibliography Options | Remove Bibliography**. Then after adding or removing the citations, return to step 7, above, to re-create an updated end reference list.

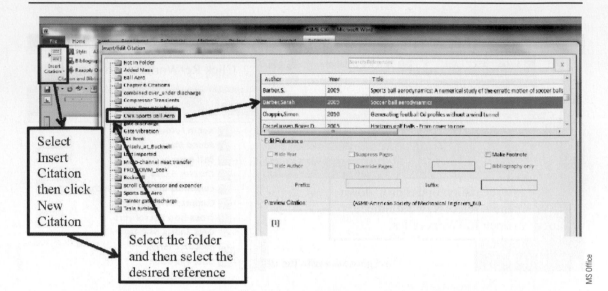

FIGURE 2.12 | To insert a citation, navigate to the folder containing the citation, select the citation, and click **OK** (bottom right corner of the dialog box). Write-N-Cite automatically inserts the in-text citation into your Word file.

In your Word document, the in-text citations are listed sequentially, and the information in the end references is in the correct order (Figure 2.13). Some minor editing may be required, but at least you will save time by not having to type a long reference list.

Here is an example of in-text citation using ASME format. The papers by Alaways et al. [1], Mehta [2], Oggiano and Stran [3], Barber and Carre [4] and Alam et al. [5] contain information on sports ball aerodynamics.

References
[1] Alaways, L. W., Mish, S. P., and Hubbard, M., 2001, "Identification of release conditions and aerodynamic forces in pitched-baseball trajectories," Journal of Applied Biomechanics, 17(1) pp. 63-76.
[2] Mehta, R.D., 1985, Aerodynamics of sports balls, Annual Review of Fluid Mechanics, 17, Annual Reviews Inc., Palo Alto, CA, USA, pp. 151-189.
[3] Oggiano, L., and Stran, L., 2010, "Aerodynamics of modern soccer balls,' 8th Conference of the International Sports Engineering Association, ISEA, July 12, 2010 - July 16, Elsevier Ltd, Vienna, Austria, 2, pp. 2473-2479.
[4] Barber, S., and Carre, M., 2009, "Soccer ball aerodynamics." Computational Fluid Dynamics for Sport Simulation, Springer Verlag, Heidelberg, Germany, pp. 83-102.
[5] Alam, F., Tio, W., Subic, A., and Watkins, S., 2008, "An experimental and computational study of tennis ball aerodynamics." Impact of Technology on Sport II, September 1, 2007, Taylor and Francis/Balkema, Singapore, pp. 437-442.

FIGURE 2.13 | In-text citation and end references using RefWorks with the ASME (C-S) format. Some editing of the reference list may be required.

Name-Year (N-Y)   In the name-year format, the in-text citation is given in the form of author and year. Remember, the number of authors determines the format of the citation:

- One author—Author's last name, Year
- Two authors—First author's last name and Second author's last name, Year
- Three or more authors—First author's last name followed by the words et al. (or *and others*) and year of publication. CSE prefers the English phrase.

The corresponding full references are listed alphabetically at the end of the paper, with all authors listed in the end reference list.

The procedure for generating a name-year in-text citation and an alphabetical list of references (as shown in Figure 2.14) is the same as that with the citation-sequence system, except parenthetical references are inserted before the sentence-ending period. It is also necessary to select a name-year citation system in the **Style** box (see Figure 2.11).

Some of the in-text citations and end references may require editing. Even so, the time saved by using RefWorks, rather than generating citations and references manually, is considerable.

Modifying RefWorks Output Style

If you are given a specific citation format that fits none of the predefined systems in RefWorks, you can modify one of the predefined systems.

Here is an example of in-text citation using ASCE format. The papers by Alaways et al. (2001), Mehta (1985), Oggiano and Stran (2010), Barber and Carre (2009) and Alam et al. (2008) contain information on sports ball aerodynamics.

References

Alam, F., Tio, W., Subic, A., and Watkins, S. (2008). "An experimental and computational study of tennis ball aerodynamics." *Impact of Technology on Sport II, September 1, 2007,* Taylor and Francis/Balkema, Singapore, 437-442.

Alaways, L. W., Mish, S. P., and Hubbard, M. (2001). "Identification of release conditions and aerodynamic forces in pitched-baseball trajectories." *Journal of Applied Biomechanics,* 17(1), 63-76.

Barber, S., and Carre, M. (2009). "Soccer ball aerodynamics." *Computational Fluid Dynamics for Sport Simulation,* Springer Verlag, Heidelberg, Germany, 83-102.

Mehta, R. D. (1985). "Aerodynamics of sports balls." Annual Review of Fluid Mechanics, 17, Annual Reviews Inc, Palo Alto, CA, USA, 151-189.

Oggiano, L., and Stran, L. (2010). "Aerodynamics of modern soccer balls." *8th Conference of the International Sports Engineering Association, ISEA, July 12, 2010 - July 16,* Elsevier Ltd, Vienna, Austria, 2473-2479.

FIGURE 2.14 | In-text citation and end references using RefWorks with the ASCE (N-Y) format. Some editing of the in-text citations and references may be required.

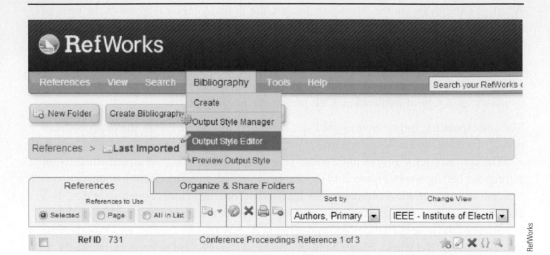

FIGURE 2.15 | To modify the in-text citation or full reference settings of a predefined style, select the style from the dropdown menu that appears after clicking **Output Style Editor**.

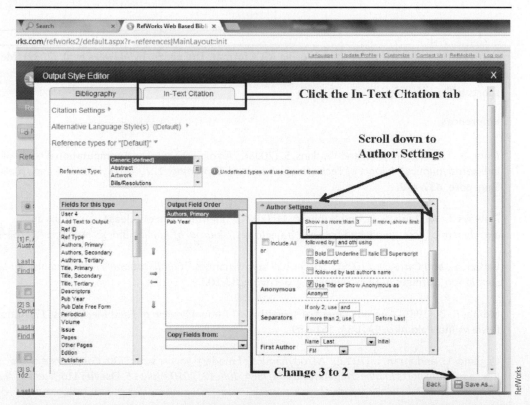

FIGURE 2.16 | Edit the in-text citation or bibliography format for the name-year system in the **Output Style Editor** dialog box in RefWorks. Select the **In-Text Citation** tab to change the in-text citation settings. Similarly, select the **Bibliography** tab to change the full reference settings. Save your changes using **Save As**.

- 7. Severn Embryonic Technologies Scheme
- 8. Next steps for Severn tidal power
- Environment results and Environmental Impact Assessment

The front matter has the expected executive summary, but it also provides instructions on how interested readers can respond to the recommendations of this feasibility study. The introductory material (Chapters 1 and 2) defines the problem for which a solution is sought. The body of the report (Chapters 3–7) presents an analysis of the constraints on building the planned facility. An assessment of how the project might move forward is given in Chapter 8. The environmental assessment data are provided in the supporting documentation at the end of the study. The Apparatus and Procedures section found in laboratory-based technical reports is not usually contained in a feasibility study, although some feasibility studies may have a separate Method of Approach section in its place.

ELEMENTS OF TECHNICAL REPORTS

In this section, we will describe the content of each element of a technical report. Knowing where to find certain kinds of information expedites a reader's acquisition of knowledge. Putting information where the audience expects to find it enables a writer to communicate information more effectively.

Title

The title is a **short, informative description of the essence of the paper**. It should contain the fewest number of words (commonly suggested in journals to be 40 characters or less, if possible) that still accurately conveys the content. Readers use the title to determine initial interest in the paper.

List of Authors

Only those people who played an active role in designing the experiment or planning the study, carrying it out, and analyzing the data are listed as authors.

Ancillary Front Matter

In a thesis or a long report, the title page may be followed by a preface, a page of acknowledgments, and the table of contents, a list of figures, and a list of tables, each on a separate page. The ancillary front material is usually numbered with italic lowercase Roman numerals. Journal articles do not generally contain front matter.

The first page of the actual manuscript may or may not repeat the title and the list of authors, but it invariably contains the Abstract or the Executive Summary. Even though this abstract page is page 1 of the manuscript, it is usually not numbered. Sequential pagination of the manuscript begins with page 2. Often the Abstract is on a separate page in a thesis or long report, but in journal articles it is followed immediately (on the same page) by the body of the paper.

Abstract (or Executive Summary)

The Abstract is a **summary of the entire report** in 250 words or less. It contains (1) an introduction (scope and purpose), (2) a short description of the methods, (3) results, and (4) conclusions. There are no literature citations or references to figures in the Abstract.

Section Headings

Sections of the report after the Abstract may be numbered or otherwise organized in a consistent hierarchy of titles and subtitles. Multilevel numbering is especially common in European publications, whereby sections are numbered with Arabic numbers and each sequentially lower subsection includes a decimal point followed by a number. Figure 3.2 shows an example of such a numbering scheme (from Singh and Nestmann, 2010).

A general rule is that if a section does not have at least two subsections, then it should not be subdivided at all. There are many variations of multilevel numbering, as you may have discovered when finding sources for the exercises in Chapter 2. Whether or not the section headings are numbered, the next section of the report after the Abstract is the Introduction.

Introduction

The Introduction concisely states what motivated the study, how it fits into the existing body of knowledge, and the objectives of the work. The Introduction consists of two primary parts:

Background or Historical Perspective on the topic
Primary journal articles and review articles, rather than other secondary sources such as textbooks and newspaper articles, are cited to provide the reader with direct access to the original work. Only the pertinent highlights are included, not a complete review of the literature. Inconsistencies, unanswered questions, or new questions that resulted from previous work set the stage for the present study. When you cite the work of others, use one of the two accepted citation systems, name-year or citation-sequence, as described in the section "Common Engineering Citation Formats" in Chapter 2.

Statement of Objectives of the Work
After the background material has been presented, the stage is set for the authors to present a clear, logical connection to what they set out to do in their study. The Introduction must contain an explanation of their goals.

Theoretical Analysis (if any)

If the study is either a numerical or an analytical study, or a combination of theory and experiment, the Theoretical Analysis section follows the Introduction. It provides the development of appropriate equations and/or numerical techniques. For analytical or numerical studies, the Solution section follows the Theoretical Analysis section and replaces the Apparatus and Procedures section. If a study contains both analytical and experimental components, the Solution section is usually combined with the Apparatus and Procedures section in one section with subheadings.

Mathematical expressions in technical reports are presented in a consistent style. Simple equations are often incorporated into the text, but more complicated mathematical

Nomenclature
1. Introduction
    1.1. Background
    1.2. Objectives and problem outline
2. The theoretical model
    2.1. Free vortex runner
    2.2. Main variables and their functional relationship
    2.3. Internal variables and their functional relationships
    2.3.1. Identifying the internal variables
    2.3.1.1. Expanding the dependent variables
    2.3.1.2. Expanding the independent variables
    2.3.1.3. Summary of the internal variables
    2.3.2. Functional relationship for the internal dependent variables
    2.3.2.1. Mean axial flow velocity
    2.3.2.2. Net tangential flow velocity
    2.3.2.3. Overall functional relationship for the internal variables
    2.3.3. Internal control variables
    2.3.3.1. System loss coefficient (k_{system})
    2.3.3.1.1. Upstream stationary loss coefficient ($k_{upstream}$)
    2.3.3.1.2. Runner loss coefficient (k_{runner})
    2.3.3.1.3. Draft tube coefficient ($k_{draft\ tube}$)
    2.3.3.2. Relative flow angle at runner exit (β_3)
    2.4. Construction of velocity triangles
    2.4.1. Inlet velocity triangle
    2.4.2. Exit velocity triangle
3. Means of solution
    3.1. Experimental test-rig
    3.2. Experimental runners
    3.3. Absolute flow conditions between guide vane exit and runner inlet
    3.4. Analysis of the internal variables
    3.4.1. Internal dependent variables
    3.4.2. Internal control variables
4. Results and discussion
    4.1. Runner A: exit tip angle changed from 85° to 77°
    4.1.1. Discussion on the behaviour of internal control variables
    4.1.1.1. Decrease of k_{system}
    4.1.1.2. Decrease of β_3
    4.1.2. Discussion on the behaviour of internal dependent variables
    4.1.2.1. Increase of c_x
    4.1.2.2. Increase of Δc_u
    4.2. Runner A: exit tip angle changed from 77° to 74°
    4.2.1. Discussion on the behaviour of internal control variables
    4.2.1.1. Decrease of k_{system} or k_{runner}
    4.2.1.2. Decrease of β_3
    4.2.2. Discussion on the behaviour of internal dependent variables
    4.2.2.1. Increase of c_x
    4.2.2.2. Increase of Δc_u
    4.3. Runner B: exit tip angle changed from 75° to 70°
    4.3.1. Discussion on the behaviour of internal control variables
    4.3.1.1. Decrease of k_{system} or k_{runner}
    4.3.1.2. Decrease of β_3
    4.3.2. Discussion on the behaviour of internal dependent variables
    4.3.2.1. Increase of c_x
    4.3.2.2. Increase of Δc_u
    4.4. Swirl effect and draft tube losses
    4.5. Comparisons with vortex runner
    4.6. Limitations of the theoretical model
    4.7. Limitations of the experimental study
5. Conclusions and recommendations

Acknowledgements
References

FIGURE 3.2 | An example of a multiple-level numbering system, extracted

Based on a technical study by Singh and Nestmann (2010)

expressions and any equation that will be subsequently referenced are usually centered on a separate line with a right-aligned sequential equation number in parentheses. The sidebar shows how to set tabs to format equations in this manner. Additional formatting conventions for mathematical expressions are shown in Table 3.3.

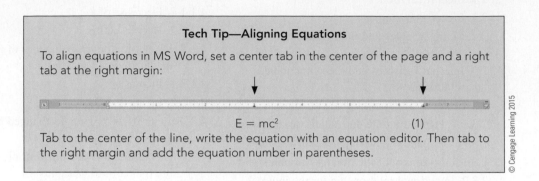

Tech Tip—Aligning Equations

To align equations in MS Word, set a center tab in the center of the page and a right tab at the right margin:

$$E = mc^2 \hspace{5cm} (1)$$

Tab to the center of the line, write the equation with an equation editor. Then tab to the right margin and add the equation number in parentheses.

TABLE 3.3 | Formatting in mathematical expressions

Expression and Examples	Usual Format	Unusual Format
Math functions sin, cos, ln, log, and exp	$\sin(\beta)$, $\cos(2\pi x/L)$ $\ln(r_o/r_i)$, $\log(v_1/v_2)$ $\exp(-t/\tau)$ $\exp(j\,\omega t)$	$\mathbf{\sin(\beta)}$, $\mathbf{\cos(2\pi x/L)}$, $\mathbf{\ln(r_o/r_i)}$, $\mathbf{\log(v_1/v_2)}$ $sin(\beta)$, $cos(2\pi x/L)$, $ln(r_o/r_i)$, $log(v_1/v_2)$ $\mathbf{\exp(-t/\tau)}$ or $exp(-t/\tau)$ $\mathbf{\exp(j\,\omega t)}$ or $exp(j\,\omega t)$
Variables and symbols a, x, y, z, t α, β, η, ξ	*italicized symbols* a, x, y, z, t α, β, η, ξ	Roman (nonitalicized) symbols a, x, y, z, t α, β, η, ξ
Vectors $\mathbf{F} = m\mathbf{a}$ or $\vec{F} = m\vec{a}$	**boldfaced vectors** $\mathbf{F} = m\mathbf{a}$ or $\vec{F} = m\vec{a}$	non-boldfaced vectors $F = ma$ (one vector component only) F = ma (symbols should be italicized)
Dimensionless parameters Reynolds number Re, Prandtl number Pr, and Strouhal number St	Re, Pr, St	*Re, Pr, St* **Re, Pr, St**
Quantities and units 10 kg or 30 sec	10 kg or 30 sec ↑ ↑ space space	*10g* or *30sec* *(italics and no space)* **10 g or 30 sec (bold)**

Apparatus and Procedures

The Apparatus and Procedures section describes, in full sentences and well-developed paragraphs, how the experiment was done. In the case of an analytical or numerical study, it would be replaced by a Solution section explaining how the analytical or numerical solution was obtained. In studies with both theoretical and experimental components, the section might be entitled Solution Methodologies with separate subsections for Theoretical Solution and Experimental Apparatus and Procedures.

In this section, regardless of its exact title, the author provides sufficient detail to allow another technically trained individual to repeat the analysis and/or experiment. A description of any specialized apparatus, associated instrumentation, data acquisition and sampling techniques, and any specialized measurement methodology are critical pieces of information that must be included. Time and location are not included unless they can be expected to affect the results (such as in a natural environment, in micro-gravity, etc.). Conventional instrumentation and measurement techniques that are common knowledge (familiar to the audience) are not explained. In some instances, authors may use references to earlier work to describe their research methods.

Results

The Results section is where the findings of the study or the experiment are summarized, without giving any explanations as to their significance (the "whys" are reserved for the Discussion section). A good Results section is expected to have two components:

• A *text* describing the results, which forms the body of this section
• One or more *visual elements (graphs or tables)* to help the reader visualize the data and process the results faster than from reading a lengthy description

Discussion and Conclusions

In the Discussion section, you will find the interpretation and possible explanations of the results. The author may:

• Summarize the results in a way that supports the conclusions
• Describe how the results relate to existing knowledge (literature sources)
• Describe inconsistencies in the data. This is preferable to concealing an anomalous result.
• Discuss possible sources of error
• Describe future extensions of the current work (not permitted in some journals)

Either in a separate section labeled Conclusions or at the end of the section labeled Discussion and Conclusions, an overview and interpretation of the impact of the study on the state of knowledge of the field is presented. This section is the "big picture" view of what your results mean.

Acknowledgments

If appropriate, the author may acknowledge people and organizations that facilitated the study. The Acknowledgments heading is usually not numbered. In longer reports, more extensive acknowledgments are placed in the front matter, as noted in the front matter description.

References

References are the outside sources that the authors consulted in preparing the paper. No one has time to return to a state of zero knowledge and rediscover known mechanisms and relationships. That is why engineers rely so heavily on information published by their colleagues. References are typically cited in the Introduction and Discussion sections of a technical report, in the Apparatus and Procedures section when procedures are modifications of earlier work, and as a basis for the Analysis section. The References heading is usually not numbered. The full references are usually listed in name-year or citation-sequence format, as described in the section "Common Engineering Citation Formats" in Chapter 2.

Appendices (if needed)

An appendix contains extra material that is *not essential* to the report but may be helpful to anyone trying to replicate the study. For example, a derivation that is outlined in the paper may be developed in detail in the appendix for the convenience of interested readers.

An appendix might also contain a tabulation of the data if it is expected that the data will be used by others as a benchmark test against which future analytical or numerical results will be compared. It is not customary to provide a table of the same data that are presented in figures in the paper.

Finally, *figures referred to in the text* are essential components of the report and so are placed in the body of the report, *not in an appendix*.

MERITS OF THE TECHNICAL REPORT FORMAT

As shown in the previous section, technical reports have a well-defined structure, and each section contains specific information. This format has benefits to both readers and writers. First, the standard format reflects the systematic approach engineers use to describe the solution to a problem. When you read technical reports, you gradually begin to assimilate the structure, which you will then adopt when you write your own reports. A systematic approach applied to the writing process often forces you to clarify what you know, and what you do not know, about the problem.

Second, the logic flow in a technical report aims to persuade readers of the validity of the procedures, results, and conclusions described in the paper. Reading, understanding, and assimilating the logic flow in technical reports will help you develop your own reasoning abilities, not just when you are required to prepare technical reports but in other areas of your life as well.

Third, observing and then developing the mental discipline required to write clear and concise technical papers will help put you on the fast track for promotion. As discussed in Chapter 1, promotions for engineers are strongly tied to their communication ability. Paraphrasing one engineering alumnus, "It has amazed me during the beginning of my career to see so many intelligent people be so unproductive because they do not know how to get their ideas across and put a plan into action."

Finally, even if your ultimate career path is not in a technical field, a logical and organized writing style is appreciated by busy people regardless of discipline.

STRATEGIES FOR READING TECHNICAL LITERATURE

Now that you have been introduced to the basic structure of a technical paper, you may have some idea where to find the information you need. Initially, however, you will probably find it difficult to read and understand technical reports because they are written for experts by experts in the field. To help you read technical reports, try following this strategy:

Determine the Topic

First, try to determine the topic of the report by reading the title and the abstract. What are the authors' objectives?

- To answer a specific question
- To explain observations
- To present a theoretical model of a process
- To determine the relationship between one or more variables or
- To accomplish something else

Acquire Background Information on the Topic

Read about the topic in your textbook. Since textbook authors generally write for a student audience, not a group of experts, your textbook is likely to be easier to read. See the subsequent section on "Strategies for Reading Your Textbook" for suggestions on reading engineering textbooks efficiently.

Read the Introduction

The Introduction is usually easier to follow than the Abstract. Skim the Introduction with the following questions in mind:

- Why did the author(s) carry out this work?
- What was previously known about the topic or problem?
- What are the objectives of the current work?

Skim the Discussion and Conclusion Sections

By skimming through the Discussion and Conclusion sections, you will gain a sense of what to look for as you read other sections of the paper. More experienced researchers possess enough background knowledge that they can assess the potential impact of the work after reading just the Discussion and Conclusion sections; some experts might read just the Conclusions. Do not worry if what you read is not readily understandable. Reading a technical paper is always a learning process, even for experienced readers.

Read the Results Section Selectively

Look at the visual elements (figures and tables) to determine what variables or parameters were studied (see the sidebar on the next page for the difference between variables and parameters). The independent parameter (the one the investigator manipulated) is plotted on the x-axis, and the dependent parameter(s) (the one that changes depending on the value of the independent parameter) is plotted on the y-axis.

What is a parameter?

Several definitions can be found online

Parameter—a computation from data values recorded—but it is not actually a data value recorded (Parameter, 2011a).

OR

pa·ram·e·ter/pə'ramitər/Noun

1. A numerical or other measurable factor forming one of a set that defines a system or sets the conditions of its operation.
2. A quantity whose value is selected for the particular circumstances and in relation to which other variable quantities may be expressed (Parameter, 2011b).

OR

One of a set of measurable factors, such as temperature and pressure, that defines a system and determines its behavior and are varied in an experiment (Parameter, 2011c).

EXAMPLE

In a drag force experiment, an engineer may vary the air velocity and measure the resulting drag force. The velocity would be identified as the independent **variable** and the drag force as a dependent **variable**. In plotting the data, the engineer may choose to present drag force as a function of velocity, as in Figure 3.3a.

However, with a bit more insight and training, the engineer may choose to represent the drag force in terms of a nondimensional drag coefficient and the velocity in terms of a nondimensional Reynolds number, as in Figure 3.3b. The drag coefficient and the Reynolds numbers are **parameters** associated with the **measured variables**. The value of this nondimensional representation is that it represents results for any geometrically similar airfoil of any size in any incompressible fluid.

The term **parameter** is used in this book to represent either a directly controlled or measured variable or a related dimensional or nondimensional representation calculated from the measured variable. In this text, parameter is the more general term that includes values calculated from a variable as well as the variable itself.

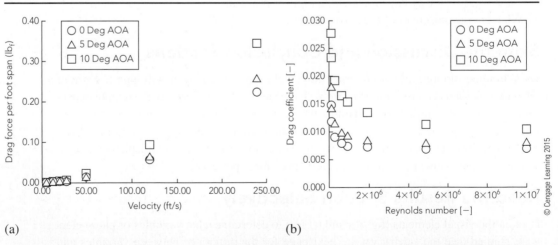

(a)   (b)

FIGURE 3.3 | (a) Drag force dependence on velocity for an NACA 0015 airfoil with a 0.5-ft chord in air at 70°F, and (b) drag coefficient for any NACA 0015 airfoil in any incompressible fluid showing dependence on Reynolds number.

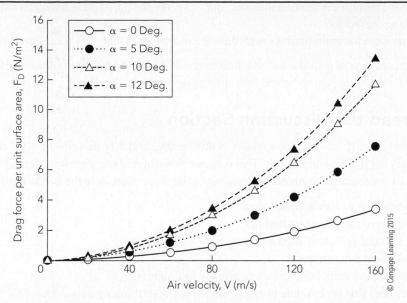

FIGURE 3.4 | Example of the drag force on an airfoil with increasing air velocity with the angle of attack (α) as a second independent controlled parameter

Data obtained with a constant value of a second independent controlled parameter may be plotted as a second series of data points, usually with a different symbol for each unique value of the second parameter. For example, drag force for an airfoil may be plotted as a function of velocity for four different angles of attack, as shown in Figure 3.4. Also look for parameters in column headings of tables.

There are two places to look for a qualitative description of each visual element:

- In the caption
- In the body of the Results section (text)

The caption states the main idea of the visual element to make it understandable for the reader. The topic sentence of the paragraph describing the visual element in the text does the same. Subsequent sentences in the paragraph provide details on what trends or findings the reader should notice in each graph or table.

Because experts in the field make up the target audience of professional journals, reports, and even handbooks, you may feel overwhelmed when you try to understand the results presented in the technical literature. You may have to read the passage several times on different days to allow your mind to digest the information. Iterative re-reading of secondary sources such as your textbook for background may be instructive.

On a first reading of the Results section, aim for a big picture overview. Look at the visual elements, read the description in the text, and then ask yourself the following questions:

- What were the independent, dependent, and controlled parameters?
- Are the parameters dimensional quantities, and, if so, which system of units (SI or U.S. Customary) is being used?

- Are the parameters nondimensional, and, if so, are the parameters ones you have seen in your textbooks?
- What were the main findings regarding the independent and dependent parameters?

If necessary, re-read the Introduction to recall the main objectives of the work. Try to understand the big picture before concerning yourself with the details.

Re-read the Discussion Section

Authors typically interpret their results in this section, and they describe how the results of the study led to their conclusions. This is where you can find out what was learned from the work. In your second, more careful reading of the discussion, note the following:

- Were the objectives accomplished?
- What were the important findings?
- Do the data presented support the authors' claims?
- Were there any surprises?
- What further work is necessary or already in progress?
- How does this paper relate to work you are doing or thinking about doing?

Skim the Apparatus and Procedures Section

Scan the subheadings (if present) and the topic sentence of each paragraph to identify the basic approach. Do not be concerned with the details.

Read the Report Several Times

Even experts must read technical reports several times before they understand the methods and the implications of the findings. Take notes the first time you read the full report, trying to sort out what is confusing. Consult your notes when you read the report again, and try to clarify what you did not understand the first time through. Each time you read the report, you will understand a little more.

STRATEGIES FOR READING TEXTBOOKS

The following strategies are based on the proposition that you cannot read a chapter in an engineering textbook just once and understand it completely. Repetition is a key ingredient in learning the material. Repetition not only provides you with multiple opportunities to be exposed to the material, but also gives you time to digest it. The basic approach is to read for organization and key concepts first, and then to fill in the details with each subsequent reading.

The two strategies described here work best with a chapter or section of text no longer than 25 to 30 pages. The first strategy is proposed by Counseling Services at the University of Victoria, Canada (Palmer-Stone, 2001).

1. Take no more than 25 minutes to
   - Read the chapter title, introduction, and summary (at the end of the chapter, if present)
   - Read the headings and subheadings

- Read the chapter title, introduction, summary, headings, and subheadings again
- Skim the topic sentence of each paragraph (usually the first sentence or the last sentence of the previous paragraph)
- Skim italicized or boldfaced words

2. Close your textbook. Take a full 30 minutes to
   - Write down everything you can remember about what you read in the chapter (make a "mind map"). Each time you come to a dead end, use memory techniques such as associating ideas from your reading to lecture notes or other life experiences; visualizing pages, pictures, or graphs; staring out the window to daydream; letting your mind go blank.
   - Figure out how all this material is related. Organize it according to what makes sense in your mind, not necessarily according to how it is organized in the textbook. Write down questions and possible contradictions to check on later.
3. Open your textbook. Fill in the blanks in your mind map with a different colored pencil.
4. Read the chapter again, this time normally. Make another mind map.

Another strategy is this:

1. Skim the chapter title, headings, and subheadings for an overview of the chapter content. Write down the headings and subheadings in the form of an outline.
2. Look at your outline and ask yourself the following questions:
   - What is the main topic of this chapter?
   - How do each of the headings relate to the topic?
   - How does each subheading relate to its heading?
3. Read each section, paying special attention to the topic sentence of each paragraph. At the end of each section, summarize the content in your own words. Answer the following questions:
   - What is the point?
   - What do I understand?
   - What is confusing?
4. If you read the assigned pages before the lecture, you can pay attention to the lecture content instead of frantically taking notes. Your instructor may provide slides for the lecture, or your textbook may come with a printed lecture notebook or a DVD with the figures. These aids allow you to spend more time listening and less time writing.
5. After the lecture, while the information is still fresh in your mind, re-read your notes on your reading. Ask yourself
   - What topics did the instructor emphasize in lecture? Fill in your lecture notes with details from your textbook.
   - What material do I understand better now?
   - What questions remain?
6. Remember that each time you read the material, you will learn a little more.

STUDY GROUPS

Even after you have read the material several times, taken notes, and listened attentively in lecture, you may still have questions. Talk to a classmate or your instructor about your questions; do not automatically assume that your questions are trivial.

Collaborative learning

Individual accountability

Small study groups are becoming increasingly popular in the technical disciplines. In the past, many instructors discouraged group learning because it was difficult to determine which members of the group actually did the work. A study led by a professor at Harvard University (Light, 2001), however, found that science students who studied in groups were more likely to stay in their science major than students who studied alone. We assume that a similar result would be obtained with engineering students.

How does studying in groups help you understand difficult subject matter? First, in most cases, the group is formed from students in the same class. The group is composed of peers who are starting from a similar knowledge base. The strengths of the individual group members make it possible for the group as a whole to work through specific questions that may stump individual students.

Second, successful technical professionals share their findings and communicate frequently with their colleagues. The personal relationships formed by these interactions not only spur technological creativity, but make their work enjoyable. Similarly, study groups build a collegial spirit in technical majors that are often regarded as highly competitive. Interactions within the group allow the group members to see that they are not alone in their struggle to understand difficult material. Camaraderie developed among a group of students helping each other learn is substantially more satisfying than struggling alone. However, you, as an individual, have certain responsibilities to yourself and to the group.

Group study is not a substitute for studying alone. You must do the reading, take notes, and figure out what you do not understand before you meet with your group. If you have not struggled with the topics yourself, you are not in a position to help a classmate. Awareness of the need to struggle and the need to overcome your limitations while learning new concepts is, in itself, one of the greatest benefits of a technical education.

MODEL PAPERS

Before writing your first technical report, visit your university library's website (or even stop by the actual building) to identify journals in your field. All professional societies have journals, and many journals are available from major publishers like Blackwell, Springer, Elsevier, and Wiley, among many others. Also note that the URLs of several reports are given in the end of chapter exercises.

On most journal websites, there is a link to a separate page labeled "Information for Authors" or "Instructions to Authors," in which specific information is conveyed regarding length of the manuscript, general format, figures, conventions, references, and so on. Skim this section to get an idea of what journal editors expect from engineers who wish to have their work published. Compare the journal requirements with the material presented in this chapter and you will see substantial uniformity in the expectations for published technical reports.

Since most students find journal articles hard to read, a sample student laboratory report is provided in Chapter 4. Read the comments in the margins as you read through the report to familiarize yourself with the basics of technical paper format and content, as well as purpose, audience, and tone.

SUMMARY

Industrial problem solving requires an understanding of the state of the art in whatever field the problem lies. The human knowledge base is expanding so rapidly that no college or university can teach you all you need to know. Learning how to acquire new knowledge as a lifelong learner is a prescribed outcome for every accredited engineering program. Embracing the lifelong learning concept will make you a valued employee. Knowing how to identify and procure the necessary background literature will help you solve engineering problems efficiently.

Reading, in general, expands your knowledge base. Reading *technical papers* has the added benefit of providing a standard template for your engineering writing assignments. Technical reports are not easy reading. However, you are likely to be rewarded for your patience and effort with better writing skills, which can accelerate your career advancement and establish your professional reputation. Steven King's advice to aspiring fiction writers applies equally to engineers: <http://www.youtube.com/watch?v=hqp7A0B7abc&feature=related>.

EXERCISES

1. Open the link on controlling the insertion of a thread into an air-jet weaving machine given as follows: <http://www.fibtex.lodz.pl/47_09_29.pdf>. First notice the country of origin of the authors and the title of the journal from which this article has been taken. Despite not having English as a native language, notice that the authors have nonetheless followed the same standard format discussed in the present chapter, supporting the universality of this report format. Read the Abstract and the Introduction and write a short summary (maximum four sentences). Can you now explain how an air-jet weaving machine operates? Examine the Conclusions (p. 5 in the downloaded report). How does the Conclusions section relate to the Introduction section? Prepare an outline of the major section headings. Comment on the formatting of the figures and tables in this report. What system of citation is used in this report?

2. Open the progress report from the Brookhaven National Laboratory, accessed at <http://www.bnl.gov/nsls2/project/progress/docs/2011/Condensed_Dec2011.pdf>. Look at the structure of this report and compare it with that presented in Table 3.1. Can you determine the topic of the project just by reading the Introduction? Write a short summary about the structure of a progress report.

3. Open the link to the wind energy feasibility study found at <http://www.bristolri.us/documents/community/EBEC_Phase_II_Final_Report.pdf>. Examine the Table of Contents and prepare a written comparison of the report structure with that outlined in the text and summarized in Table 3.1.

4. Open the link to the feasibility study concerning a DME project in Iceland at <http://www.nea.is/media/eldsneyti/A-DME-feasibility-study-in-Iceland-summary-report.pdf>. Examine the Table of Contents and prepare a written comparison of the report structure with that outlined in the text and summarized in Table 3.1.

5. In the Chapter 2 exercises, you were asked to find primary sources on various topics. Choose one of those journal articles (or another that interests you) and fill in the following template.

Write the full reference for the paper in name-year or citation-sequence format.

Read the Abstract and record the following information:

1. What motivated the study?
2. How was the study done (experimental, theoretical, numerical, or some combination of these)?
3. What were the results?

Read the Introduction

1. How long have engineers been studying this problem? In other words, what is the date of the oldest reference?
2. Is there a recent review article in this field, which would provide detailed background information?
3. Why was the author interested in this topic?
4. What are the objectives of the study?
5. What citation style is used?

Skim the Discussion and Conclusion sections
    List the main conclusions

Skim the remainder of the paper and note major section headings below:

1. Abstract
2. Keywords?
3. Introduction
4. _____
5. _____
6. _____
7. _____ (as many as you need)

Look at the Results section.

1. What parameters are presented graphically?
2. Are they dimensional or nondimensional?
3. Make a sketch of the most significant figure in the Results section.

Read the Discussion section.

1. Were the objectives accomplished?
2. What were the important findings?
3. Do the data presented support the authors' claims?
4. Were there any surprises?
5. What further work is necessary or already in progress?
6. How does this paper relate to work you are doing or might consider doing?

Read the Materials and Methods/Analysis/Apparatus and Procedures section.

1. What measurement techniques were used?
2. What type of analysis was done?
3. Was any kind of special equipment used?
4. Do you have any experience with the equipment used?

Now read the full paper and note any specific questions that arise as you read.

1.
2.
3.

Are all end references cited in the text?

6. Open the report on computing airfoil drag found at the following link: <http://www
.nasa-usa.de/centers/dryden/pdf/88778main_H-2551.pdf>. Read the Abstract and suggest how the writing might be improved. Next read the Introduction and write a short
summary (maximum 4 sentences). Examine the Conclusions (p. 20 in the downloaded
report). How does the Conclusions section relate to the Introduction section? Prepare
an outline of the major section headings. Comment on the formatting of the figures and
tables in this report. What system of citation is used in this report?

7. Open the report on reservoir-induced seismicity found at the following link:
<http://goo.gl/MKmox>. Read the Abstract. Do you know what reservoir-induced
seismicity is after reading this brief section? Next, read the Introduction and write a
short summary (maximum four sentences). Can you now explain reservoir-induced
seismicity? Examine the Conclusions (p. 6 in the downloaded report). How does the
Conclusions section relate to the Introduction section? Prepare an outline of the major
section headings. Comment on the formatting of the figures and tables in this report.
What system of citation is used in this report? What is the affiliation of the author? Is
this an engineering report?

8. Open the report on airfoil drag and lift found at the following link: <http://smiller
.sbyers.com/temp/AE510_03%20NACA%200015%20Airfoil.pdf>. Read the Abstract
and suggest how the writing might be improved. Next read the Introduction and write a
short summary (maximum 4 sentences). Examine the Conclusions (p. 20 in the downloaded report). How does the Conclusions section relate to the Introduction section?
Prepare an outline of the major section headings. Comment on the formatting of the
figures and tables in this report. What system of citation is used in this report?

9. Open the report on concentrated solar power by clicking the "download
publication" link found at the following website: <http://www.iea.org/publications
/freepublications/publication/name,3903,en.html> or directly from the following
link: <http://www.iea.org/publications/freepublications/publication/csp_roadmap.
pdf>. Who do you think wrote this report? Read the Key Findings section. How does
this section differ from an abstract? Now look at the Table of Contents. How does the
format of this report differ from the standard format of technical reports? Next read
the Introduction and write a short summary (maximum four sentences). Examine
the Conclusions (p. 41 in the downloaded report). How does the Conclusions section
relate to the Introduction section? Prepare an outline of the major section headings.
Comment on the formatting of the figures and tables in this report. What system of
citation is used in this report?

10. Go to <http://web.mit.edu/ceepr/www/publications/index.html>, click on the picture
of the reports on the future of various energy sources (coal, nuclear, natural gas, and
geothermal). Examine the long reports on the future of various forms of energy. Do not
try to read the reports page by page (they are all more than 100 pages long), but look at
the formatting of the reports. For each energy source, prepare an outline using the major section headings. Where are the captions located for the graphs and tables in each
report? What is the punctuation and style of the captions ("Figure 1", "Fig. 1:", "Fig-
ure1." and so on). What is the most frequently used type of graph (x-y plots, bar charts,
or pie charts)? What system (citation-sequence or name-year) is used for citing references? What is the departmental affiliation of the authors? Do you notice any differences in format, which might be related to the departmental affiliation of the authors?

REFERENCES

Future of Natural Gas. 2011. The future of natural gas, an interdisciplinary MIT study. Accessed 12 January 2013 at <http://mitei.mit.edu/publications/reports-studies/future-natural-gas>.

Light, R.J. 2001. Making the most of college: Students speak their minds. Cambridge, MA: Harvard University Press.

Palmer-Stone, D. 2001. How to read university texts or journal articles. University of Victoria Counseling Service, Victoria (BC), Canada. Updated 2008. Accessed 9 July 2011 at <http://www.coun.uvic.ca/learning/reading-skills/texts.html>.

Parameter. 2011a. Accessed 8 July 2013 at <http://en.wikipedia.org/wiki/Parameter>.

Parameter. 2011b. Accessed 25 April 2013 at <http://goo.gl/th6Ip>.

Parameter. 2011c. Accessed 8 July 2013 at <http://www.thefreedictionary.com/parameter>.

Severn Tidal Power. 2010. Severn Tidal Power Feasibility Study Conclusions and Summary Report. HM Government, Southwest RDA. Accessed 12 January 2013 at <http://www.decc.gov.uk/assets/decc/what%20we%20do/uk%20energy%20supply/energy%20mix/renewable%20energy/severn-tp/621-severn-tidal-power-feasibility-study-conclusions-a.pdf >.

Singh, P., and Nestmann, F. 2010. Exit blade geometry and part-load performance of small axial flow propeller turbines: An experimental investigation. Experimental Thermal and Fluid Science 34 (6): 798–811. Accessed 26 April 2013 at <http://dx.doi.org/10.1016/j.expthermflusci.2010.01.009>.

Part 2
PREPARING TECHNICAL REPORTS

©iStockphoto.com/melhi

The paragraph describing the important result shown by Figure 4.1 has a number of notable characteristics:

- The opening sentence of the paragraph emphasizes the take-home message.
- The figure number is cited parenthetically (enclosed in parentheses following the opening sentence), directing the reader to the data.
- The subsequent sentences in the paragraph provide specific details about the data displayed in the figure.
- Past tense is used to indicate that the author is referring specifically to his or her own results and is not suggesting that these results are generally valid. The use of present tense to indicate that a statement has been accepted as true by the technical community is an idiosyncrasy of technical writing. Clearly, it would be inappropriate to use present tense for one's own work, when that work has not yet been published.

In-text References to Figures, Tables, and Equations

The caption to Figure 4.1 is presented in three different styles. If the caption style has not been specified by your instructor or by your organization, select a style that appeals to you and use it consistently throughout your report.

The format of the figure reference in the text should be the same as that in the caption, either *Figure 4.1* or *Fig. 4.1*. The reference to a specific figure is usually considered to be a proper noun and is thus always capitalized, regardless of its placement in the sentence.

Similarly, when referring to a specific equation, use the word *Equation* or *Eq.* followed by the equation number.

When referring to a specific table, always write out the word *Table* followed by the table number; do not abbreviate the reference as *Tab*.

Alternatively, an explicit reference to the graphical data could be made:

> The pump curve (head versus flow rate curve) in Figure 4.1 moved up and to the right, indicating higher heads and larger flow rates with increasing pump speed. At 2000 RPM, the shut-off head (the head when the flow rate approaches zero) was about 11 m. At 2500 RPM, the shut-off head was about 17.50 m, and at 3000 RPM, the highest speed tested, the shut-off head was 25 m. The shut-off head varied in proportion to the square of the RPM.

The explicit citation of figures is commonly used in engineering, while the parenthetical citation is more common in technical disciplines related to the natural sciences. It is acceptable to use both explicit and parenthetical citation of figures in technical reports.

The following sentence is vague, uninformative, and unsuitable:

> The following figures show the results.

This sentence is unsuitable because it provides no guidance concerning *what* results are important and *which* figures are being referenced, and it is unclear *how* the data in the figures support the findings.

In professional publications, visual elements typically are positioned at either the top or the bottom of a page. Therefore, it is good practice, though not essential, to put tables and figures at the top or bottom of the page whenever possible in all of your technical writing projects.

MAKING CONNECTIONS

Now that you have written (1) what you did and how you did it (the Apparatus and Procedures section), (2) the theoretical basis for your study (the Theory section), and (3) an organized summary of your results (the Results section), you are ready to make connections. The connections are made in the Discussion section, the Conclusions section, and the Introduction section.

Write the Discussion

The Discussion section gives you the opportunity to *interpret your results, relate them to published findings, and explain why they are important*. The structure of the discussion is specific to broad, as illustrated by the triangle in Figure 4.2.

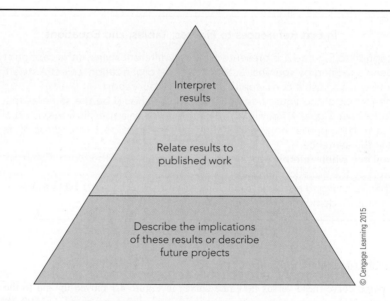

FIGURE 4.2 | Structure of a Discussion section

Start the discussion by *restating the objectives* of the current work. Then *recap each result* and *try to explain it*. Interpret the results so that the reader understands how you arrived at your conclusions. Ideally, there will be a one-to-one correspondence between the important points you address in the Discussion section with the problems or questions you stated in the Introduction section.

Were the results as expected from your theoretical analysis? If not, consider these possible reasons:

- Resolution of instruments was too imprecise.
- Calibration of instruments was incorrect.

- Human error (failure to follow the procedure, failure to use the equipment properly, failure to calibrate the measuring instrument accurately, variability when multiple lab partners measure the same thing, and simple arithmetic errors). If you suspect that human error may have influenced your results, it is important to acknowledge its contribution.
- Numerical values were entered incorrectly in the computer plotting program.
- Sample size was too small.
- Variability was too great to draw any conclusions.

If you can rule out these possibilities, discuss your results with your lab partner, teaching assistant, or instructor. If there is an obvious error, an "outsider" may be able to spot it immediately. Furthermore, informal discussions may help you clarify what you know and what you do not know about the topic.

Next, compare your results with those in the literature. Do they agree with the findings in previously published reports? Finally, if warranted, add a conclusion about the significance of your work. You may describe how your results apply to a related field or possible future work that focuses on an interesting observation.

An extract from the discussion of the pump characteristics data in Figure 4.1, for example, might be written as follows:

> The pump curves in Figure 4.1 demonstrate behavior consistent with the predictions from the extended Bernoulli Equation (Equation 1 in the Theory section). The extended Bernoulli Equation predicts that the head loss in the pump varies with the square of the mean velocity, which is, in turn, proportional to the volumetric flow rate, given as $\dot{V}$. In Figure 4.1, the head loss is the vertical distance from a horizontal line passing through the shut-off head (when $\dot{V} = 0$) to the pump curve. The curves fitted to the pump curve data in Figure 4.1 are second-order polynomials. The conclusion is that the head loss does indeed vary with the square of the volumetric flow rate, and the data appear to be consistent with theoretical expectations.

Write the Introduction

After having written drafts of the Apparatus and Procedures, Theory (if applicable), Results, and Discussion sections, you should be intimately familiar with the procedure, the data, and the discussion of what the data mean. Now you are in a position to put your study into perspective:

- What was already known about the topic before you carried out your study?
- Were there any deficiencies, inconsistencies, or unanswered questions?
- What was the purpose of your study?

The Introduction consists of two primary parts:

- Background information with references to published literature or in-house reports
- Objectives of the current work

Background Information   To locate background information in the literature, enter keywords in the search box of a database such as Web of Knowledge or Engineering Village (see Chapter 2). Evaluate the titles in the "hits" critically to make sure the sources are relevant to your work.

Objectives of the Current Work   If you designed your own experiment, you will have already stated the purpose of your study in your lab notebook. If this investigation was a laboratory exercise prepared by your instructor, re-read the exercise. Put the key goals of the lab into your own words and incorporate them in the Introduction section.

Structure of the Introduction Section   The structure of the Introduction is comparable to an upside-down pyramid—broad concepts to specific details (Figure 4.3). The opening sentence or two usually states a general observation or result familiar to the reader. Subsequent sentences narrow down the topic to the specific focus of the current study. Subsequent paragraphs then provide background information from published sources and describe unanswered questions or research extensions. The objectives of the current work are usually stated in the last paragraph of the Introduction section. It is *not* customary to give a preview of the results of the current work in the Introduction.

One way to structure an introduction to a pump characteristics lab, for example, is as follows:

It is estimated that pumps consume about 20% of the electricity produced in the US (Frenning, 2001). To improve the efficiency of pump installations, engineers need a good understanding of the fluid dynamics of pumps and their interaction with the piping systems they service. There are multiple types of pumps, classified according to the flow direction relative to the drive axis. Axial flow pumps generally provide large volumetric flow rates with relatively small heads, while centrifugal (also called radial outflow) pumps produce smaller volumetric flow rates with higher heads (see Fox et al., 2008, or Cengel and Cimbala, 2009, for example, for a more complete discussion of these pump types).

Centrifugal pumps are commonly used in water supply systems, waste water systems, as well as in many other residential (heating systems and swimming pools, for example) and industrial systems. The performance of a centrifugal pump depends upon the operating speed (shaft RPM) and the system resistance. Ideally, the pump's operating speed can be adjusted to keep the pump operating near the pump's best efficiency point for the system head against which it is pumping. Knowledge of pump performance characteristics permits the design of pumping systems for minimum energy consumption.

The objective of the experiments contained in this report was to determine experimentally the operating characteristics of a centrifugal pump. The pump was operated at three rotational speeds with the system head being adjusted by the position of a valve downstream of the pump. The flow rate and pump head were measured. In addition, the power input to the pump as well as the shaft RPM were also measured. From the recorded data, the head-flow rate characteristic and the efficiency were calculated for each shaft speed.

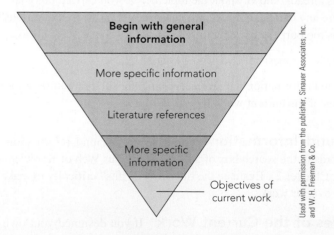

FIGURE 4.3 | Structure of an Introduction section

Write the Conclusions

In some disciplines, the Conclusions section is the final paragraph or two in the Discussion section. In many fields, however, the Conclusions section is a separate and final section of the report. The conclusions refer back to the objectives stated in the introduction and show how they have been achieved. The significant results of the study are summarized, and any broader implications of the work are described.

EFFECTIVE ADVERTISING—SELLING YOUR WORK

With any writing project, it is prudent to take a moment while you are writing to reflect on what you are trying to accomplish. Beyond communicating your knowledge, what motivated you to write this report and who is the intended audience? In industrial settings, as in academia, technical personnel and managers typically have very little time. They cannot possibly read every document published in their area of expertise. Therefore, to make sure that *your* work is given serious consideration, you need a strategy to "sell" potential readers on its merit. This is where effective advertising comes in. Recall from the material presented in Chapter 3 that experienced readers may initially read just the title, the abstract, and the Discussion and Conclusions sections of a report to decide whether the report is worth more detailed reading. These three sections collectively are thus the initial "deal breaker" and have to be your best writing—precise as well as concise.

Write the Abstract

The abstract is a *summary of the entire paper* in 250 words or less. It contains

- An introduction (scope and purpose)
- A short description of the methods
- The results
- The conclusions

Usually, there are no literature citations or references to figures in the abstract.

After the title, the abstract is the most important part of the technical paper used by the audience to determine initial interest in the author's work. Abstracts are indexed in databases such as Web of Knowledge and Engineering Village, which catalogue technical publications. If an abstract suggests that an author's work may be relevant to your own work, you will probably want to read the whole paper. On the other hand, if an abstract is vague or essential information is missing, you will probably decide that the paper is not worth reading. Writing abstracts is an excellent opportunity to develop and polish your professional voice. The question is: How can you effectively summarize your work so that the most important elements catch the eye of busy readers and entice them to read your entire paper?

One strategy for writing an abstract is to list the key points of each section of the paper, as though you were taking notes on your own work. Then write the key points in full sentences. Revise the draft for clarity and conciseness, employing strategies such as appropriate use of the active voice, combining choppy sentences with connecting words, rewording run-on sentences, and eliminating redundancy. With each revision, look for ways to shorten the text so that the resulting abstract is a concise and accurate summary of your work.

Compose the Title

The title is a short, informative description of the essence of the paper. You may choose a working title when you begin to write your paper, but revise the title after subsequent drafts. Remember that readers use the title to determine initial interest in the paper, so accuracy is essential.

Some examples of vague and undescriptive titles are as follows:

FAULTY: Pump Heads [too simplistic]

FAULTY: Variation of Pump Head as a Function of Flow Rate [one of the variables is missing]

FAULTY: Pump Heads for Varying Speeds [one of the variables is missing]

REVISION: Experimental Determination of Pump Characteristics for Three Pump Speeds

Review and Revise the Conclusions

After writing the abstract and fine-tuning the title so that it is concise and accurate, it is prudent to re-read your conclusions from the perspective of a reader who has just read the title and the abstract. Does your Conclusions section clearly, concisely, and accurately show that you achieved the objectives? Were there any contradictions in your data? Do your data align with those in the published literature? What are the broader implications of the work? Does your work suggest additional potentially fruitful research topics? If your Conclusions section provides well-written answers to these questions, the interested reader will probably be motivated to read your entire paper for the details.

REFERENCE AND CITATION FORMAT

References typically are cited in the Introduction, Theory, and Discussion sections of a technical paper, and the procedures given in Apparatus and Procedures are often modifications of those described in previous work. The citation and reference format in engineering differs from that in the liberal arts in three important ways:

- It is not customary to use direct quotations in technical papers. Therefore, when you describe another person's work, put the information into your own words and cite the source.
- In the in-text citation, it is not customary to give the page number(s) of the source; instead, use either the citation-sequence or the name-year systems explained in Chapter 2.
- In the references at the end of the report, give the page numbers for the entire paper (inclusive page numbers), not just the page(s) from which you extracted the information.

The References section is also called Literature Cited, because it includes only the published sources you cited in your paper. It is not a bibliography, which is a list of all the works you consulted to learn more about the topic. Properly citing your sources helps avoid issues related to plagiarism.

PLAGIARISM

Plagiarism is using someone else's ideas or work without acknowledging the source. Plagiarism is ethically wrong and demonstrates a lack of respect for members of your academic community (faculty and fellow students) and the engineering community in general. For more details on how to avoid plagiarism, see Chapter 2.

CITATION SYSTEMS

The citation-sequence system and the name-year system are two common systems for citing sources. For details on these systems, see "Common Engineering Citation Formats" in Chapter 2. Familiarize yourself with the order of the elements in each of these systems because you will use them often in your career.

ENGINEERING LAB REPORT CHECKLIST

The checklist in Table 4.3 provides an outline of the components of a technical report. Refer to this checklist when you complete your first draft of a lab report and also during the revision process. The checklist can also be downloaded from <www.CENGAGE/Knisely>.

TABLE 4.3 | Engineering lab report checklist

TITLE PAGE (if omitted, type the title and authors' names on the same page as the abstract)

☐ Title is descriptive and concise

☐ Your name and lab partners' names are listed, first name followed by surname

☐ Course and instructor

☐ Date experiment/study was performed

☐ Date report was submitted

ABSTRACT

☐ Begins a new page after the title page

☐ Contains an introduction (scope and purpose of your experiment/study)

☐ Contains a brief description of methods

☐ Contains results

☐ Contains your conclusions

INTRODUCTION

☐ Section comes immediately after the abstract

☐ Starts with a general introduction to the topic

☐ Contains a question or unresolved problem

(Continued)

TABLE 4.3 | Continued

- ☐ References support the background information (*primary* sources are preferred)
- ☐ Selected references are directly relevant to your study
- ☐ Information from the literature is paraphrased, and the source is cited. Direct quotations are not used.
- ☐ Citation format is correct (name-year or citation-sequence system)
- ☐ Objectives of the study are clearly stated in the final paragraph

THEORY (if appropriate)

- ☐ Fundamental concepts are expressed as equations
- ☐ Derivation of equations is logical and accurate
- ☐ Equations are formatted consistently and appropriately

APPARATUS AND PROCEDURES
(or METHODS, SOLUTION, or similar heading for computational or theoretical studies)

- ☐ Materials are not listed separately
- ☐ Contains all relevant information to enable the reader to repeat the procedure
- ☐ Routine procedures are not explained
- ☐ Written in paragraph form (not a numbered list) and in complete sentences
- ☐ Written in past tense
- ☐ Written in passive voice (active voice is allowed in some disciplines)
- ☐ No preview is given of how the data will be organized or interpreted

RESULTS

- ☐ Consists of a body (text) and visuals (tables and graphs)
- ☐ Text describes the trend, rather than listing the actual numbers
- ☐ Reference is made to each table and figure by number, either explicitly (*Figure 1 shows…*) or in parentheses at the end of the first sentence in which that visual is described *(Figure 1)*
- ☐ No explanation is given for the results (explanations belong in the Discussion section)
- ☐ Visuals are a summary of the raw data
- ☐ For any particular data set, include *either* a figure *or* a table, *not both*
- ☐ Visuals are positioned immediately after the paragraph in which they were first described
- ☐ Figure and table titles are informative and can be understood apart from the text
- ☐ Figure captions are placed *below* the figure, table captions *above* the table

DISCUSSION

- ☐ Results are briefly restated
- ☐ Results are explained and interpreted
- ☐ Results are related to published work

TABLE 4.3 | Continued

☐ Errors and inconsistencies are pointed out

☐ Implications of the results are described

REFERENCES

☐ References consist of primary sources, textbooks, and Internet sources as allowed by instructor

☐ Reference format is correct and complete (name-year or citation-sequence system)

☐ All references have been cited in the text

☐ All in-text citations have been included in the references list

APPENDICES (if applicable)

REVISION

☐ All questions from the laboratory exercise have been answered

☐ Calculations and statistics have been double-checked

☐ Overall structure

☐ Figures and tables

☐ Sections, paragraphs, sentences, and words

☐ Word usage

☐ Grammar

☐ Spelling

☐ Punctuation

☐ Standard abbreviations

☐ Numbers

☐ Format

Used with permission from the publisher, Sinauer Associates, Inc. and W. H. Freeman & Co.

SAMPLE LAB REPORT

To provide additional guidance in the preparation of laboratory reports, this chapter presents an annotated sample report. This report was originally prepared by a student in a junior-level mechanical engineering laboratory course. The student has asked not to be identified, but has given us permission to revise his report and to include it in this book. The course number, dates, and lab partners' names have also been changed.

The sample report has many positive characteristics. Notice the tone and the style of the writing, as well as the format of the report. The comments and annotation in the margins alert you to important points to keep in mind as you prepare your own reports.

The presentation here has been typeset to fit the format of the book and to accommodate the annotations. Your report should be formatted to fit on standard 8.5 × 11 inch paper. Follow your instructor's instructions, or use the formatting guidelines in Table 4.2. See "The Home Tab" and "The Page Layout Tab" sections in Appendix I for details on formatting documents in MS Word 2010.

Title is concise and informative.

Course

Instructor

Author

List lab partners' names alphabetically.

Include date on which the lab was performed.

Include submission date.

Do not number the title page.

Vibration of a Cantilever Beam with End Mass

MECH 4435

Prof. Knisely

Christopher Davidson

Lab Partners:
Nick Bainborough
Ross Carlson
Bob Kownurko

Lab Performed: September 23, 20XX

Report Submitted: September 30, 20XX

Abstract

The vibration characteristics of an aluminum cantilever beam were investigated experimentally. The equivalent spring constant for one degree-of-freedom vibration was determined from static tip deflection measurements with increasing static tip load. The damped natural frequencies were determined using a beam instrumented with strain gages to measure the dynamic response of the beam released from an initial displacement. The reduced amplitudes of forced vibrations for three different damping ratios were measured using a stroboscope and an engineer's rule to determine the steady state peak-to-peak amplitude for selected forcing frequencies in the range from 10 to 24 Hz. The non-dimensional vibration amplitudes (tip amplitude/base excitation amplitude) were compared with the equivalent one degree-of-freedom theoretical predictions. The experimental trends were qualitatively consistent with the theoretical predictions. However, the quantitative deviations of the experimental forced vibration amplitude values from the theoretical predictions exceeded the experimental error.

Introduction

Distributed mechanical systems can vibrate in multiple modes of vibration. Frequently, the lowest frequency mode can be approximated by a simpler one degree-of-freedom analysis. As an example of a relatively simple distributed system, consider a cantilever beam with a concentrated mass mounted very close to the free end of the beam. Many vibration textbooks, from Timoshenko (1937) to Inman (2001) and Palm (2007) provide analyses of a cantilever beam as a one degree-of-freedom system.

In viscously damped one degree-of-freedom systems, the vibration characteristics are determined by three parameters: the system mass, the damping coefficient and the spring constant. The equivalent values of mass and spring constant can be evaluated based on the analyses found in Timoshenko (1937) and Palm (2007). The damping coefficient can be determined from the dynamic response of the beam after being released from an initial displacement.

The experiments documented in this report were undertaken to determine the validity of the approximations made in the equivalent one-degree-of-freedom analysis of a cantilever beam with an end mass.

Label sections of the report clearly.

Introduce the topic.

Describe the procedure.

Limit abstract to 250 words maximum.

Do not cite references in the abstract.

Describe the results without reference to tables or figures.

State your conclusions.

Provide background information.

Use proper citation format (e.g., name-year system).

Clearly state the purpose of the experiments.

In an actual report, page 1 would not be numbered. Begin page numbering with page 2.

If there is a theoretical underpinning to your experiment, present it in the Theoretical Analysis.

If there is no relevant theory, omit this section.

Do not use direct quotations. Paraphrase and cite the source text.

Italicize variables.

Center equations, with right-aligned consecutive numbering.

Here no punctuation is included in the equation. Some organizations require punctuation in equations. Check for guidelines before writing your report.

Use an equation editor to insert symbols, subscripts, and superscripts.

In an actual report, remember to insert page numbers.

Theoretical Analysis

Treating the cantilever beam as a single degree-of-freedom system, it can be modeled as a mass-spring-damper system. Using Rayleigh's energy method, Timoshenko (1937) showed that the vibration of a cantilever beam with an end mass can be analyzed as an equivalent one degree-of-freedom system with an equivalent mass of 33/140 times the cantilever beam mass plus the concentrated end mass. Timoshenko (1937) notes that even when the end mass is removed, the equivalent one degree-of-freedom analysis predicts a frequency of vibration within 1% of the exact value.

In the following analysis, the vertical tip displacement is denoted as x, the mass as m (assumed to be the equivalent mass), the damping coefficient as c and the spring constant as k. With an assumed initial beam displacement, the sum of the forces acting on the mass-spring-damper system results in the following second order differential equation with constant coefficients:

$$m\ddot{x} + c\dot{x} + kx = 0 \tag{1}$$

Following the development in Inman (2001), Equation 1 can be re-written in terms of the natural frequency $\omega_n = \sqrt{k/m}$ and the damping ratio $\zeta = c/(2m\omega_n) = c/(2\sqrt{mk})$ as follows:

$$\ddot{x} + 2\omega_n \zeta \dot{x} + \omega_n^2 x = 0 \tag{2}$$

The solution to Equation 2 is assumed to be of the form $x = Ke^{rt}$ which, when substituted into Equation 2, yields the corresponding characteristic equation

$$r^2 + 2\omega_n \zeta r + \omega_n^2 = 0 \tag{3}$$

The roots of Equation 3 can be written as

$$r_{1,2} = -\zeta\omega_n \pm i\omega_n\sqrt{1 - \zeta^2} \tag{4}$$

The imaginary part of Equation 4 is called the damped natural frequency $\omega_d = \omega_n\sqrt{1 - \zeta^2}$ and would be the frequency of vibration of the damped system.

The two roots given by Equation 4 can be (a) two real distinct values when $\zeta > 1.0$ (called over-damped), (b) two real repeated roots when $\zeta = 1$ (called critically damped), or (c) two complex conjugate roots when $\zeta < 1.0$ (called under-damped). Only the under-damped case with two complex roots leads to oscillatory behavior of the system. The remainder of this analysis will focus on the oscillatory case with two complex conjugate roots.

With the characteristics roots known, and then substituted back into Equation 2, the solution of the equation of motion, Equation 2, can be written as

$$x = e^{-\zeta\omega_n t}(A \cos \omega_d t + B \sin \omega_d t) \tag{5}$$

in which the constants A and B must be determined from the initial conditions. In the case where there is an initial displacement with no initial velocity, the value of B is zero and A is equal to the initial displacement.

The rate of decay of the oscillatory amplitude is governed by the value of $\zeta\omega_n$ in the exponential term in Equation 5. In experimental studies it is often very convenient to use an alternative form of damping, called the logarithmic decrement, δ_{ln}. The logarithmic decrement can be defined in terms of two adjacent peak amplitude values or in terms of two amplitudes n cycle apart as

$$\delta_{ln} = ln\left(\frac{A_1}{A_2}\right) = \frac{1}{n} ln\left(\frac{A_1}{A_{1+n}}\right) \tag{6}$$

Palm (2007) shows the relationship between the logarithmic decrement and the damping ratio:

$$\zeta = \frac{\delta}{\sqrt{4\pi^2 + \delta^2}} \tag{7}$$

When an equivalent one degree-of-freedom system is driven by a force P_o at a selected forcing frequency ω_f, the steady state system response will be at the forcing frequency after the initial transient has died out. The differential equation of motion for the forced vibration of a one degree-of-freedom is

$$m\ddot{x} + c\dot{x} + kx = P_o \sin(\omega_f t) \tag{8}$$

The steady state solution to Equation 8 is given as

$$x_f(t) = \frac{P_o}{\sqrt{(k - m\omega_f^2)^2 + (c\omega_f)^2}} \sin(\omega_f t - \psi) \tag{9}$$

or, after factoring k out of the denominator and recalling $\omega_n^2 = k/m$

$$x_f(t) = \frac{P_o/k}{\sqrt{(1 - \omega_f^2/\omega_n^2)^2 + (2\zeta\omega_f/\omega_n)^2}} \sin(\omega_f t - \psi) \tag{10}$$

The driving force P_o divided by the system spring constant k can be interpreted as the static deflection of the system exposed to a steady force P_o. By setting $x_{st} = P_o/k$ it is possible to represent the dimensionless response amplitude as

$$|x_f| / x_{st} = \frac{1}{\sqrt{(1 - \omega_f^2/\omega_n^2)^2(2\zeta\omega_f/\omega_n)^2}} \tag{11}$$

Reference equations using a format parallel to that used for figures. If you write Fig. #, use Eq. #. If you write out Figure #, then use Equation #. Some organizations require parentheses around any equation-number citation, such as ". . . back into Equation (2). . ." Check for guidelines before writing your report.

The phase angle ψ can be written as

$$\psi = \tan^{-1}\left(\frac{2\zeta(\omega_f/\omega_n)}{1 - \left(\dfrac{\omega_f}{\omega_n}\right)^2} \right) \quad (12)$$

Thus, by knowing the frequency ratio ω_f/ω_n and the damping ratio ζ it is possible to determine both the amplitude and the phase of the system response to a driving oscillatory force.

Experimental Apparatus and Procedures

The experiment was conducted using two identical 2024-T4 aluminum beams, one with strain gages mounted on the beam and one without. Each beam was approximately 0.5" wide, 0.25" high, and 13.35" long. The length of the cantilever extending from the base was 9.75" as shown in Figure 1. An end mass made from a slice of a carbon steel rod, flattened on one side, was attached with a threaded eyebolt though the end of the beam. An optional damping dashpot could be used by attaching the damper disk to the eyebolt and placing the disk in the dashpot. The dashpot could be used with air, light oil, or water filling the dashpot. As the perforated disk oscillated in the fluid in the dashpot with either free or forced vibration of the beam, additional viscous damping was produced.

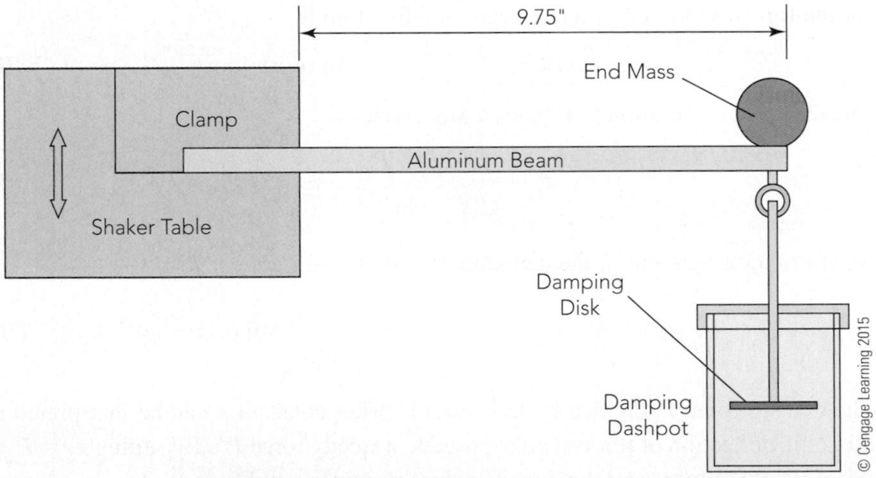

Figure 1 Simplified schematic of the experimental apparatus for forced vibration of an aluminum cantilever beam with end mass

© Cengage Learning 2015

Write the Apparatus and Procedures section in the past tense.

Provide sufficient detail to permit the reader to repeat the experiment.

Include figures in the Apparatus and Procedures section if they help clarify the setup and the methods.

Any figure included in the report must be discussed in the text of the report.

The precise dimensions and mass of the beam were measured and recorded. The mass of the concentrated tip mass and of the damping disk were also recorded. The beam was then clamped as a cantilever with a 9.75" free length extending from the rigid support. A static load was applied in sequential five pound force (lbf) increments to a maximum of 35 lbf. For each static load, the deflection at the tip of the beam was measured using a point gage with a dial scale. The static load and the corresponding beam deflection were recorded. From these static force-deflection data, the equivalent spring constant, the modulus of elasticity, and the approximate undamped natural vibration frequency were calculated.

After removing the un-instrumented test beam from the cantilever mounting apparatus, a second beam with strain gages mounted on it was clamped as a cantilever. The same concentrated end weight and damping disk were attached. The beam tip was statically deflected in increments of about 1/16" to a maximum of 0.25". From this deflection, the calibration factor for the strain gages was entered into the dynamic signal analyzer so the scale read directly in inches of tip deflection. The tip was then deflected about 0.25" and released from rest. The strain gage output was low-pass filtered at 45 Hz with a Krohn-Hite filter and then digitized at 500 Hz on the dynamic signal analyzer (HP 35665A). The trigger settings on the dynamic signal analyzer permitted the capture of 1024 data points at a sampling frequency of 500 Hz, producing about 2 seconds worth of data.

The digitized signals were converted to Excel files for plotting and further analysis. Transient responses were recorded in succession for three damping conditions: the undamped system, the system connected to an air dashpot, and the system connected to a water-filled dashpot. From these transient data, the natural frequency and damping constant for each damping condition were calculated. Concern for breaking the strain gages dictated using the un-instrumented beam for the subsequent forced vibration test.

The un-instrumented beam was used in the forced vibration study. The beam was mounted in the cantilever jig as shown in Figure 1. With small amplitude forced vibration of the shaker table at a set shaker frequency, the peak-to-peak amplitude of the beam tip was measured using a stroboscope and an engineering rule, readable to about ± 0.01 in. Shaker table forcing frequencies in the range of 10 to 24 Hz were used. Adjusting the strobe light frequency to freeze the visual image of the vibrating beam permitted the measurement of the beam frequency to within ± 0.1 Hz.

The original un-instrumented beam (with the concentrated end weight and the damping disk re-attached) was then once again placed in the cantilever mounting apparatus on the shaker table. The shaker table was turned on with a small driving amplitude of approximately 0.035" at a selected frequency between 10 and 24 Hz. The peak-to-peak amplitude of the beam tip was measured using a stroboscope and an engineering rule, readable to about ± 0.01 in. By adjusting the strobe light frequency to freeze the visual image of the vibrating beam, the beam frequency could be measured to within ± 0.1 Hz. The peak-to-peak vibration amplitude and corresponding excitation frequency were recorded.

Report the precision of the instruments for later reference when determining the expected experimental error.

Then for the case of the undamped system (no attached dashpot), the forcing frequency was set close to the undamped natural frequency and the time until failure was measured using a stop watch.

Results

The precise measured beam dimensions were 0.501" wide by 0.250" high and 13.35" long. The total beam mass was 89.6 grams. Table 1 provides a summary of the mass properties of the beams, as well as the mass of the concentrated end mass and the damping disk. The system masses were then calculated as

$$m_{sys} = \frac{33}{140} m_{cant} + m_{canc} + [m_{disk}] \tag{13}$$

The damping disk mass in the square brackets was included only when the dashpot was used (*i.e.*, for the "air-damped" and "water-damped" cases).

The un-instrumented beam was mounted as a cantilever and mass loaded in five lb_m increments. Figure 2 shows the linear relationship between beam loading and tip deflection. The tip deflection was measured with a point gage with a dial scale with an estimated accuracy of $\pm$ 0.001". Note that even though the beam loading was the independent parameter, it is purposely plotted on the *y*-axis so the slope of the best-fit line could be interpreted as the spring constant. For a linear spring

$$\Delta F_{spring} = k\, \Delta x \tag{14}$$

Thus, the spring constant is the slope ($k = \Delta F/\Delta x$) of the force-displacement line in Figure 2. For the present case it took on the value $k = 18.30\ lb_f/in.$

Table 1 Measured and calculated component masses

Item	Mass [g]	Mass [lb_m]
Aluminum beam (0.501 × 0.250 × 13.35")	89.6	0.198
Cantilever beam mass (0.501 × 0.250 × 9.75")	65.4	0.144
Equivalent beam mass (33/140 × cantilever mass)	15.4	0.034
Concentrated end mass	303.7	0.670
Damping disk mass	40.0	0.088
"Undamped" system mass	319.1	0.704
"Air-damped" system mass	359.1	0.792
"Water-damped" system mass	359.1	0.792

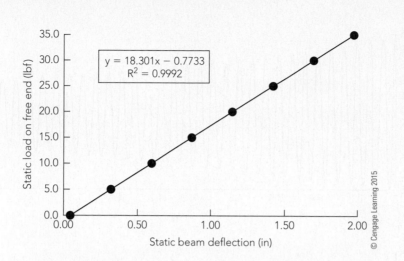

© Cengage Learning 2015

Figure 2 Equivalent spring characteristic of the cantilever beam
determined from the best-fit slope of the load versus static
deflection line. Equivalent spring constant was found to be
18.30 lb$_f$/in. Deflection error was ±0.001".

Knowing the spring constant of the system, it was then possible to calculate the modulus of elasticity for the material, making use of the formulation for the equivalent spring force for a cantilever beam (see Palm, 2007) of rectangular cross-section of width b and height h, given as follows:

$$E = \frac{kL^3}{3I} = \frac{4kL^3}{bh^3} \qquad (15)$$

From the spring constant, $k = 18.30$ lb$_f$/in, a value of $E = 8.67 \times 10^6$ psi was calculated.

Figure 3 shows the time history traces for the case of no dashpot damping, air-dashpot damping and water dashpot damping. From these data traces, which are plotted to approximate the original analog signal (even though the data is actually discrete data points), it was possible to determine the peak amplitudes at an initial location and then count the number of peaks to a reduced peak amplitude to calculate the logarithmic decrement for the equivalent mass-spring-damper system. The theoretical vibration frequencies were calculated using $\omega_d = \sqrt{k/m}\sqrt{1 - \zeta^2}$. Table 2 provides a summary of the theoretical frequency and experimental values for frequency, logarithmic decrement, damping coefficient, and damping ratio.

Place Figure 2 at the top of the next page after its initial mention since Table 1 occupies the bottom of the preceding page.

On axis labels, write the variable followed by the units in parentheses

Include sufficient detail in the figure caption so that the figure can be understood apart from the text.

Describe the figures in order.

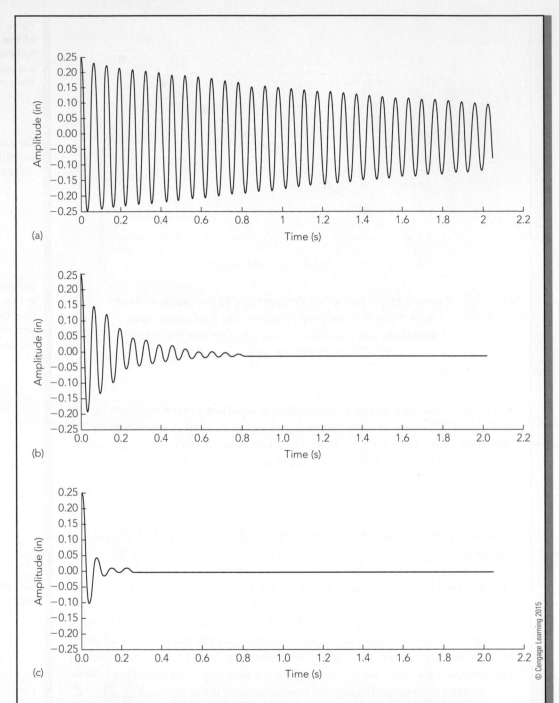

Position the figure caption below the figure.

Figure 3 Free vibration of (a) the undamped system (no dashpot), (b) the air-dashpot damped system, and (c) the water-dashpot damped system

Table 2 Characteristics of measured free vibrations

Damping	Th. Freq. [Hz]	Peak Count n	Δt [s]	Freq [Hz]	Initial Ampl. A_1 [in]	(n+1)st Ampl. A_{n+1} [in]	Log Decr [−]	Damp Coef [lb$_f$-s/ft]	Damp. Ratio [−]
No Dashpot	15.95	32	2.026	15.8	0.249	0.1118	0.025	0.00144	0.004
Air Dashpot	14.99	8	0.532	15.0	0.250	0.0103	0.399	0.0246	0.063
Water Dashpot	14.47	3	0.224	13.4	0.248	0.0012	1.773	0.0938	0.271

© Cengage Learning 2015

The un-instrumented beam was then subjected to forced vibration. The data for the forced vibration tests, shown in Table 3, includes the forced vibration frequency, the amplitude of the excitation of the shaker table and the beam tip peak-to-peak displacement amplitude. The theoretical systems' responses were calculated using Equation 11. The experimentally measured tip peak-to-peak amplitudes were first divided by 2 to get the peak amplitude and then normalized using the corresponding shaker table amplitude (labeled as x_b in Table 3). The experimental and theoretical amplitudes are shown as functions of the reduced excitation frequency (ω_f/ω_n) in Figure 4. The final part of this laboratory exercise was to explore the time to failure when the beam was driven at a frequency close to its natural frequency. The shaker was set at 15.5 Hz; after 5 minutes and 15 seconds the beam failed. At 15.5 Hz, the beam underwent 4882 cycles before failure.

Assuming the endurance stress for 2024-T4 aluminum is $SE = 20 \times 10^3$ psi (ASM Material Data Sheet, 2010a), the allowable bending moment can be determined from

$$M_O = \frac{(SE)\,I}{b/2} = (20 \times 10^3 \text{ psi})(6.52 \times 10^{-4} \text{ in}^4)/(0.125 \text{ in}) = 104.2 \text{ in} \cdot \text{lb}_f \qquad (16)$$

The end load producing this bending moment can be calculated as

$$F_{tip} = M_O/L = 104.2 \text{ in} \cdot \text{lb}_f / 9.75 \text{ in} = 10.68 \text{ lb}_f \qquad (17)$$

Position the table caption above the table.

Table 3 Forced vibration frequency and amplitude data

No Dashpot Damping $x_b = 0.35"$		Air dashpot damping $x_b = 0.35"$		Water dashpot damping $x_b = 0.35"$	
Freq [Hz]	Peak-to-peak Ampl. [in]	Freq [Hz]	Peak-to-peak Ampl. [in]	Freq [Hz]	Peak-to-peak Ampl. [in]
10.4	0.15	10.2	0.10	10.2	0.13
12.1	0.20	11.7	0.15	12.0	0.19
15.0	0.65	13.7	0.30	13.7	0.16
14.5	0.45	14.5	0.60	15.3	0.14
17.4	0.52	14.8	0.72	18.1	0.06
17.9	0.40	16.3	0.48		
20.0	0.16	18.3	0.15		

© Cengage Learning 2015

Use black symbols and lines on a white background for optimal contrast when this page is printed in black and white.

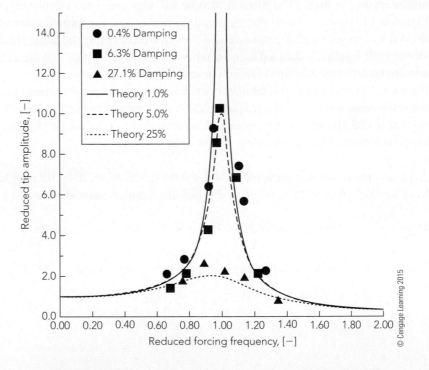

© Cengage Learning 2015

Make the figure title descriptive. Never use "*y*-axis label vs. *x*-axis label" as the figure title.

Figure 4 Forced vibration response of an aluminum cantilever beam. The reduced amplitude is the tip amplitude divided by the shaker amplitude; the reduced frequency is the ratio of the forcing frequency to the natural frequency of the system.

From the force-displacement diagram, Figure 2, an end load of 10.68 lb_f produced a deflection of 0.626 in. To maintain tip deflections less than 0.626" for the "undamped" beam with $\zeta = 0.004$ and $f_n = 15.8$ Hz, the equivalent non-dimensional amplitude (using a base excitation amplitude of 0.035") is 17.89. From Figure 4, for deflections less than 0.63", the two ranges of frequencies that would be safe from fatigue failure are $f_f < 15.3$ Hz and $f_f > 15.95$ Hz. This result is consistent with the fatigue failure after 4882 cycles due to small amplitude excitation at 15.5 Hz. If the material is not actually 2024-T4 (as suspected from the low E value), or if it is a bad batch, then the endurance stress may be lower than that listed for 2024-T4. For example, 6061-T6 aluminum has an endurance stress $SE = 14.0 \times 10^3$ psi (ASM Material Data Sheet, 2010b). Repeating the above analyses using Equations 16 and 17, with $SE = 14.0 \times 10^3$ psi, results in the safe frequency limits of $f_f < 15.1$ and $f_f > 16.4$ Hz. This frequency range would also be consistent with fatigue failure at a forcing frequency of 15.5 Hz. There is no apparent inconsistency with the observed fatigue failure occurring for any value of SE between 14×10^3 and 20×10^3 psi.

For the air-damped and the water-damped cases, with an assumed endurance stress of 20×10^3 psi, all operating frequencies would be safe from fatigue failure. For an assumed SE of 14×10^3 psi, the air-damped system would be safe from fatigue if $f_f < 14.5$ or $f_f > 15.3$ Hz. The water-damped case appears to be safe from fatigue for all frequencies even when $SE = 14 \times 10^3$ psi.

Discussion

In general, the measurements taken in this study support the one degree-of-freedom analysis of a cantilever beam. There are a few discrepancies that need closer examination.

The first issue is the apparent error in the measurement of the modulus of elasticity. Recall that from the Results section, a value of $E = 8.67 \times 10^6$ psi was calculated using Equation 15. The ASM Material Data Sheet (2010a) for 2024-T4 aluminum alloy lists a value of $E = 10.6 \times 10^6$ psi. The 18.2% error suggests the beam may not have been 2024-T4 or that it came from a bad batch of the alloy. A difference of this magnitude, while quite large (18.2%), is indeed statistically possible. The listed handbook value is the statistically most probable value, but the data are distributed about the mean in what is usually described as a Gaussian distribution. Variations of the modulus of elasticity of $\pm 30\%$ for 304 stainless steel, of $\pm 10\%$ for 1090 HR steel, and of $\pm 25\%$ for 1018 CF steel have been reported (Kieboam, 2001). The observed 18.2% variation in the modulus would appear to be within manufacturing tolerances based on the reported variations for other alloys. For experimental data in this study to agree with the handbook value, the tip deflection at the maximum static load would need to be reduced by about 0.10", which is 100 times greater than the estimated 0.001" error. It is concluded that the measured modulus would be accurate to within $\pm 1.0\%$ even if the error in the measured deflection was 10 times the estimated value.

Not all experiments provide perfect data. Address discrepancies and explore possible explanations for the differences between measured data and theoretical predictions.

Cite the source of a personal communication in the text, but do not include the source in the References section.

A second discrepancy was the substantial reduction in the natural frequency of the cantilever beam with the dashpot damper filled with water. With the measured damping ratio ($\zeta = 0.271$), the predicted natural damped frequency should have been about 14.47 Hz as given in Table 2. The measured frequency was lower, at 13.4 Hz, suggesting either the system spring constant was reduced or the system mass was increased, since the damped natural frequency can be written as $v_d = \sqrt{k/m}\,\sqrt{1 - \zeta^2}$. The spring constant is unaffected by the damper, leaving the mass as the only variable that changed with the addition of the water-damping. From a conversation concerning these results (C.W. Knisely, personal communication, 26 Sept 2010), it was learned that a system oscillating in a dense fluid (such as water) also experiences an added mass effect as well as the added damping. The added mass is the amount of fluid that is accelerated with each change of direction of the system. To estimate the added mass, the known values of frequency (13.4 Hz), spring constant (18.30 in-lb$_f$), and damping ratio (0.271) were substituted into the above equation for the damped natural frequency. The mass term was assumed to consist of the system mass plus added water mass. The equation then becomes

$$13.4 \cdot 2\pi = \sqrt{\frac{18.3 \cdot 12 \cdot 32.17}{0.792 + m_a}}\sqrt{1 - 0.271^2} \tag{18}$$

Equation 18 for m_a yields a value of 0.13 lb$_m$ for the added mass. To check if this value would be reasonable, the water mass in the dashpot was estimated. The dashpot had an inside diameter of about 1.5" and was about 3" high. It was not completely filled with water, but if it had been, the water mass would have been 0.19 lb$_m$. With a maximum physical mass about 1.7 times that of the calculated added water mass, it would appear feasible to attribute the frequency reduction for the water-damped case to the increased system mass.

Give possible explanations for the results.

On the left side of the resonance curve in Figure 4, the data for the 0.4% damping (the "undamped" case) and for 6.3% damping (the "air-damped" case) are above and below the theoretical curve for values of reduced frequencies between 0.6 and 0.9. The amplitudes for these points (circles and squares) that straddle the theoretical curve differ by 0.01", the available resolution of the measurement system. Thus, it is concluded that the data in this frequency range agree with the theoretical predictions, within the precision of the instrument. For the water-damped system, the amplitude data (triangles) are on or above the theoretical curve for 25% damping, except for the highest frequency point. The experimental data with 27% damping was expected to lie below the dotted theoretical curve. The difference in the measured amplitude relative to the predicted amplitude is about 2 times the amplitude resolution. If, for any reason, there was a different water mass in the dashpot for the forced vibration test than was used for the free vibration test, the system damping could have been different from that measured in the free vibration test. The changed water mass could also have possibly changed the system natural frequency, which is in the denominator of the reduced frequency. The water-damped tests should be repeated with a constant water mass in the dashpot, to avoid introducing artifacts into the damping and frequency of the system.

On the right side of the resonance peak (reduced frequencies greater than 1.0), there appears to be a systemic error (particularly for the 0.4% damping data). Magnitude differences between experiment and theory are greater by more than the precision of the instruments used; however, the trend in the experimental data follows very closely the theoretical trend. Another possibility is that the amplitude is correct, but the measured frequency was too high. However, the frequency decrease needed to put these points on the theory line would be five to seven times the frequency resolution. The frequency calibration of the stroboscope should be checked to determine if there was a systemic frequency error. Additional testing to minimize the error statistically would also have been helpful, but was not possible in the time allotted for this experiment.

Conclusions

Other than the anomalous frequency for the water-damped system, which was probably due to the added water mass, and the slight discrepancy on the right side of the resonance peak, the experimental results in Figure 4 strongly support the theoretical development of the one degree-of-freedom analysis for the cantilever beam with an end mass.

This experiment demonstrated that even for small input amplitudes, fatigue failure can occur if the driving frequency is close to the natural frequency and damping is low. It was very instructive to see how a small input amplitude near the natural frequency could produce fatigue failure. Adding damping to avoid sustained large amplitude vibration and the resulting fatigue failure is a useful concept that can be used in any mechanical system.

State the conclusion in light of the objectives.

References

ASM Material Data Sheet (2010a) ASM Material Data Sheet for Aluminum 2024-T4; 2024-T351 <http://asm.matweb.com/search/SpecificMaterial .asp?bassnum=MA2024T4>, accessed Sept 25, 2010.

ASM MDS (2010b) ASM Material Data Sheet for Aluminum 6061-T6; 6061-T651 <http://asm.matweb.com/search/SpecificMaterial.asp?bassnum=MA6061T6>, accessed Sept 25, 2010.

Inman, D. J. (2001) *Engineering Vibrations* (2nd Ed.), Prentice Hall, Upper Saddle River, NJ.

Kieboam, J. (2001) "Georgia Pacific's bolt material justification report for the 2001 Waferizer's blade holder," <http://web.vtc.edu/mt/114sum0/TTsum01/Kieboam /JoalsReport.html>, accessed Sept 25, 2010.

Palm, W. J. (2007) *Mechanical Vibration*, John Wiley & Sons, Hoboken, NJ.

Timoshenko, S.P. (1937) *Vibration Problems in Engineering* (2nd Ed.), D. Van Nostrand Co., NY.

Include all cited references.

For the name-year system, list references in alphabetical order.

Include the URL and date accessed when citing Internet sources.

SUMMARY

Engineering lab reports have a standard structure that allows readers to find information quickly. Lab reports are usually divided into sections. While the section headings may be different for different assignments, the body of a lab report usually consists of Introduction, Procedures, Results, and Discussion sections. Most writers do not write the sections in this order; instead, they write the sections in chronological order: Procedures, then Results and Discussion, and finally the Introduction. This writing strategy allows the writer to review the methods, analyze the data, summarize the data in tables, plot the results in figures, explain and interpret the results, and then put the findings into perspective by relating the current work to the published body of knowledge. The most difficult parts of the lab report to write are also the shortest: the abstract and the title. These parts are usually written when the body of the lab report is close to its final form. In journal articles and technical reports that are indexed in engineering databases, the title and abstract are used by readers to determine initial interest in the paper. These elements must convey the objectives and the most important results clearly and concisely, because most readers have only a few minutes in which to make their decision.

Writing lab reports in college is good practice for writing technical reports, one of the most common communications you will write in your career, because lab reports and technical reports have a similar style and structure. Use each writing assignment to expand your arsenal of tools with which to craft your communications. Remember that you are investing in your future; good communication skills are one of the top skills desired by employers and are essential for career advancement. Practice the strategies presented in this chapter to develop your professional voice.

EXERCISES

1. Go to the associated website for this book and download one of the "Reports in need of revision." Read the report at least twice. Consult the Engineering Lab Report Checklist in this chapter and the Revision Checklist in Chapter 5 if you are unsure of what might need editing. Write an improved version of the report. Submit both the original and the revised lab report to the course instructor.

2. Revise a lab report that you wrote for one of your engineering courses using the strategies presented in this chapter.

REFERENCES

Cengel, Y. and Cimbala, J. 2009. Fluid mechanics with student resources DVD, 2nd ed. New York: McGraw-Hill.

Fox, R.W., Pritchard P.J., and McDonald, A.T. 2008. Introduction to fluid mechanics, 7th ed. Hoboken, NJ: John Wiley, and Sons.

Frenning, L. (ed.). 2001. Pump life cycle costs: A guide to LCC analysis for pumping systems. Parsippany, NJ: Europump and Hydraulic Institute.

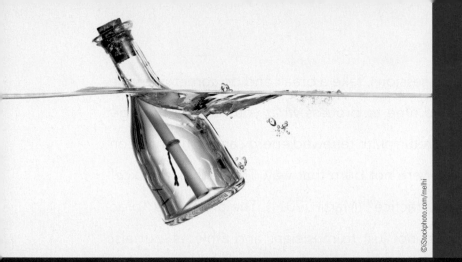

Chapter

5

REVISING REPORTS AND REVIEWING GRAMMAR

Revision—that is, reading your report and making corrections and improvements, is an important task that seldom gets the attention it deserves. Too many students and working engineers write the first draft of their report at the last minute, just before it is due. Often the submitted hardcopy is still warm from the printer and has not even been proofread. Most writers, and especially inexperienced writers, cannot produce a clear, concise, and error-free product on the first try. It may take a writer several rounds of revision, often with feedback from a reviewer, to be able to convey the objectives of the report, the important findings, and the conclusions in the "tightly wound" braid discussed at the end of Chapter 1.

Marathon work sessions, whether writing the first draft or reviewing and rewriting, are not productive. Instead, you should break up the process into short, focused sessions in which you give the task your undivided

Parts of this chapter have been adapted from *A Student Handbook for Writing in Biology*, Chapter 5. Used with permission from the publishers, Sinauer Associates, Inc. and W. H. Freeman & Co.

attention. Between writing sessions, take a break and do something else. The break gives your mind time to process what you have written, get help if necessary, and then return with renewed energy and concentration.

Most excellent writers were not born that way. They achieved excellence through "deliberate practice" (Martin, 2011). The old adage "practice makes perfect" applies not just to musicians and athletes, but also to you in your quest to develop your professional voice in engineering. So if writing does not come naturally to you, take heart. Writing becomes easier with practice, especially if you learn from your mistakes.

Many writers argue that they simply do not have time to go through the review and revision processes. Indeed, McKee (2012) notes that workplace writing often cannot follow the established "write and revise" processes practiced at the university because of other work commitments. To prepare themselves for these workplace constraints, students must practice and learn from revision in their college classes. By learning to read their writing critically and making a conscious effort not to repeat their mistakes, students will enter the workforce with skills that should reduce the time needed for both writing and revision. Further, employers value engineers who can plan ahead and who have developed the discipline needed to meet deadlines, both in their writing and in their technical assignments.

By virtue of having grown up in an English-speaking country, native English speakers usually write intuitively, without thinking deeply about the parts of speech or the reason for a particular grammatical construction. Native speakers, however, also have different levels of grammatical proficiency and may benefit from a refresher on writing essentials. For this reason, the sidebars included in this chapter address grammatical issues that may be helpful for multilingual writers and native speakers alike.

GETTING READY TO REVISE

You may notice that as you write your draft, the report evolves. The report may become more involved in some parts, requiring added detail; you may remove unneeded detail in other parts. Making small edits along the way as you write the draft certainly makes revising the final draft much less arduous.

Now your draft is done, and you are anxious to cross it off your to-do list. You are about ready to begin your revisions, when you remember a critical piece of advice: Give your writing time and distance before you revise.

Take a Break

The first step in revision is *not* to do it immediately after you have completed the first draft. You need to distance yourself from the paper to gain the objectivity needed to read the paper critically. So take a break, go for a run, socialize with friends, or get a good night's sleep.

Slow Down and Concentrate

When you are rested and ready to revise, find a quiet room where you will not be disturbed. Do not read your paper the same way you wrote it. Instead, **change its appearance** by increasing the zoom level on the computer monitor or by converting the paper to a different format such as a pdf (LR Communication Systems, 1999; Corbett, 2011). Reading a printed copy is even more effective. Next, **slow down your reading speed** by getting some of your other senses involved. For example, reading aloud, or having someone read the paper back to you, involves the sense of hearing. Pointing to each word with your finger as you read adds the sense of touch. Listen for discontinuities in the flow of words, where you or the reader need to stop and reflect or look ahead to know what was meant by the written words. These discontinuities are the areas that probably require revision. Finally, **concentrate on one type of error at a time**. Use multiple passes through your document to check for overall structure, content within individual sections, consecutive numbering of tables and figures, one-to-one correspondence between in-text citations and end references, and other issues.

Think of Your Audience

Remember for whom you are preparing your written or oral communications. Focus on what the audience wants to know, and present this information clearly. Use language and visual aids that are familiar to your audience. When communicating in person, show enthusiasm for the topic, be confident without being cocky, and work on your interpersonal skills. Revise your communications with the goal of meeting or exceeding the expectations of your audience.

Assess Readability

How do you know if your writing will be comprehensible to your target audience? One tool for assessing readability is the Flesch score. This score is computed from an algorithm that takes into account the number of syllables per word and the number of words per sentence in a document. As shown in Table 5.1, the lower the score, the harder the text is to read. To provide some standard for comparison, Table 5.2 lists selected publications commonly available in the United States along with their Flesch score.

TABLE 5.5 | Examples of redundancy

Redundant	Revised
It is absolutely essential…	It is essential…
mutual cooperation	cooperation
basic fundamental concept	basic concept or fundamental concept
totally unique	unique
The aqueous solution was obtained and poured…	The aqueous solution was poured…

Redundancy

Redundancy means using two or more words that mean the same thing. This problem is easily corrected by eliminating one of the redundant words (Table 5.5). Along with empty phrases, redundancy is a source of wordiness, using too many words to convey an idea.

Some people think that using a lot of words makes them sound important. Actually, using an excessive number of words in technical reports or laboratory reports detracts from the content. For student writers whose reports are evaluated by instructors, lack of clarity translates into a low grade. Researchers and faculty members, whose reputation depends on the number and quality of their publications, simply cannot afford not to write clearly because poorly written papers may be equated with sloppy methods.

Empty Phrases

Replace empty phrases with a concise alternative (Table 5.6). Put yourself in your reader's shoes. Which of the following two sentences, inspired by Van Alstyne (2005), would you rather read?

> **WORDY:** It is absolutely essential that you use a minimum number of words in view of the fact that your reader has numerous other tasks to complete at the present time.

> **REVISION:** Write concisely, because your reader is busy.

Frequently Confused Word Pairs

When you use the right words in the right situations, readers have confidence in your work. Use a standard dictionary whenever you are not sure about word usage. Consult your textbook or another technical reference for proper spelling and use of technical terms. The following word pairs are frequently confused in students' lab reports.

affect, effect

Affect is a verb that means "to influence." Affect is never used as a noun. *Temperature affects thermal conductivity. Effect* can be used either as a noun or a verb. When used as a noun, *effect* means "the result, the outcome, the influence." *We examined the effect of temperature on specific heats.* When used as a verb, *effect* means "to cause, to bring about." *High temperatures effected an increase in the specific heat values.*

TABLE 5.6 | Examples of empty phrases

Empty	Concise	Empty	Concise
a downward trend	a decrease	functions to	*omit this phrase*
a great deal of	much	higher degree of	higher, more
a majority of	most	in a manner similar to	like or such as
accounted for the fact that	because	in the amount of	of
as a result	so, therefore	in the vicinity of	near, around
as a result of	because	is dependent upon	depends on
as soon as	when	is situated in	is in
at which time	when	it is interesting to note that, it is worth pointing out that	*omit these kinds of unnecessary introductions*
at all times	always	it is recommended	I (we) recommend
at a much greater rate than	faster	on account of	because, due to
at the present time, at this time	now, currently	prior to	before
based on the fact that	because	provided that	if
brief in duration	short, quick	referred to as	called
by means of	by	so as to	to
came to the conclusion	concluded	through the use of	by, with
despite the fact that, in spite of the fact that	although, though	with regard to	on, about
due to the fact that, in view of the fact that	because	with the exception of	except
for this reason	so	with the result that	so that
in fact	*omit*		

all ready, already

All ready is an adjective that means "completely prepared." *When the instruments were finally all ready to use, the lab period was over. Already is an adverb that involves time. By the time we had set up the instruments, the lab period was already over.*

all right, alright

All right is an adjective or adverb used to indicate that something is satisfactory. The alternate, less formal spelling, *alright*, should not be used in academic or technical writing. *My professor asked me if I was all right after I accidentally let the positive and negative battery cables touch.*

all together, altogether

All together refers to a group gathered in one place. *The graduation ceremony was the last time the 45 mechanical engineering students were all together. Altogether means "wholly, completely, or entirely." Altogether, 45 mechanical engineering students graduated last year.*

amount, number
Use *amount* when the quantity cannot be counted. *The output from a PV cell depends on the amount of sunlight.* Use *number* if you can count individual pieces. *The maximum output of an array of PV cells depends on the number of cells.*

analysis, analyses   See plurals in this list.

anyone, any one
Anyone is a pronoun that means "any person at all." *The instructions are so easy to understand that anyone can follow them. Any one* is "any single person or thing." *Any one of those textbooks will provide a good introduction to heat transfer.*

anyway, any way
Anyway is an adverb that means "nevertheless" or "in any case." *Her father told her she didn't need to go to college to become a mechanic, but she decided to study engineering anyway. Any way* means "no matter which way." *It seemed to the student that any way she tried to solve the problem, the professor was not satisfied.*

complementary, complimentary
Complementary means "something needed to complete" or "matching." *The male and female electrical connectors have complementary profiles that allow them to be connected quickly. Complimentary* means "given free as a courtesy." *The brochures at the Visitor's Center are complimentary.*

continual, continuous
Continual means "going on repeatedly and frequently over a period of time." *The continual chatter of a group of inconsiderate students during the lecture annoyed me. Continuous* means "going on without interruption over a period of time." *The single crystals of bismuth were grown continuously for 24 hr.*

create, generate, prepare, produce
Create is to cause to come into existence out of nothing. *The artist used wood and plastic to create this sculpture.* Another example: *Energy is neither created nor destroyed. Generate* means to bring into existence from another source. *The conversation generated new ideas. The power plant generated 500 MW of electricity. Prepare* means "to make ready." *The tensile test samples were prepared by our machinist. Produce* means to make or manufacture. *The students produced a working prototype in eight weeks.*

datum, data   See plurals.

different, differing
Differing is the participle of the verb "to differ" and can usually be replaced by the words *divergent, contradictory, contrary, contrasting,* or *opposing.* As is frequently the case for participles, *differing* can serve as an adjective. *Differing opinions on matters of faith are quite common. Different* is an adjective that means "not alike" or "distinct from." It may imply the items being considered are distinct from each other, or have certain characteristics that are not the same. Alternatively, "different" may also imply that the items considered are not the same as the usual assortment of such items. *Often the opinions of scientists are different from those of creationists.*

The black pearls and pink pearls are different. [Context may suggest "from white pearls" or "from each other." It would be awkward to say black pearls and pink pearls have differing colors. The colors are not opposing, even though they are different.]

effect, affect   See affect, effect.

everyone, every one   See anyone, any one.

fewer, less

Use *fewer* when the quantity can be counted. *Today there are far fewer passenger trains than 100 years ago.* Use *less* when the quantity cannot be counted. *The weight of this sample was less than I expected.*

formula, formulas, formulae   See plurals.

generate   See create, generate, prepare, produce.

hypothesis, hypotheses   See plurals.

its, it's

Its is a possessive pronoun meaning "belonging to it." *The RTD is preferred over thermocouples because of its greater precision.* It's is a contraction of "it is." *The RTD is preferred over thermocouples because it's more precise.* Note: Contractions should not be used in formal writing.

lead, led

Lead is a heavy metallic element having the symbol Pb. *I was covered with a lead apron while I had dental x-rays taken.* Led is the past tense and past participle form of the verb "to lead." *The results of the test led the engineers to conclude that the bridge design was flawed.*

less, fewer   See fewer, less.

maybe/may be

Maybe is an adverb meaning "perhaps." *If I get all my studying done in time, maybe I can go out with my friends.* May be expresses the likelihood of something taking place. *This may be my last chance to go out with my friends before final exams.*

phenomenon, phenomena   See plurals.

plurals

The plural and singular forms of some words used in engineering are shown in Table 5.7.

TABLE 5.7 | Singular and plural forms of words frequently encountered in technical writing

Singular	Plural	Singular	Plural
analysis	analyses	hypothesis	hypotheses
criterion	criteria	phenomenon	phenomena
datum	data*	radius	radii
formula	formulas, formulae	ratio	ratios
index	indexes, indices	vortex	vortices

* Some disciplines treat the word *data* as singular, but scientists and engineers typically use *data* in the plural sense. *The data show…* [not *shows*] is correct.

Used with permission from the publisher, Sinauer Associates, Inc. and W. H. Freeman & Co.

prepare   See create, generate, prepare, produce.

produce   See create, generate, prepare, produce.

ratio, ration
Ratio is a proportion or quotient. *The ratio of heat rejected to heat added is used to determine the efficiency of a heat engine. Ration* is a fixed portion of food. *Rations for the refugees were distributed by the Red Cross.*

that, which
Use *that* only with restrictive clauses. A restrictive clause limits the reference to a certain group. Use *which* only with nonrestrictive clauses. A nonrestrictive clause does not limit the reference, but rather provides additional information. Commas are used to set off nonrestrictive clauses but not restrictive clauses. Consider the following examples:

EXAMPLE 1:   The Carnot cycle, which is one of several reversible heat engine cycles, serves as a standard for the maximum possible efficiency of a heat engine operating between two thermal reservoirs. [*Which* begins a phrase that provides additional information, but is not essential to complete the rest of the sentence.]

EXAMPLE 2:   The thermocouples that were used were chromel-alumel (Type K). [*That* refers specifically to *The thermocouples*.]

than, then
Than is an expression used to compare two things. *Collisions between molecules occur more frequently at high temperatures than at low temperatures. Then* means "next in time." *First the sample was weighed and then it was placed in a beaker of water, displacing a measurable volume of water equal to the sample's volume.*

various, varying
Various is an adjective meaning "different." *Various hypotheses were proposed to explain the observations. Varying* is a verb meaning "changing." *By varying the wind speed in the wind tunnel, drag could be measured over a range of Reynolds numbers.*

Jargon and Scientific Terminology
Jargon refers to words and abbreviations used by specialists. Always define terms that may be unfamiliar to your audience and always write out the full expression when first using an abbreviation. Technical words that you learned in class are not jargon and should be used in your writing. When you use technical terminology correctly, your readers have confidence in your knowledge.

Clichés and Slang
Clichés and slang should not be used in technical writing. **Clichés** are tired, worn out expressions that have lost their impact as a result of overuse. **Slang** is colloquial language that detracts from the seriousness of a formal, written communication. Furthermore, many clichés and slang are based on idiomatic expressions, which can be confusing for a non-native audience.

Gender-Neutral Language
Years ago, it was customary, for the sake of simplicity, to use masculine pronouns to refer to antecedents that could be masculine or feminine, but that use of language is no longer accepted.

SEXIST: The clarity with which an engineering student writes *his* lab reports affects *his* grade.

This practice is no longer considered appropriate. One solution that preserves equality, but makes sentences unnecessarily complex, is to include both masculine and feminine pronouns, as in the following example.

EQUAL BUT AWKWARD:	The clarity with which an engineering student writes *his or her* lab reports affects *his or her* grade.

Two better alternatives are to make the antecedent plural (revision 1) or to rewrite the sentence to avoid the gender issue altogether (revision 2).

REVISION 1:	The clarity with which engineering students write *their* lab reports affects *their* grades.
REVISION 2:	Writing clearly has a positive effect on an engineering student's grade. (Change the subject from *an engineering student* to *writing clearly*.)

Figures and Tables

Visuals often make the difference in how well you convey your meaning to your readers or listeners. Make sure every visual serves a purpose, because unnecessary visuals only dilute the significance of your message. Make sure you use the appropriate visual for the data (see the "Visual Elements" section in Chapter 6). Check that the visuals are positioned in the right order and that each visual is described accurately in the text.

PROOFREADING: THE HOME STRETCH

Proofreading is the last stage of revision, which involves correcting errors in *spelling, punctuation, grammar, and overall format.* Like editing, proofreading requires intense focus and slow, careful reading. It is beyond the scope of this book to cover punctuation and grammar in detail. Instead, we recommend Hacker and Sommers (2011 and 2012), and Lunsford (2010, 2011, 2013) to readers who would like a concise, easy-to-read reference for grammar and punctuation. Each of these books has a chapter especially for writers whose native language is not English. The companion websites for these books and the Purdue Online Writing Lab (OWL) provide study guides along with many exercises.

Grammatically Correct Sentences

The "Tips for Multilingual Writers—VII" sidebar shows how simple sentences may be formed from one noun and one verb. Sentences that contain adjectives, adverbs, and other structural elements add complexity and provide more information.

Subject–Verb Agreement   Early on in our formal education we are taught to make the verb agree with the subject. Most of us know that *the steel sample was* …, but that *the steel samples were*… Most errors with subject–verb agreement occur when there are words between the subject and the verb, as in the following example.

EXAMPLE:	The steel *samples* placed in the oven *were* [not *was*] heated to 975°C for two hours.

Tips for Multilingual Writers—VII

Building Sentences

The simplest structure for a sentence is a two-word sequence: a subject noun and an intransitive verb. For example,

NOUN   VERB
Temperature changes.

Complexity increases when the verb is a transitive verb. A transitive verb requires at least one object. For example,

SUBJECT   VERB   OBJECT
Temperature improves performance.

Further, adjectives and adverbs can be added to modify the subject, verb, and object. The sentence then becomes

3 ADJs   SUBJECT   ADV   VERB   2 ADJs   OBJECT
Cold ambient air temperature actually improves steam power-plant performance.

Often even more details are needed to convey the precise details of the situation. These details may require adding a subordinate (dependent) clause. A subordinate clause has a subject and a verb, but cannot form a complete sentence by itself.

SUBORDINATE CLAUSE   INDEPENDENT CLAUSE
According to well-developed theory, cold ambient air temperature actually improves steam power-plant performance.

An independent clause can form a complete sentence by itself. To join two independent clauses in one sentence, use a comma and a coordinating conjunction (CC) such as *and, or, but, nor, yet, for,* or *so.* Alternatively, use a semicolon.

INDEPENDENT CLAUSE   CC
Cold air temperature improves steam power-plant performance, so

INDEPENDENT CLAUSE
more electric power can be generated per unit mass of fuel.

When you write more complicated sentences, ask yourself what the subject of the sentence is. Look for the verb that goes with that subject, and then mentally remove the words in between. Make the subject and its verb agree.

A second situation in which subject–verb agreement becomes confusing is when there are two subjects joined by *and*, as in the following example.

EXAMPLE: The enthalpy value and that for internal energy *are* [not *is*] tabulated as functions of temperature and pressure in the appendix.

Compound subjects joined by *and* are almost always plural.

A third situation involving subject–verb agreement involves numbers. When numbers are used in conjunction with units, the quantity is considered to be singular, not plural.

EXAMPLE: In the engine cycle, 1500 KJ/kg of heat *was* [not *were*] added to the system during the constant pressure process.

When the subject is an indefinite pronoun, verb agreement depends on the nature of the pronoun. Some indefinite pronouns are always singular, some are always plural, and others can be either singular or plural.

The indefinite pronouns *each, either, neither,* and *one* are always singular and always take a singular verb.

> **EXAMPLES:**   *Each* thermocouple mounted in the test channel *is* read sequentially using the digital readout.
>
>   *Each* first-year student *is* expected to attend the meeting.

The indefinite pronouns *both, few, many, others*, and *several* are always plural and always take a plural verb.

> **EXAMPLES:**   *Both* thermocouples mounted in the test channel *are* read sequentially using the digital readout.
>
>   *Several* students *were* sick and *were* excused from the quiz.

The indefinite pronouns *all, any, enough, more, most, none,* and *some* can be singular or plural depending on the context of the sentence.

> **EXAMPLES:**   *All* [plural] students from last year's class *are* [plural] gainfully employed.
>
>   *All* my time *is* already scheduled.
>
>   *Any* writing practice *is* expected to lead to improvement.
>
>   *Any* tests done using uncalibrated instruments *are* of dubious value.

Finally, a group of words known as collective nouns—words such as *team, group, department, class, committee,* and *family,* among others, which refer to a group—can be used in either the singular or the plural sense. British writers tend to use the plural verb form, with the logic that the verb corresponds to the individual members of the group. American writers, on the other hand, prefer the singular verb form, which refers to the group as a whole.

EXAMPLES:		
	The team [implied *members of the team*] of students *work* well together.	**British usage**
	The team of students *works* well together.	**American usage**
	The department [implied *members of the department*] vote on all curricular changes.	**British usage**
	The department *votes* on all curricular changes.	**American usage**

Ambiguous Pronouns (him, her, he, she, they, it, its)   Beware of ambiguous pronoun references when two or more antecedents are present in the sentence. Consider the following examples and observe how the initial ambiguity has been eliminated by specifying the antecedent.

> **INCORRECT 1:**   The disagreement between the *client* and the *consultant* resulted in *his* seeking a new job. [Who was seeking the new job?]
>
> **REVISION 1A:**   The consultant was forced to find a new job because of a disagreement with the client.

Tips for Multilingual Writers—VIII

Verb Conjugation

For **regular verbs** such as *talk* and *help*, the present tense is simply the infinitive (*to* + verb) without the word *to*, except in the third person singular (he, she, or it). The past tense of regular verbs is always the base verb + *–ed*.

Infinitive	Present Tense	Past Tense
to talk (regular verb)	I talk	I talked
	you talk	you talked
	he, she, or it *talks*	he, she, or it talked
	we talk	we talked
	they talk	they talked

For **irregular verbs** such as *to go*, *to do*, and *to have*, the present tense is formed just like regular verbs, but the past tense is formed differently (not by adding *–ed*). Consult a dictionary to determine if a verb is regular or irregular. The verb *to be* is a special case because both the present and past tense are formed differently.

Infinitive	Present Tense	Past Tense
to go (irregular verb)	I go	I went
	you go	you went
	he, she, or it *goes*	he, she, or it went
	we go	we went
	they go	they went

Infinitive	Present Tense	Past Tense
to be (special irregular verb)	I *am*	I *was*
	you are	you were
	he, she, or it *is*	he, she, or it *was*
	we are	we were
	they are	they were

REVISION 1B: The client sought a new consultant after they disagreed on an important matter.

INCORRECT 2: The *coaches* did not like meeting the *players* in the locker room after games because *they* smelled. [Who smelled? The *coaches* or the *players*?]

REVISION 2A: Because the players smelled, the coaches did not like meeting them in the locker room after the games.

REVISION 2B: Because the locker room smelled, the coaches did not like meeting the players there after the games.

Tips for Multilingual Writers—IX

Adjectives

Adjectives modify nouns and pronouns and answer questions such as

- Which one?
- What color?
- What size?
- How many?
- How does it feel?

An adjective usually precedes the noun it modifies, or it may also follow the verb.

	ADJ NOUN
BEFORE THE NOUN:	Brian was tired after the long trip.

	VERB ADJ
AFTER THE VERB:	Brian's trip was long.

When two or more adjectives modify the noun, they can be classified as *coordinate adjectives* or *cumulative adjectives*. **Coordinate adjectives** can be connected with a comma or the word *and*, as shown in the following example.

	ADJ ADJ NOUN
ADJ SEPARATED BY *AND*:	Brian just returned from a long and difficult trip.

	ADJ ADJ NOUN
ADJ SEPARATED BY COMMA:	Brian just returned from a long, difficult trip.

Cumulative adjectives are adjectives that cannot be connected with a comma or the word *and*. These kinds of adjectives frequently cause problems for non-native English speakers because the adjectives have to come in a particular order ahead of the noun they modify.

Order	Adjective Describes	Examples
1st	An article, a demonstrative pronoun, a possessive pronoun, or a possessive noun	*a, the, an, my, his, those, Sheila's*
2nd	Order	*first, second, last, initial, final*
3rd	Quantity	*five, all, some, each*
4th	An opinion or observation	*wonderful, beautiful, difficult, ugly*
5th	Shape or size	*long, square, big, enormous*
6th	Color	*red, blue, yellow, black*
7th	Nationality	*American, Japanese, German, Serbian*
8th	Religion	*Buddhist, Christian, Jewish, Muslim*
9th	Material	*metallic, wooden, oak, cotton*

	ART QTY SIZE MATERIAL
CUMULATIVE:	The five 8-inch long stainless steel bolts failed under the load induced by the vibration.

Adjectives can be further modified by adverbs. See the next sidebar, "Tips for Multilingual Writers—X," for a short discussion of adverbs.

Vague Demonstrative Pronouns (it, this, that, these, those, which) Ambiguity also results from the misuse of demonstrative pronouns. Demonstrative pronouns should not be used by themselves when there is more than one antecedent.

Consider the following pairs of examples. In the incorrect sentences, the reader cannot easily determine to which antecedent the italicized pronoun refers. In the revisions, a specific antecedent follows the demonstrative pronoun to remove the ambiguity, or the demonstrative pronoun has been eliminated.

INCORRECT 1:	Lift acting on a wing is important for aircraft flight. With an increase in *its* angle of attack, *it* increases up to a certain angle. Then *it* separates. [Does *it* refer to lift, wing, angle of attack, or something as yet unnamed?]
REVISION 1:	Lift acting on a wing is important for aircraft flight. With increasing angle of attack of the wing, the lift on the wing increases up to a certain angle. At this angle, known as the stall angle, and beyond, the flow on the upper surface separates.
INCORRECT 2:	The pollution resulting from the combustion of DME is less than *that* due to petroleum-based fuels. [Does the word *that* refer to *pollution* or *combustion*?]
REVISION 2:	The combustion of DME produces less pollution than the combustion of petroleum-based fuels.
INCORRECT 3:	I have to read two chapters in economics, write a paper in philosophy, and work six calculus problems. I need to get these done for tomorrow. [What needs to be done tomorrow: *the calculus problems* or *all of the assignments*?]
REVISION 3A:	I need to get all of *these assignments* done for tomorrow.
REVISION 3B:	I need to get *these calculus problems* done for tomorrow.

In summary, to avoid ambiguity, examine every pronoun to be sure the antecedent cannot be mistaken. If there is any doubt, replace the pronoun with the appropriate noun phrase.

Run-on Sentences Run-on (fused) sentences consist of two or more independent clauses joined without proper punctuation. Each independent clause could stand alone as a complete sentence. Run-on sentences are common in first drafts, where your main objective is to write down your ideas. When revising drafts, however, use one of the following strategies to eliminate run-on sentences:

- Make two separate sentences.
- Use a semicolon to signal a close relationship between the two clauses.
- Insert a comma and a coordinating conjunction (*and*, *but*, *or*, *nor*, *for*, *so*, or *yet*).
- Rewrite the sentence.

These strategies have been applied to the examples below:

INCORRECT 1:	When you are writing, ideas often come faster than you can type, you write the way you think, without necessarily paying attention to grammar.
REVISION 1A:	When you are writing, ideas often come faster than you can type. You write the way you think, without necessarily paying attention to grammar. [Make two separate sentences.]

Tips for Multilingual Writers—X

Adverbs

Adverbs modify verbs, other adverbs, or adjectives. Adverbs answer the questions *why*, *when*, *where*, and *how*. In a sentence, adverbs precede the verb in the simple present and simple past tenses, except in the case of the verb *to be*, when they follow the verb.

ADV VERB

WITH MOST VERBS: The professor *usually* posts her lectures online.

VERB   ADV

WITH THE VERB TO BE: The supervisor is *almost always* at work by 7 AM.

Examples of positive adverbs include *already*, *always*, *almost always*, *ever*, *finally*, *frequently*, *generally*, *just*, *probably*, *occasionally*, *often*, *sometimes*, and *usually*. Examples of negative adverbs include *almost never*, *hardly*, *hardly ever*, *never*, *not ever*, *rarely*, and *seldom*. A negative adverb cannot be used with a negative verb.

NEG V NEG ADV

INCORRECT: The professor *does not never* post her lectures online.

NEG V POS ADV

CORRECT: The professor *does not ever* post her lectures online.
(*ever* always follows the negative verb)

POS ADV NEG VERB

CORRECT: The professor *usually does not* post her lectures online.
**(adverbs other than *ever* always precede the
negative verb)**

In the future and progressive tenses, adverbs are placed between a helping verb (HV) and the main verb (MV).

HV ADV MV

FUTURE: He *will finally finish* the problem tomorrow.

HV ADV   MV

PROGRESSIVE: She *is probably eating* lunch now.

HV ADV   MV

PROGRESSIVE: The system *has finally reached* equilibrium.

When the adverb is used in a question, the adverb is placed directly after the subject.

HV   SUBJ ADV MV

ADV USED IN A QUESTION: Did the system *finally reach* equilibrium?

REVISION 1B: When you are writing, ideas often come faster than you can type; you write the way you think, without necessarily paying attention to grammar. [Replace the comma with a semi-colon to show related thoughts, one following the other.]

REVISION 1C: When you are writing, ideas often come faster than you can type, and you write the way you think, without necessarily paying attention to grammar. [Leave the comma in place and add a coordinating conjunction.]

INCORRECT 2:　　The temperature distribution showed a logarithmic decline with increasing radial distance, this distribution is predicted by the conduction equation in radial coordinates, which combines Fourier's Law of Conduction with the conservation of energy.

REVISION 2A:　　The temperature distribution showed a logarithmic decline with increasing radial distance. This distribution is predicted by the conduction equation in radial coordinates, which combines Fourier's Law of Conduction with the conservation of energy. [Make two separate sentences.]

REVISION 2B:　　The temperature distribution showed a logarithmic decline with increasing radial distance; this distribution is predicted by the conduction equation in radial coordinates, which combines Fourier's Law of Conduction with the conservation of energy. [Use a semicolon to separate the two clauses.]

INCORRECT 3:　　An increase in the heat flux increased the magnitude of the temperature gradient as did a decrease in the thermal conductivity of the specimen, so the temperature gradient in the planar sample can be determined from the quotient of heat flux divided by the thermal conductivity.

REVISION 3:　　The magnitude of the temperature gradient was found to be proportional to the heat flux and inversely proportional to the thermal conductivity of the planar specimen. [Rewrite the sentence]

Tips for Multilingual Writers—XI

Prepositions and Idiomatic Expressions

Prepositions frequently describe the position of a person or thing in time and space. Some examples of commonly used prepositions include *on*, *in*, *at*, *by*, *to*, and *from*. Prepositions are especially challenging for non-native speakers because the direct translation of a preposition from their native language is often not the correct one to use in English in that particular situation.

Some prepositions are used in combination with adjectives and verbs to form standard expressions such as *different from*, *interested in*, *similar to*, *made of*, *consist of*, and *refer to*. However, some expressions involving prepositions are idiomatic; in other words, their meaning cannot be taken literally. For example, different prepositions after the verb *cut* produce expressions with different meanings:

a cut above	better than
cut down	to decrease in size, to humiliate, to kill
cut in	to interrupt
cut off	to separate from or disinherit
cut out	to remove something or to leave a place
cut through	to pierce
cut up	to cut into pieces

The best way to learn prepositions and idiomatic expressions is to read (newspapers, magazines, fiction, technical material—any genre you enjoy), listen to the radio, watch movies, and ask for clarification from native speakers with good communication skills. Make a list of idiomatic expressions you hear frequently, and practice using them. Consult references such as the Collins Cobuild English Dictionary (1995), the Longman Dictionary of American English (2008), Cambridge Dictionary of American English (2000), and online resources such as the Corpus of Contemporary American English (Davies, 2013).

Write in Complete Sentences A complete sentence consists of a subject and a verb. If the sentence starts with a subordinate word or words such as *after*, *although*, *because*, *before*, *but*, *if*, *so that*, *that*, *though*, *unless*, *until*, *when*, *where*, *who*, or *which*, however, another clause must complete the sentence.

INCORRECT 1: Fatigue failure occurs in materials subjected to a cyclic load. Which usually requires a large number of cycles. [The second sentence is a fragment.]

REVISION 1: Fatigue failure occurs in materials subjected to a cyclic load and usually requires a large number of cycles. [Use a coordinating conjunction such as *and* to combine the fragment with the previous sentence.]

INCORRECT 2: The signal was sampled at 100 Hz. Allowing a maximum frequency of 50 Hz in its spectrum. [The second sentence is a fragment.]

REVISION 2: The signal was sampled at 100 Hz, but only had a maximum frequency of 50 Hz in its spectrum. [Combine the two sentences using a subordinating conjunction]

Spelling

Spell checkers in word processing programs are so easy to use that there is really no excuse for *not* using them. Remember, however, that spell checkers may not know technical terminology, so consult your textbook or another technical reference for correct spelling. In some cases, the spell checker may even try to get you to change a properly used technical word to an inappropriate word that happens to be in its database (for example, *vorticity* to *vortices* or *radiosity* to *grandiosity*, *adiposity*, or *graciosity*).

The following poem is an example of how indiscriminate use of the spell checker can produce garbage:

Wrest a Spell

Eye halve a spelling check her
It came with my pea sea
It plainly Marques four my revue
Miss steaks eye kin knot sea.

Eye strike a key and type a word
And weight four it two say
Weather eye am wrong oar write
It shows me strait a weigh.

As soon as a mist ache is maid
It nose bee fore two long
And eye can put the error rite
Its rare lea ever wrong.

Eye have run this poem threw it
I am shore your pleased two no
Its letter perfect awl the weigh
My check her tolled me sew.

—*Sauce unknown*

Spell checkers will also not catch mistakes of usage, for example, "form" if you really meant "from." Print out your document and proofread the hard copy carefully.

Tech Tip—Spelling and Grammar Check

In MS Word, words that are possibly mis-spelled are underlined with a wavy red line. Phrases that may contain a gram-matical error are underlined with a wavy green line. Do not ignore these visual cues. Check spelling and grammar using an authoritative reference and revise in-correct phrases. Consider adding correctly spelled technical words to Word's diction-ary by right-clicking the word and select-ing **Add to Dictionary**.

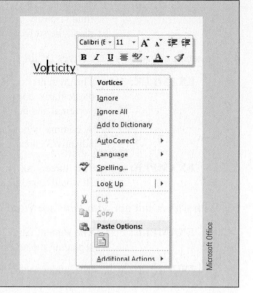

Punctuation

The purpose of punctuation marks is to divide sentences and sentence segments to make the meaning clear. A few of the most common punctuation marks and their uses in techni-cal writing are described in the following section. For a more comprehensive, but still con-cise, treatment of punctuation, see Hacker and Sommers (2011, 2012) or Lunsford (2010, 2011, 2013).

The Comma   The comma inserts a pause in the sentence in order to avoid confusion. Note the ambiguity in the following sentence:

> **CONFUSING:**   While the sample was heating the students set up a spreadsheet for their data.

> **CLEARER:**   While the sample was heating, the students set up a spreadsheet for their data.

A comma should be used in the following situations:

1. To connect two independent clauses that are joined by *and, but, or, nor, for, so,* or *yet.* An independent clause contains a subject and a verb, and can stand alone as a sentence.

> **EXAMPLE:**   Feel free to call me at home, but don't call after 9 p.m.

2. After an introductory clause, to separate the clause from the main body of the sentence.

> **EXAMPLE:**   Although she spent many hours writing her lab report, she earned a low grade because she forgot to answer the questions in the laboratory exercise.
>
> A comma is not needed if the clause is short.

> **EXAMPLE:**   Suddenly the power went out.

3. Between items in a series, including the last two.

EXAMPLE: The boundary layer thickness is affected by the air velocity, the fluid viscosity, the distance from the leading edge, the pressure gradient, the surface roughness, and the freestream turbulence level.

4. Between coordinate adjectives (if the adjectives can be connected with *and*)

EXAMPLE: The students' original, humorous remarks made my class today particularly enjoyable. [Replacing the comma with *and* as in "original and humorous remarks" makes sense.]

A comma is not needed if the adjectives are cumulative (if the adjectives cannot be connected with *and*).

EXAMPLE: The three, tall, muscular students look like football players. [It would sound strange to say three and tall and muscular students.]

5. With *which*, but not with *that* (see Word usage: that, which)

EXAMPLE: The student damaged the milling machine, which resulted in us replacing the machine.

INCORRECT: The student so badly damaged the milling machine, that it had to be replaced.

6. To set off conjunctive adverbs such as *however, therefore, moreover, consequently, instead, likewise, nevertheless, similarly, subsequently, accordingly*, and *finally*.

EXAMPLE: Instructors expect students to hand in their work on time; however, illness and personal emergencies are acceptable excuses.

7. To set off transitional expressions such as *for example, as a result, in conclusion, in other words, on the contrary*, and *on the other hand*.

EXAMPLE: Curt participates in many extracurricular activities in college. As a result, he rarely gets enough sleep.

8. To set off parenthetical expressions. Parenthetical expressions are statements that provide additional information; however, they interrupt the flow of the sentence.

EXAMPLE: Fluency in a foreign language, as we all know, requires years of instruction and practice.   *or*

Fluency in a foreign language (as we all know) requires years of instruction.

A comma should *not* be used in the following situations:

1. After *that*, when *that* is used in an introductory clause

INCORRECT: The student could not believe that he lost points on his laboratory report because of a few spelling mistakes.

2. Between cumulative adjectives, which are adjectives that would not make sense if separated by the word *and* (see Item 4 above)

The Semicolon

The semicolon inserts a stop between two independent clauses not joined by a coordinating conjunction (*and, but, or, nor, for, so*, or *yet*). Each independent clause (one that contains a subject and a verb) could stand alone as a sentence, but the semicolon indicates a closer relationship between the clauses than if they were written as separate sentences.

EXAMPLE: Outstanding student-athletes use their time wisely; this trait makes them highly sought-after by many employers.

A semicolon is also used to separate items in a series in which the items are already separated by commas.

> **EXAMPLE:**   Participating in sports has many advantages. First, you are doing something good for your health; second, you enjoy the camaraderie of people having a common interest; third, you learn discipline, which helps you make effective use of your time.

The Colon

The colon is used to call attention to the words that follow it. Some conventional uses of a colon are shown in the following examples.

Dear Sir or Madam:

5:30 p.m.
2:1 (ratio)

In references, to separate the place of publication and the publisher, as in
> Pacific Grove (CA): Brooks/Cole Publishing Company

A colon is often used to set off a list, as in the following example.

> **EXAMPLE:**   The College of Engineering has the following accredited programs: aeronautical engineering, biomedical engineering, chemical engineering, civil engineering, computer engineering, electrical engineering, environmental engineering, mechanical engineering, engineering mechanics, and zymological engineering.

A colon should *not* be used when the list follows the words *are*, *consist of*, *such as*, *including*, or *for example*.

> **INCORRECT:**   Doctors encourage us to eat plenty of vegetables such as: spinach, lettuce, onions, carrots, and broccoli.

The Period

The period is used to end all sentences except questions and exclamations. It is also used in some American English abbreviations, for example, *Mr.*, *Ms.*, *Dr.*, *Ph.D.*, *i.e.*, and *e.g.*

The Dash

The dash may be used to set off a list or parenthetical material. However, dashes are not usually used in technical writing, because they interrupt the flow of the sentence. When used to set off a list, the dash should be replaced with a colon.

> **EXAMPLE:**   Doctors encourage us to eat plenty of vegetables—spinach, lettuce, onions, carrots, and broccoli.

> **PREFERRED:**   Doctors encourage us to eat plenty of vegetables: spinach, lettuce, onions, carrots, and broccoli.   **Colon is preferred**

When used to set off parenthetical material that already contains commas, the dash should *not* be replaced with a comma.

> **EXAMPLE:**   The instruments that she plays—oboe, guitar, and piano—are not traditionally used in the marching band.   **Dashes may not be replaced with commas**

Parentheses

Parentheses are used mainly in two situations in technical writing: to enclose supplemental material and to enclose references for citations. Use parentheses sparingly because they interrupt the flow of the sentence.

EXAMPLE:	Human error (failure to read scales correctly; arithmetic errors; failure to correct for latitude, elevation, and temperature; and failure to zero the barometer) was the main reason for the unexpected results.
REFERENCE TO A TABLE OR FIGURE:	Atmospheric carbon dioxide increased at Mauna Loa from 2007 to 2012 (Figure 6.7).
CITATION-SEQUENCE SYSTEM:	Friction coefficients in laminar flows are unaffected by surface roughness (1). **Also see square brackets**.
NAME-YEAR SYSTEM:	Friction coefficients in laminar flows are unaffected by surface roughness (Schlichting, 2004).

Square Brackets   Square brackets are used in the following ways:

- To indicate an addition/correction to a direct quote
- To add a parenthetical comment inside a higher level parenthetical comment that is already using parentheses
- To cite a reference in the text in the citation-sequence system
- To present a dimensional analysis, whereby the dimensions are usually enclosed in square brackets

EXAMPLE:	The project manager of NASA's Jet Propulsion Laboratory commented, "Preparations for launching Curiosity [a robotic rover carrying scientific equipment to Mars] are on track."
EXAMPLE:	Most engineering faculty members at my university belong to an engineering professional society (American Society of Mechanical Engineers [ASME], American Institute of Chemical Engineers [AIChE], American Society of Civil Engineers [ASCE], among others).
CITATION-SEQUENCE SYSTEM:	Friction coefficients in laminar flows are unaffected by surface roughness **[1]**.
EXAMPLE:	The Reynolds number is a dimensionless parameter formed from the product of a characteristic velocity [L/T] and a reference length [L] divided by the kinematic viscosity [L^2/T], where [L] is a length dimension and [T] is a time dimension.

Standard Notation and Units

In each engineering discipline, accepted symbols are used to denote physical properties and dimensions. Some common technical abbreviations are given in the next chapter in Table 6.3. Pay attention to the use of spacing, case (capital or lowercase letters), and punctuation. Make sure that your usage is consistent with that in your textbook or that used by your instructor. With very few exceptions, unit abbreviations are the same for singular and plural units (for example, 30 min, *not* 30 mins).

To reduce the tediousness of formatting mathematical expressions with Greek letters, sub- or superscripts, and italics, take advantage of the AutoCorrect feature in MS Word. As shown in the "Tech Tip—AutoCorrect Simplifies Formatting" sidebar, AutoCorrect can be programmed to replace complicated technical expressions with a simple keystroke combination.

Tech Tip—AutoCorrect Simplifies Formatting

Mathematical expressions such as freestream velocity, V_∞, take time to format. To save yourself the trouble of inserting a symbol (∞), subscripting it, and italicizing the entire expression each time you type it, program the expression in AutoCorrect as follows:

1. First type the expression without format (e.g., V∞). The infinity symbol is inserted by clicking **Insert** (→Symbols) | **Symbol** and selecting ∞ from the dialog box.
2. Select the infinity symbol and subscript it by clicking the **Subscript** button on **Home** (→Font) (or use **Ctrl+=**). The expression then becomes V_∞.
3. To italicize the entire expression, select it and click **Home** (→Font) | **Italic** (or use **Ctrl+I**).
4. Now select the entire formatted expression and click **File | Options | Proofing | AutoCorrect Options** (see Figure A1.5 in Appendix I).
5. On the **AutoCorrect** tab, you will notice "V∞" already entered in the **With** text box. Click the **Formatted text** option button to add the subscripting and italics.
6. Type a mnemonic abbreviation of your choosing in the **Replace** textbox. For example, let ";vi" stand for "<u>v</u> <u>i</u>nfinity."
7. Click the **Add** button.
8. Click **OK** to exit the **AutoCorrect** dialog box.
9. Click **OK** to exit the **Word Options** box.

After AutoCorrect has been programmed in this way, every time you type ";vi" and hit the space bar, Word automatically changes the abbreviation to V_∞.

Numbers

Numbers are used for quantitative measurements. Traditional technical writing guidelines, still in use in the majority of technical and scientific textbooks, specify that numbers less than 10 (some argue less than 12) are spelled out, and larger numbers are written as numerals. The modern scientific number style recommended in the *CSE Manual* (2006) aims for a more consistent usage of numbers. The new rules are as follows:

1. Use numerals to express any *quantity*. This form increases their visibility in technical writing and emphasizes their importance.
   - Cardinal numbers, for example, 3 observations, 5 samples, 2 times
   - In conjunction with a unit, for example, 5 kg, 0.5 m/s, 37°C, 50%. Pay attention to spacing, capitalization, and punctuation.
   - Mathematical relationships, for example, 1:5 ratio, 20x amplification, 10-fold
2. Spell out numbers in the following cases:
   - When the number begins a sentence, for example, *Fifty ml of water* was (not *were*) *placed in the beaker"* rather than *"50 ml of water was placed in the beaker."* Alternatively, restructure the sentence so that the number does not begin the sentence. Notice that when numbers are used in conjunction with units, the quantity is considered to be singular, not plural.
   - When there are two adjacent numbers, retain the numeral that goes with the unit, and spell out the other one. An example of this is *The solution was divided into four 250-mL flasks.*

- When the number is used in a nonquantitative sense, for example, *one of the treatments.*
- When the number is an ordinal number less than ten, and when the number expresses rank rather than quantity, for example, *the second time*, or *was first discovered.*
- When the number is a fraction used in running text, for example, *one-half of the mixture*, or *nearly three-quarters of the sample.* When the precise value of a fraction is required, however, use the decimal form, for example, *0.5 L* rather than *one-half liter.*

3. Use scientific notation for very large or very small numbers. For the number 5,000,000, write 5×10^6, not *5 million.* For the number 0.000005, write 5×10^{-6}.
4. For decimal numbers less than one, always mark the ones column with a zero. For example, write *0.05*, not *.05*.

Format

Most university writing centers and professional editors recommend proofreading your paper in multiple "passes," looking for one kind of mistake in each pass (Writing Center, 2012; Cook Counseling Center, 2009; Every, 2012). This strategy works particularly well for finding formatting errors, which are much easier to detect on printed pages than on the computer screen (Table 5.8).

The Styles feature of MS Word is a handy tool for ensuring that each heading level is formatted consistently (see the "Tech Tip—Styles" sidebar). Using this feature has two additional benefits:

- Long documents can be navigated efficiently (see Appendix I, "Navigation Pane")
- The headings can be used to generate a table of contents (see Appendix I, "Table of Contents")

TABLE 5.8 | Checklist for proofreading format

Category	Check for…
Section headings	Correct order, format consistent, not separated from section body
Lists (bulleted or numbered)	Sequential numbering and consistent style
	Parallelism in sentence structure
	Indentation consistent for each level
Figures and tables	Sequential numbering in the order they are described
In-text references to figure and table numbers	Correspondence with the actual figures and tables
In-text citations	One-to-one correspondence with the end references
	Formatted correctly

TABLE 5.8 | Continued

Category	Check for...
End references	One-to-one correspondence with the in-text citations
	Complete and formatted correctly
Headers and footers **(if needed)**	Correct position on each page
Page numbers	Sequential (check especially after section breaks in Microsoft Word)
Typography	Consistent typeface, font size, and spacing

© Cengage Learning 2015

Tech Tip—Styles

Apply consistent format to headings by following these steps:

1. Select the appropriate heading level from **Home** (→Styles). Then type the heading and hit **Enter**.

2. To change a built-in style, right-click the style and select **Modify**.

3. In the **Modify Style** dialog box, click the **Format** button and make the desired changes.

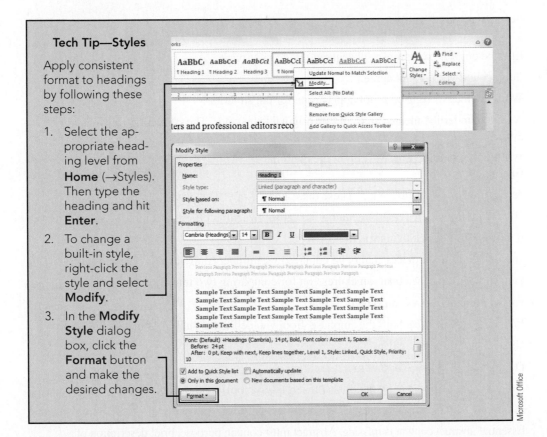

Microsoft Office

FINAL CHECK

The Laboratory Report Mistakes Checklist (Table 5.9) illustrates some of the common mistakes made by students when they write lab reports. It may be beneficial for you to scan this list of mistakes when you self-evaluate your writing or peer review someone else's work. This checklist can also be downloaded from <www.CENGAGE.com/Knisely>.

TABLE 5.9 | Laboratory report mistakes (adapted from Knisely, 2013)

Mistake	Explanation
lc	Use lowercase letter
CAP or uc	Use capital (uppercase) letter
∧	Insert text
♂	Delete text
⌒ or #	Leave space between the two characters
⌒⌣	Close up
¶	Start a new paragraph
Page break	End the current page; move subsequent text to next page.
agr	Subject and verb do not agree.
wc	Word choice. Word used is not appropriate for the situation.
....	A dotted underline means "stet," or "let original text stand." The correction was made in error.
1.	Word usage incorrect. Look it up in the dictionary. Examples:
d.	data = plural of "datum"; e.g., "These data show…" not "This data shows…"
e.	effect (noun) vs. effect (verb meaning "to cause") vs. affect (verb meaning "to influence")
i.	it's (it is) vs. its (belonging to "it")
m.	amount (individual pieces cannot be counted) vs. number (individual pieces can be counted)
t.	then (next in time) vs. than (an expression for comparing two things)
v.	varying (changing) vs. various (different)
2.	Write in passive voice. Shift the emphasis from yourself to the subject of the action.
3.	Use punctuation correctly.
4.	Do not write awkward or incomplete sentences.
5.	Do not write symbols in by hand. Use Insert (→Symbol) on the Ribbon.
6.	Make proper superscripts and subscripts.
7.	Essential abstract content is missing. Abstract must contain purpose, brief description of methods, results, and conclusions or possible explanation for the results.
8.	Too much detail for abstract.
9.	Cite sources when providing background.
10.	State purpose of the current experiment in the Introduction.
11.	Use proper citation format (name-year or citation-sequence system). See Chapter 2 for specific examples.
12.	Do not use direct quotations in technical writing. Paraphrase and cite the source.

TABLE 5.9 | Continued

Mistake	Explanation
13.	Write apparatus and procedures in paragraph form. Do not make a numbered list.
14.	Do not list apparatus separately from procedures, unless it is necessary to emphasize that the apparatus is not commercially available.
15.	Do not describe routine procedures in detail.
16.	In the Apparatus and Procedures section, do not give a "preview" of how data will be organized in the Results section.
17.	Essential details are missing in the Apparatus and Procedures section. Provide enough detail to enable the reader to repeat the experiment.
18.	Do not include raw data in the Results section; instead, summarize and organize the data.
19.	Include a text that describes the results shown in the figures and tables.
20.	Describe each table and each figure individually and sequentially.
21.	Refer to each figure/table either explicitly or parenthetically by number as you describe the results.
22.	Do not include a table when the figure(s) shows the same data.
23.	Include a number and a title in each figure and table caption.
24.	Make the figure/table title self-explanatory so that it can be understood without referring to the text.
25.	Do not write uninformative figure titles such as "y-axis label vs. x-axis label." See Chapter 4 for examples.
26.	Position the figure caption *below* the figure. Position the table caption *above* the table.
27.	Do not include a title above the figure. When you make the figure in Excel, leave the "Chart title" space blank.
28.	Use proper spacing on the axis. Choose "XY Scatter" plot in Excel (see Appendix 2).
29.	Give each data set its own geometrically distinct symbols or lines. Choose a dark color for both the line and the symbol. Do not rely on color alone to distinguish data or lines.
30.	Use scientific notation when numbers are very large or very small.
31.	Include units.
32.	Turn figures and tables in landscape orientation so that the bottom of the graphic is on the right side of the page.
33.	Use correct number format. Do not start sentences with a numeral (write out the number). Make sure decimal numbers less than one start with zero, e.g., 0.1 m/s, not .1 m/s.
34.	Use past tense to describe your *own* results; use present tense for general knowledge or published results.
35.	Try to explain the results in the Discussion section.
36.	Compare your results to published results or what is expected from theory.
37.	Use proper reference format. See Chapter 2 for specific examples.
38.	List all in-text citations in the References section.
39.	Cite all end references in the text.

Adapted from Knisely, 2013. Used with permission from the publisher, Sinauer Associates, Inc. and W. H. Freeman & Co.

GET FEEDBACK

When we are engrossed in our work, we may fail to recognize that what is obvious to us is not obvious to an "outsider." That is where feedback from someone who is familiar with the subject matter is very beneficial. If your instructor allows it, ask your lab partner, another classmate, or your teaching assistant to review your paper. Return the favor by reviewing someone else's. You may also get valuable feedback from a writing counselor at the writing center at your university.

First, ask yourself the following questions, which are listed on the self-assessment form in Table 5.10.

- What are the goals and objectives of this paper? Do I make these objectives clear to the reader?
- What questions or concerns do I have with this draft? What are some ways I can begin to address these concerns?
- On what parts would I like feedback? For example, *I am not sure if my meaning is clear*, or *I am not sure if I correctly understood this concept*, among other issues.
- What do I like about the paper? What are its strengths, and what has gone well?

TABLE 5.10 | Draft self-assessment form

DRAFT SELF-ASSESSMENT

Writer's name _____

Title of paper _____

1. What are the goals and objectives of this paper? Have they been sufficiently emphasized? Do I make these objectives clear to the reader?

2. Is every sentence in my draft complete, clear, and understandable?

3. What questions and concerns do I have with this draft? What are some ways I can begin to address these concerns?

4. What parts would I like feedback on (not sure if my meaning is clear, not sure if I correctly understood this concept, and other issues)?

5. What do I like about this paper? What are its strengths and what has gone well?

TABLE 5.11 | Peer review form

PEER REVIEW

Writer's name _____

Title of paper _____

Reviewer's name _____

1. Do I know what the writer is trying to accomplish with this paper? Is the purpose clear?

2. What questions or concerns do I have about this paper? Are there sections that are difficult to follow? Is the report wordy? Can it be shortened without losing meaning? Are the organization, content, flow, and level appropriate for the intended audience?

3. What suggestions can I offer to help the writer clarify the intended meaning?

4. What do I like about the paper? What are its strengths? What has gone well?

© Cengage Learning 2015

The questions your reviewer will focus on are listed below and on the peer review form in Table 5.11.

- Do I know what the writer is trying to accomplish with this paper? Is the purpose clear?
- What questions or concerns do I have about this paper? Are there sections that were difficult to follow? Are the organization, content, flow, and level appropriate for the intended audience?
- What suggestions can I offer the writer to help him or her clarify the intended meaning?
- What do I like about the paper? What are its strengths?

In addition to answering these questions, your reviewer may also mark up the draft you supply. The markup can be done on paper or electronically (see "Tech Tip—Tracking Changes Electronically" sidebar). Think of the peer review process as a team sport: The reviewer is not challenging the writer's right to be on the team. The two are working together to get the best possible result.

Copies of the self-assessment form and the peer review form are available as pdf files on the companion website (<www.CENGAGE.com/Knisely>).

Tips for Being a Good Peer Reviewer

Two issues with which you may struggle when asked to review someone's paper are: (1) I am not confident that I know the "right" answer or know enough about the writing process to give good suggestions, and (2) I do not want to hurt the writer's (my friend's) feelings. These are valid concerns, and resolving them will require, first, a willingness to learn as much about technical writing as possible, and second, the attitude that if something is unclear to you, it will most likely also be unclear to other readers. With each paper you review, you will gain more confidence in your ability to give constructive feedback. In the meantime, however, a good rule of thumb is to give the kinds of suggestions and consideration that you would like to receive on your paper.

Here are some concrete tips for being a good peer reviewer:

- Read the writer's draft self-assessment prior to reading the paper.
- Use the Engineering Lab Report Checklist (Table 4.3) for content.
- Use the Laboratory Mistakes Checklist (Table 5.9) to familiarize yourself with common types of mistakes.
- Use proofreader's marks to point out awkward sentences, spelling and punctuation mistakes, and formatting errors. Alternatively, use the Track Changes feature of MS Word when reviewing electronic files (see "Tech Tip—Tracking Changes Electronically" sidebar). Do not feel you have to rewrite individual sentences—that is the writer's job.
- Ask questions. Let the writer know where you cannot follow the paper's logic, where you need more examples, where you expect a more detailed analysis, and so on.
- Do not be embarrassed about making many comments, because the author does not have to accept your suggestions. On the other hand, if you say only good things about the paper, how will the writer know whether the paper is accomplishing the desired objectives?

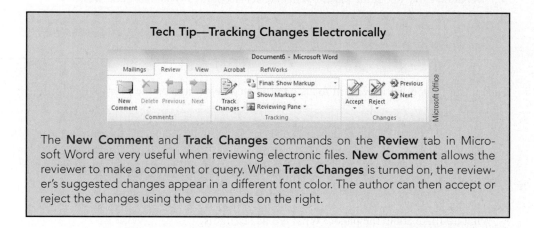

Tech Tip—Tracking Changes Electronically

The **New Comment** and **Track Changes** commands on the **Review** tab in Microsoft Word are very useful when reviewing electronic files. **New Comment** allows the reviewer to make a comment or query. When **Track Changes** is turned on, the reviewer's suggested changes appear in a different font color. The author can then accept or reject the changes using the commands on the right.

You can fine-tune your proofreading skills on any text. You may recognize some of your own problems in other people's writing, and with persistence and practice, you will find creative solutions to correct these problems. **Keep a log of the problems that recur in your writing**, and review them from time to time. Repetition builds awareness, which will help you achieve greater clarity in your writing. Making a habit of learning from your errors will make you a better writer and accelerate the development of your professional voice.

Have an Informal Discussion with Your Peer Reviewer

Sometimes the comments made by the peer reviewer are self-explanatory. Other times, however, the peer reviewer cannot respond to certain parts of the paper because more information is required. Under these circumstances, an informal discussion between the writer and the reviewer is helpful. There are two important rules for this discussion:

- First, the writer talks and the reviewer listens. The objective is to help the writer express exactly what he or she wants to say in the paper.
- Second, the reviewer talks, in nonjudgmental terms, about which parts of the paper were readily understandable and which parts were confusing. The reviewer does not have to be an experienced writer to do this—no two people have exactly the same life experiences, and you can always learn something positive from looking at your writing from someone else's perspective.

SUMMARY

The revision process requires concentration, familiarity with the overall structure and content of different types of technical communications, and attention to detail. Because it is hard to detect errors in your own writing, seek feedback from experienced writers, and keep a writing log to avoid repeating your mistakes. When you are a student, view your writing assignments as an opportunity to develop your professional voice. Pay attention to your instructor's comments regarding format, content, grammar, and conventions of the discipline. It is normal to make mistakes during the learning process. However, if you continue to make these kinds of mistakes in the workplace, your reputation will suffer. Take pride not only in your technical competence, but in your ability to communicate your knowledge to others.

EXERCISES

1. Develop a short outline on one of the topics below. Choose an organizational structure for the topic. Explain why you chose that structure.

> History of motive power
> History of fluid mechanics
> History of computing machines
> History of aviation
> History of commercial airlines
> The Fukushima nuclear disaster
> The space shuttle Columbia accident
> The Silver Lake Dam failure
> The Folsom Dam gate failure
> The Chernobyl nuclear disaster
> The 2010 Christchurch earthquake
> Types of personal vehicles (include full-size sedan, compact sedan, station wagon, sports car, van, pick-up, SUV, scooter, motorcycle)
> Renting or purchasing a home after college
> Options for post-fossil fuel power

2. Edit the sentences below, removing unnecessary lead-ins such as "In fact, . . .", "In order . . ." For the case of . . . ," and similar unnecessary words that contain no new information.

(a) In order to comply with FAA regulations, pilots must complete the pre-flight protocols.

(b) Due to the fact that we had a severe snowstorm, it was necessary to cancel the planned field testing.

(c) We attempted to comply with the specified testing procedures, but, in fact, the procedures were inconsistent and contradictory.

(d) However, please keep in mind the fact that the calculated, theoretical phase shift cannot be detected without a reference signal.

(e) Please compare the results of the tabulated cost factors with the corresponding anticipated costs in the proposal.

(f) For the case of a brass propeller in seawater, it is necessary to supply a replaceable, sacrificial zinc anode to avoid corrosion of the propeller.

(g) In order to calculate the power spectrum of a sampled signal, a fast Fourier transform (FFT) algorithm must be used.

(h) Changing the damping ratio in a single degree-of-freedom system changes the nature of the system response curve.

(i) Another important consideration is the fact that we have two progress reports due next week.

(j) The report issued by the president stressed the point that we need careful planning for large expenditures in the next quarter.

(k) To prevent the possibility of electromagnetic interference, all cables connecting machinery on the factory floor must be shielded.

3. Edit the sentences below, selecting the appropriate word.

(a) The (affect/effect) of the high velocity exiting the duct was a loud noise.

(b) To (affect/effect) change you must be persistent and untiring.

(c) Smoking has been shown to (affect/effect) adversely a person's health.

(d) (Affective/Effective) advertising increases sales.

(e) The discussion focused on how to help (fewer/more) students make (fewer/less) writing errors.

(f) We have had (fewer/less) snow accumulation because there were (fewer/less) storms this year.

(g) (Fewer/less) disposable income means you have (fewer/less) dollars for free-time activities.

(h) If (its/it's) broken, fix it. If (its/it's) not, leave it alone.

(i) In the course of life, (its/it's) one's choice how to deal with (its/it's) ups and downs.

(j) She (lead/led) the discussion on mass-spring-damper systems.

(k) The usual material in a sailboat keel is (lead/led).

(l) If it is later (than/then) 3:00 p.m., (than/then) we are late for our meeting.

(m) If the sum of two digits is greater (than/then) their product, (then/than) the two digits are still not uniquely defined, right?

(n) (Various/varying) the velocity by a factor of two changes the kinetic energy by what factor?

(o) (Various/varying) combinations of integers satisfy the Pythagorean theorem.

4. Edit the following sentences, removing words such as "proper," "suitable," and "appropriate," which needlessly establish suitability, when suitability can be inferred.

(a) Insertion of a suitable pin permitted the proper alignment of the mating surfaces.

(b) The zinc anode was suitably located to minimize the corrosion of the brass propeller.

(c) The signal was sampled at an appropriate rate that was sufficiently fast to resolve the highest frequency component.

(d) Correct identification of suitable materials was not possible.

(e) The wind tunnel model length cannot exceed 18" in length because our currently existing machining capabilities cannot produce longer components.

5. Go to the companion website for this book and download one of the "Reports in Need of Revision." Read the report at least twice, paying attention to the kinds of mistakes listed in the Laboratory Mistakes Checklist (Table 5.9). Write an improved version of the report. Submit both the original and the revised lab report to your instructor.

6. Edit the sentences below, removing unnecessary words such as "type," "condition," "basis," "nature," and similar expressions.

(a) We replaced the readout with a newer, inexpensive, and reliable type of readout.

(b) The sophomores attend lab on an as-required basis.

(c) The laptop computer was used primarily in a backup capacity.

(d) The first production run of 400 units indicated no substantial problem areas.

(e) She works primarily in the areas of failure analysis and forensic engineering.

(f) Transient thermocouple signals are of short duration in nature.

(g) Start recording data when the light changes to the green condition.

(h) Sort the items for the appendix into alphabetical order.

(i) The sample-and-hold circuit is shown in schematic form in Figure 4.

(j) The backup system operates in exactly the same manner.

(k) The proposed system has a relatively high reliability factor with a moderate cost factor.

(l) The cause of failure appeared to be more of a mechanical nature than of an electrical nature.

7. Revise a lab report that you wrote for one of your engineering courses using the strategies presented in this chapter.

8. Edit the sentences below, removing superfluous words such as "actual," "current," "real," "active," "existing," and similar expressions.

(a) The existing lab equipment was actually built by students.

(b) The research group has been actively studying active control of bow wave generation for the past two years.

(c) Is the currently existing schedule of courses for next year accurate?

(d) What is the real dollar value of an engineering degree?

(e) The presence of an error signal will cause the red light to blink intermittently.

(f) The department is continually and actively pursuing international exchange agreements with other universities.

(g) The architects will provide the actual design of the building next week.

(h) The primary purpose of the new CO alarm system is to notify occupants of carbon monoxide danger.

(i) The variable speed AC drives can provide real dollar savings over the life of the project.

(j) The new robotic floor sweeper automatically reverses direction if the input polarity actually reverses.

REFERENCES

Cambridge Dictionary of American English. 2000. Cambridge Dictionary of American English (Landau, S.I., ed.). New York: Cambridge University Press.

Collins Cobuild English Dictionary. 1995. Collins Cobuild English Dictionary (Sinclair, J.M., ed.). London: Harper Collins.

Cook Counseling Center. 2009. Proofreading [Internet]. Cook Counseling Center–Virginia Tech: Blacksburg, VA. Accessed 1 November 2012 at <http://www.ucc.vt.edu/stdysk/proofing.html>.

Corbett, P.B. 2011. The reader's lament [Internet].The New York Times. Accessed 31 October 2012 at <http://afterdeadline.blogs.nytimes.com/2011/10/04/the-readers-lament/>.

CSE Manual. 2006. Scientific style and format: The CSE Manual for authors, editors, and publishers, 7th ed. Reston, VA: Council of Science Editors.

Davies, M. 2013. Corpus of contemporary American English. Accessed 25 January 2013 at <http://corpus.byu.edu/coca/>.

Every, B. 2012. My 15 best proofreading tips, c2006–2012 [Internet]. BioMedical Editor. Accessed 1 November 2012 at <http://www.biomedicaleditor.com/proofreading-tips.html>.

Flesch, R. 1979. How to write plain English, Chapter 2 excerpt reproduced at <http://pages.stern.nyu.edu/~wstarbuc/Writing/Flesch.htm>. Accessed 1 February 2013.

Hacker, D., and Sommers, N. 2011. A writer's reference, 7th ed. Boston: Bedford/St. Martin's.

Hacker, D., and Sommers, N. 2012. A pocket style manual, 6th ed. Boston: Bedford/St. Martin's.

Hacker and Sommers books companion website. Accessed 12 February 2013 at <http://bcs.bedford-stmartins.com/writersref7e/default.asp#t_612701____>.

Hofmann, A.H. 2010. Scientific writing and communication: papers, proposals, and presentations. New York: Oxford University Press.

Hooks, I.F. 1993. "Writing good requirements (A Requirements Working Group information report)," Proceedings of the Third International Symposium of the NCOSE, Volume 2, 1993. Accessed 30 April 2013 at <http://www.incose.org/rwg/writing.html>.

Knisely, K. 2013. A student handbook for writing in biology, 4th ed. Sunderland, MA: Sinauer Associates, Inc.

Longman Dictionary of American English. 2008. Longman Dictionary of American English, 4th ed. White Plains, NY: Longman.

LR Communication Systems, Inc. 1999. Proofreading and editing tips: A compilation of advice from experienced proofreaders and editors, c1999 [Internet]. Berkeley Heights, NJ. Accessed 31 October 2012 at <http://www.lrcom.com/tips/proofreading_editing.htm>.

Lunsford, A.A. 2010. Easy writer, 4th ed. Boston: Bedford/St. Martin's.

Lunsford, A.A. 2011. The new St. Martin's handbook, 7th ed. Boston: Bedford/St. Martin's.

Lunsford, A.A. 2013. The everyday writer, 5th ed. Boston: Bedford/St. Martin's.

Lunsford books companion websites:
<http://bcs.bedfordstmartins.com/easywriter4e/#t_518364____>,
<http://bcs.bedfordstmartins.com/everydaywriter5e/default.asp#t_798016____>, and
<http://bcs.bedfordstmartins.com/smhandbook7e/#t_623495____>

Martin, B. 2011. Doing good things better. Ed (Sweden): Irene Publishing. Accessed 2 February 2013 at <http://www.bmartin.cc/pubs/11gt/>.

McKee, C.D. 2012. Information design: A new approach to teaching technical writing service courses, Ph.D. Dissertation, Oklahoma State University, Stillwater, OK 74078.

Purdue Online Writing Lab (OWL). 2013. West Lafayette, IN: The Writing Lab & The OWL at Purdue and Purdue University, c1995–2013. Accessed 1 February 2013 at <http://owl.english.purdue.edu/owl/>.

Romanin, V.D., and Carey, V.P. 2011. An integral perturbation model of flow and momentum transport in rotating microchannels with smooth or microstructured wall surfaces, Phys. Fluids 23 (8), 082003. Accessed 1 February 2013 at <http://dx.doi.org/10.1063/1.3624599>.

Stockmeyer, N.O. 2009. Using Microsoft Word's readability program. Michigan Bar Journal (January): 46–47.

Van Alstyne, J.S. 2005. Professional and technical writing strategies, 6th ed. New York: Pearson Longman.

Writing Center. 2012. Revising drafts, The Writing Center at UNC Chapel Hill, c2010–2012. [Internet]. University of North Carolina: Chapel Hill. Accessed 31 October 2012 at <http://writing-center.unc.edu/handouts/revising-drafts/>.

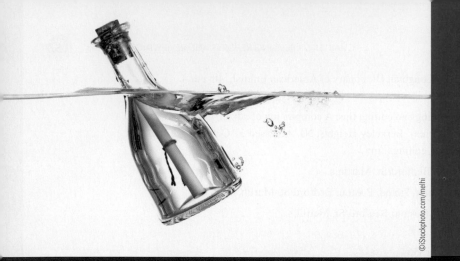

Chapter
6

ENGINEERING TOOLBOX AND VISUAL ELEMENTS

An engineering report consists of both written and visual material. The purpose of this chapter is to acquaint you with the tools you will need to analyze and prepare visual materials for your reports. In particular, this chapter will provide an overview of the following topics:

- Standard engineering symbols, units, and dimensions

- Rules for significant figures and rounding

- Types of visual elements (and when to use them)

- Guidelines for formatting figures

- Graphs of standard functions

- Graphical analysis

- Software and tools for preparing illustrations and editing images

Parts of this chapter have been adapted from *A Student Handbook for Writing in Biology*, Chapter 4. Used with permission from the publishers, Sinauer Associates, Inc. and W. H. Freeman & Co.

SYMBOLS AND CONVENTIONS

This section covers three components of an engineer's toolbox: nomenclature, units of measurement, and significant figures. **Nomenclature** is the representation of physical parameters by symbols. Because there are more parameters than letters in the English alphabet, some parameters are represented by Greek letters. Standard **units of measurement** define the magnitude of quantities such as length, mass, and time. There are two main systems of measurement: the international (metric) system (called SI units, from the French *système international d'unités*) and the U.S. customary units system. When making measurements, engineers record data with a certain precision, expressing uncertainty in terms of **significant figures**. When these measured values are used in subsequent calculations, the calculated value cannot be more precise than the least precise measurement used in its calculation. The goal of this section is to help engineering students and entry-level engineers develop fluency in reading the symbols in their fields, applying the appropriate units, and using significant figures to present data honestly.

Basic Nomenclature

Each engineering discipline and, in some cases, each subdiscipline, has a set of symbols that represents mathematically the physical parameters that are the subject of study. For example, in electrical engineering, the quantities voltage, current, resistance, capacitance, inductance, and power are commonly denoted by V (or E), I, R, C, L, and P, respectively. When you use commonly accepted symbols, your audience of fellow engineers is more likely to understand your writing than if you define your own nomenclature. Most engineering students have already learned that the relationship between voltage, current, and resistance is

$$V = IR \qquad\qquad (6.1)$$

Power, denoted as P, is the product of volts and current, and can be expressed in one of the following forms:

$$P = VI = I^2R = V^2/R \qquad\qquad (6.2)$$

Many physical parameters share a common symbol. For example, the symbol V, identified above as voltage, may also variously be used to represent volume or velocity. Often, the same symbol might appear in a different font face to denote a different parameter, such as V for voltage, **V** for velocity, and $\mathbb{V}$ for volume. To avoid confusion, all parameters should be defined in a given manuscript. Commonly used mechanical and electrical engineering nomenclature is shown in Table 6.1. Make a list of the nomenclature in your engineering field to use as a reference when you write your lab reports and other communications.

TABLE 6.1 | Commonly (*but not universally*) used engineering symbols

Symbol	Meaning	Symbol	Meaning
A	area, Helmholtz function (thermodynamics)	a	acceleration, speed of sound, specific Helmholtz function (thermodynamics)
B	breadth (linear dimension)	b	breadth (linear dimension)
C	specific heat, drag or lift coefficient, molar concentration, heat capacity, friction coefficient, capacitance	c	specific heat, speed of sound, damping coefficient
D	distance, diameter, substantial derivative, diffusion coefficient	d	diameter, distance, derivative
E	total energy, total internal energy, emissive power	e	specific energy, specific internal energy
F	force, function (math)	f	function (math)
G	center of gravity, function (math), Gibbs free energy (thermodynamics)	g	acceleration of gravity, function (math), specific Gibbs free energy
H	total enthalpy	h	specific enthalpy, convective heat transfer coefficient, elevation
I	moment of inertia, total enthalpy, total irreversibility (thermodynamics), modified Bessel function of the first kind (math), current	i	square root of −1, current, specific enthalpy, specific irreversibility (thermodynamics)
J	polar moment of inertia, radiosity (heat transfer), Bessel function of the first kind	j	diffusive mass flux, Colburn j-factor
K	modified Bessel function of the second kind (math)	k	ratio of specific heats, thermal conductivity, spring constant
L	length, inductance	l	length
M	molar mass, mass, moment, torque	m	mass
N	index, number of moles, number of teeth, fins, etc.	n	polytropic exponent, index
O	origin of coordinate system	o	origin of coordinates
P	pressure, power	p	pressure, perimeter
Q	total heat transfer, volumetric flow rate	q	heat transfer, specific heat transfer
R	radius, ideal gas constant, electrical resistance, thermal resistance	r	radius, radial coordinate, compression ratio
S	total entropy, area, conduction shape factor	s	specific entropy
T	temperature, torque	t	time, thickness
U	total internal energy, velocity, overall heat transfer coefficient	u	specific internal energy, velocity component in x-direction

TABLE 6.1 | Continued

Symbol	Meaning	Symbol	Meaning
V	voltage, volume, velocity	v	specific volume, velocity in y-direction
W	work, weight	w	velocity in z-direction
X	coordinate direction (Cartesian), exergy (thermodynamics)	x	coordinate direction (Cartesian), quality (thermodynamics), specific exergy (thermodynamics)
Y	coordinate direction (Cartesian), Bessel function of the second kind	y	coordinate direction (Cartesian), mole fraction, extraction ratio (thermodynamics)
Z	coordinate direction (Cartesian), compressibility factor (thermodynamics)	z	coordinate direction (Cartesian), elevation

The nomenclature used in the United States may differ from that used in other countries. When preparing a written communication, use the nomenclature specified in the "Instructions to Authors." When giving oral presentations in a different language or in a different country, check with your host about the "local" nomenclature. By considering the needs of your audience, you will not only avoid confusion, but show that you are sensitive to regional differences.

Greek Letters

There are more parameters in engineering than there are letters of the English alphabet. For this reason, Greek letters are customarily used to represent certain quantities. It is valuable to learn the Greek alphabet, given in Table 6.2, so that when your professor or co-worker refers to *alpha*, *nu*, *eta*, or *xi*, you will know that the corresponding symbols are α, ν, η, and ξ, respectively.

TABLE 6.2 | Greek alphabet and common engineering uses for selected Greek letters

Name	Uppercase Symbol	Uses	Lowercase Symbol	Uses
Alpha	A		α	angle, angular acceleration, absorptivity, thermal diffusivity
Beta	B		β	angle, bulk modulus
Gamma	Γ	gamma function (math), circulation (fluids)	γ	specific weight, ratio of specific heats
Delta	Δ	change in a point function	δ	angle, inexact differential boundary layer thickness
Epsilon	E		ε	small quantity, emissivity

TABLE 6.2 | Continued

Name	Uppercase Symbol	Uses	Lowercase Symbol	Uses
Zeta	Z		ζ	damping ratio
Eta	H		η	transformed y-coordinate
Theta	Θ	angle	θ	dimensionless temperature, momentum thickness, angle
Iota	I		ι	
Kappa	K		κ	
Lambda	Λ	pressure gradient function (fluids)	λ	wavelength, eigenvalue
Mu	M		μ	coefficient of friction, dynamic viscosity
Nu	N		ν	kinematic viscosity, frequency
Omicron	O		o	
Pi	Π	product function (math)	π	circumference/diameter (3.14159...)
Xi (or ksi)	Ξ		ξ	transformed x-coordinate
Rho	P		ρ	density
Sigma	Σ	summation function (math)	σ	normal stress, Stefan-Boltzmann constant
Tau	T		τ	torque, transmissivity, shear stress
Upsilon	Y		υ	
Phi	Φ	relative humidity	φ	equipotential function, relative humidity
Chi	X		χ	
Psi	Ψ		ψ	stream function
Omega	Ω	angular velocity, frequency, vorticity	ω	angular velocity, frequency, specific humidity, vorticity

To illustrate the advantage of using Greek letters, consider the quantities time, thickness, temperature, and torque, which all start with the letter t. By convention, the lowercase t is most commonly used for *time* and is often also used for *thickness* when there is no chance of confusing it with time. The uppercase T is commonly used for *temperature*. Nondimensionalized temperatures are often given by theta, θ. Torque may also be given by T, but is more commonly (but by no means universally) denoted by the Greek letter tau, τ, or on occasion by M (for the turning moment of a force acting through a distance). Similarly, the parameters density, distance, diameter, and derivative all start with the letter d. To distinguish *density* from these other parameters, the Greek letter rho (ρ) is commonly used.

TABLE 6.3 | SI and U.S. customary units

Quantity	SI Unit	U.S. Customary Unit
Length	meter, m	foot, ft
Mass	kilogram, kg	pound mass, lb_m
Time	second, s	second, s
Force	Newton, $N = kg \bullet m/s^2$ (derived unit)	pound force, lb_f
Temperature	Kelvin, K	Rankine, R
Electric current	Ampere, A	Ampere, A
Amount of matter	mole, mol	pound mole, lbmol
Intensity of light	candela, cd	candela, cd

Units and Dimensions

The international system of units (abbreviated SI) consists of seven base units of measurement and many derived units that are combinations of the base units. Table 6.3 presents the base SI units and the corresponding U.S. customary units. It is common practice to provide SI units in published papers or at least to give the SI equivalent in parentheses after the customary U.S. units. Most engineering textbooks and handbooks have unit conversion tables. One convenient online unit converter, among many, can be found at <http://www.engineeringtoolbox.com/unit-converter-d_185.html#Force>.

In many engineering subdisciplines, results may be presented in nondimensional form to eliminate the dependence on any specific set of units. Nondimensional parameters are often used to guide experimental testing to confirm theoretical results. Finding appropriate nondimensional groups and functional relationships may reduce the required number of experimental validation tests by a factor of 10 to 100. Nondimensional analysis is covered in many upper-level engineering textbooks (for example, see Fox et al, 2008 and Bergman et al, 2011).

When writing derived units, use the "•" (middle dot) to separate the component units that are multiplied and a "/" to represent "per." However, use only one "/" in any given unit. For example, in Table 6.3, the force unit in SI is a derived quantity written as $kg \bullet m/s^2$. While this quantity might be spoken as "kilograms times meters per second per second," by convention, only one "/" is used when writing the expression. Thus it would be strange to see force written as $kg \bullet m/s/s$.

Significant Figures

Measurements always have a degree of uncertainty, which may come from the measuring instrument, the object being measured, the environment, or another source. Let us consider the uncertainty due to the measuring instrument. Ideally, the measuring instrument chosen for a particular task is both precise and accurate. In engineering, precision and accuracy are expressed in terms of significant figures.

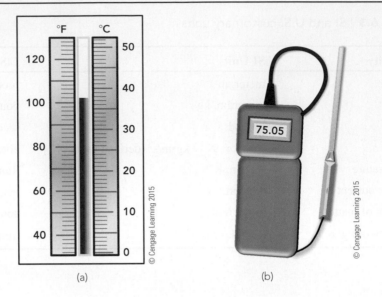

FIGURE 6.1 | Measurement of temperature using (**a**) a thermometer, and (**b**) a resistance temperature device (RTD) with a digital readout

As an example, let us compare the display on the two temperature-measuring instruments shown in Figure 6.1. The thermometer in Figure 6.1a has divisions of 2° on the Fahrenheit scale. The alcohol level indicates that the temperature is 102°F, but if the level fell between two divisions, you should be able to estimate the temperature to half of one division, or 1°F in this case. A general rule for reporting measurements is to include all of the digits you are sure of plus an estimate of the next smaller digit. Stated more formally, the number of significant digits is the number of digits known accurately plus one that is uncertain. Thus, the temperature on the thermometer in Figure 6.1a should be recorded as 102.°F, not 102.0°F, because our best estimate is to the closest 1°F, not 0.1°F. On the other hand, the resistance temperature device (RTD) in Figure 6.1b has a digital readout that displays a temperature to ±0.01°; however, the manufacturer gives the precision of the RTD as ±0.05°. Thus, the temperature registered on the display should be recorded as 75.05°F.

To determine the number of significant digits, start counting with the first nonzero digit and stop after counting the first doubtful digit (see Table 6.4).

THERMOMETER: 102. has three significant figures.

RTD: 75.05 has four significant figures.

If we use the thermometer to measure the temperature at one location and the RTD at another, any values we calculate using these measurements cannot have more significant figures than the number of significant figures in the least precise temperature measurement. For example, let $T_1 = 102.°F$ (from the thermometer) and $T_2 = 75.05°F$ (from the RTD). The temperature difference is then

$$\Delta T = T_2 - T_1 = 75.05 - 102. = -26.95°F \qquad (6.3)$$

TABLE 6.4 | Rules for determining the number of significant figures

Type of Number	Number of Significant Figures	Examples
Integer	Infinite	A gear with 8 teeth has precisely 8.00000… teeth.
Scientific notation	All digits	$3.400 \cdot 10^3$ has four significant digits. $4.7 \cdot 10^{-6}$ has two significant digits.
Number less than one	Zero is significant only when it is between two nonzero digits or it is the last digit.	0.0045 has two significant digits. 0.407 has three significant digits. 0.4070 has four significant digits.
Number greater than one	Zero after the decimal point is significant. If the number is an integer (no decimal point) that ends in zero, rewrite in scientific notation to avoid confusion.	300.10 has five significant digits. $300 \Rightarrow 3 \cdot 10^2$ (one significant digit) or $300 \Rightarrow 3.00 \cdot 10^2$ (three significant digits) $300 \Rightarrow$ see integer rule (infinite significant digits)

However, the temperature difference should be reported as $-27.0°F$ because the least precise temperature measurement has only three significant digits. The rules for rounding are given in Table 6.5. The take-away lesson is that the *number of significant figures of any calculated result cannot be greater than the number of significant figures of the values from which it was calculated.*

TABLE 6.5 | Rules for rounding

Number to Be Dropped	Change to Preceding Number	Examples
Is less than 5	No change	54.93 is rounded to 54.9
Is greater than 5	Rounded up	76.38 is rounded to 76.4
Is 5 and the preceding number is		
• Odd	• Rounded up	• 85.75 is rounded to 85.8 • 34.95 becomes 35.0
• Even	• None, just drop the 5	• 85.65 rounds to 85.6 • 34.85 rounds to 34.8

When performing operations with numbers, the following rules apply:

1. The result of **addition or subtraction** is no more accurate than the least accurate of the inputs.

   EXAMPLES: 28.5 (three significant figures) + 18.634 (five significant figures) = 47.1 (three significant figures)

   396,000 (three significant figures) − 145,596.1 (seven significant figures) = $2.50 \cdot 10^3$ (three significant figures)

2. The result of any **multiplication or division** is no more accurate than the least accurate of the inputs.

   EXAMPLES: 3.14159 (six significant figures) • 7. (one significant figure) = $2 \cdot 10^1$ (one significant figure)

   3.14159 (six significant figures) • 7.0 (two significant figures) = 22 (two significant figures)

   569,000 (three significant figures) / 436.123 (six significant figures) = $1.30 \cdot 10^3$ (three significant figures)

VISUAL ELEMENTS

Practicing engineers do not have to be convinced of the truth in the proverb "A picture is worth a thousand words." Much of their work involves tabulating data and other numerical values; plotting numerical data as graphs; creating and using sketches, drawings, and schematics; and explaining the graphics to their audience. Information is easier to process when a visual element supplements a verbal description.

Tables and figures are probably the most common visual elements encountered in technical communications. They are very separate and distinct entities.

- A **table** is defined by Webster's dictionary as "a systematic arrangement of data usually in rows and columns for ready reference."
- A **figure** is any visual element that is not a table. Thus, line (or *x–y*) graphs, bar graphs, pie graphs (also called pie charts), drawings, sketches, photographs, clip art images, and schematics are all called *figures* in engineering reports.

Although equations do indeed have a visual impact, they are technically part of the text. In this chapter, we discuss equations in the context of visual elements because they often provide the theory that guides an engineer's decision on how to plot and interpret the data.

The type of visual element you select depends on the objectives of your study, the training of your intended audience, and the nature of the data.

Use a **table** when

- The exact numbers are more important than the trend
- Statistics such as sample size, standard error, and P-values are used to support your conclusions
- The number of data pairs is relatively small (less than 6 to 10 data pairs)
- A summary of the categorical parameters and other nonquantitative information makes it easier to interpret the results

Use a **graph** to show relationships between or among parameters or variables (for more on variables and parameters, see the "What Is a Parameter" sidebar in Chapter 3).

The most appropriate type of graph for the given data is often dictated by the nature of the parameters—categorical or quantitative.

Categorical parameters are groups or categories that have no units of measurement (treatment groups, age groups, habitat, etc.). Bar graphs and pie graphs are commonly used to display results involving categorical parameters.

Quantitative parameters, on the other hand, have numerical values with units. Line graphs (*x*–*y* plots, called scatterplots in Excel) display relationships between quantitative parameters and are probably the most common type of graph used in engineering reports. Table 6.6 provides a summary of graphs frequently encountered in technical papers. These graphs are described individually in the following sections.

Not every result requires a visual element. If the result can be stated in a sentence, then **no visual element** is needed (see Example 1 on p. 198).

> A **parameter** is either
> - a directly controlled or measured variable
>
> or
> - a related dimensional or nondimensional representation calculated from the measured variable

TABLE 6.6 | Types of graphs and their purpose

Graph Type	Purpose	Examples
Histogram	To show the frequency of occurrence of a quantitative variable.	Statistical distribution of material properties, students' grades, and many other types of data; the *x*-axis shows the value of the numerical property, and the *y*-axis shows the number of occurrences.
x–y plots Known by many names: • scatterplot • scattergram • line graph All may have axes that are: • linear (Cartesian) • logarithmic • semilogarithmic	To show the relationship between two paired sets of quantitative data, which may come from experimental measurements or from theoretical calculations. When possible, use the theoretical relationship to guide selection of the appropriate type of trendline. The correlation between a trendline and the experimental data, the R^2 value, indicates the strength of the relationship.	Relationship between force and acceleration (linear); between aerodynamic forces and air velocity (F goes as V^2); between boundary work and volume change for a constant pressure process (linear); between heat transfer rates and velocities (usually a noninteger power law); or any natural law where one parameter is a function of at least one other parameter.
x–t plots (also known as a time series) Variant of x–y with time as the independent parameter along the abscissa	To show time-dependent measured or calculated data in vibrations, electrical oscillations, control systems, transient aerodynamics and transient heat transfer, among other fields.	Decaying damped vibration signal to determine damping; transient exponential decay or growth.

TABLE 6.6 | Continued

Graph Type	Purpose	Examples
Area graph (also known as a surface graph)	To show the behavior of multiple related or similar components changing relative to a common parameter (usually time). Illustrates the contribution of components to the whole as the independent parameter varies.	Time history of utilization of energy sources; time history of energy consumption by sector; growth of corporate assets by division or product line over time.
Bar graph	To show the distribution of a categorical (nonquantitative) variable.	Consumption of resources. One axis shows the resource category, and the other shows the numerical value of consumption.
Pie graph	To show the distribution of a categorical (nonquantitative) variable in relation to the whole. All categories must be included so that the pie wedges total 100%.	Distribution of energy consumption by energy source. Each wedge represents the percentage of power from the energy source. Sources with low representation may be combined into an "Other" wedge to complete the pie.

Equations

Equations are neither tables nor figures; they are a part of the text, but they do have a highly visual character. Equations should be set off from the rest of the text on a separate line. In technical writing, common practice is to number equations sequentially so that you can refer to them later unambiguously (this is how the equations are formatted in this chapter). This guideline may be violated for relatively simple equations that you do not plan to reference later in your text. These simple, unreferenced equations may be included in the body of the text.

Do not type equations as normal running text. Use an equation editor. Many equation editors have standard equations that can be inserted with just a mouse click. Alternatively, modify one of the equation structures. Symbols, Greek letters, and operators can be inserted into the equations, and elements can be subscripted and superscripted. Appendix I provides details on typing equations with the equation editor in MS Word.

To produce a centered equation with a right-aligned equation number, set a centered tab stop at the midpoint of the line, and then a right-aligned tab stop at the right margin of the line (see "Tech Tip—Aligning Equations" in Chapter 3). Once you set the tabs, type your equation in the equation editor (in MS Word, **Insert** (→Symbols) **| Equation**), exit the equation editor, insert a tab, and then type the right-aligned equation number. When preparing your manuscript for publication, check with the journal editor on how equations are to be incorporated into the text, as some typesetting programs have difficulty translating tab stops from word processing files.

The inclusion or exclusion of punctuation in equations on a separate line from the text appears to vary. One school of thought is not to include anything in an equation line that is not part of the equation. On the other hand, other sources argue that the equation is part of a sentence and should be punctuated as needed. We prefer no punctuation in a separate equation line.

Tables

Tables are used to display large quantities of numbers and other information that would be tedious to read in prose. Tables are formatted for simplicity to focus the reader's attention on the message, with minimal distractions from lines and extraneous text.

In engineering reports, do not include a table when a graph shows the same data. Make *either* a table *or* a graph—not both—to present a given data set. If the recipient of your report wants to see the raw data, you may be asked to attach the tabulated results as an appendix.

Tables can be constructed in either Microsoft Word (see "The Insert Tab" section in Appendix I) or Microsoft Excel (see Figure A2.7 in Appendix II).

Table Format   To make an effective table, follow these guidelines:

- Arrange the categories vertically rather than horizontally, as this arrangement is easier for the reader to follow (see, for example, Table 1.2 in Figure 6.2).

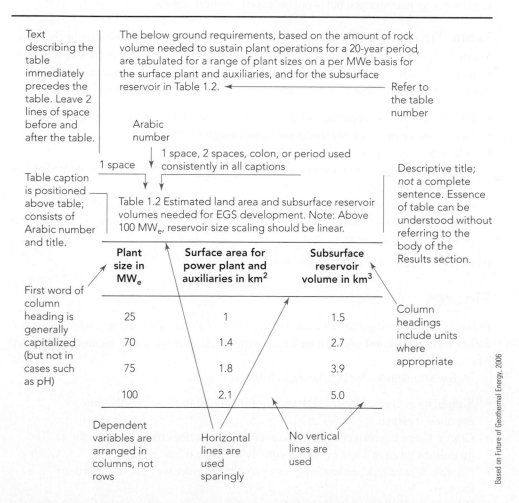

Text describing the table immediately precedes the table. Leave 2 lines of space before and after the table.

The below ground requirements, based on the amount of rock volume needed to sustain plant operations for a 20-year period, are tabulated for a range of plant sizes on a per MWe basis for the surface plant and auxiliaries, and for the subsurface reservoir in Table 1.2.

Refer to the table number

Arabic number

1 space

1 space, 2 spaces, colon, or period used consistently in all captions

Table caption is positioned above table; consists of Arabic number and title.

Table 1.2 Estimated land area and subsurface reservoir volumes needed for EGS development. Note: Above 100 MW$_e$, reservoir size scaling should be linear.

Descriptive title; *not* a complete sentence. Essence of table can be understood without referring to the body of the Results section.

First word of column heading is generally capitalized (but not in cases such as pH)

Plant size in MW$_e$	Surface area for power plant and auxiliaries in km^2	Subsurface reservoir volume in km^3
25	1	1.5
70	1.4	2.7
75	1.8	3.9
100	2.1	5.0

Column headings include units where appropriate

Dependent variables are arranged in columns, not rows

Horizontal lines are used sparingly

No vertical lines are used

Based on Future of Geothermal Energy, 2006

FIGURE 6.2 | Excerpt from a Results section showing a properly formatted table preceded by the text that describes the data in the table

- List the items in a logical order (e.g., sequential, alphabetical, or increasing or decreasing value).
- Include the units in each column heading, not after each number entry in the table.
- Limit the number of lines. Use horizontal lines to separate the table caption from the column headings, the headings from the data, and the data from any footnotes. Do not include an external border or use vertical lines to separate the columns.

Table Captions   Each table in an engineering report has a consistently-formatted caption that includes a **number** and a **title**. The caption is either centered or aligned on the left margin *above* the table, as shown in the example in Figure 6.2. In this example, the authors of the cited study left aligned their table caption. In this book, captions for both tables and figures are left aligned (with a special format used by the publisher).

Tables are numbered so that they can be referred to unambiguously in the body (text) of the report. Use Arabic numbers and number the tables consecutively in the order they are discussed in the text. Number tables independently of figures. Notice that in this book, the table and figure numbers are preceded by the chapter number. This system helps orient the reader in long manuscripts, but is not necessary in short reports.

Table Titles   The title succinctly describes the information presented in the table. When composing a table title, put yourself in your reader's shoes: Aim for maximum information content with the fewest number of words. While brevity and accuracy are the two most important considerations, other attributes of an effective title include:

- Use of a precise noun phrase rather than a full sentence for simple table titles
- Use of one or more full sentences for more complex tables
- Adherence to the rules of English grammar:
  » Do not capitalize *common* nouns (*general* classes of people, places, or things) unless they begin the phrase or sentence.
  » Capitalize *proper* nouns (names of *specific* people, places, or things).
  » Do not capitalize words that start with a lowercase letter (for example, pH), even if they begin a sentence.

Examples of faulty and revised titles are shown in Table 6.7.

Figures

Figures include all elements that are not text or tables. Labels such as *Photograph 1, Sketch 3*, and *Drawing 5* are **not** used. In technical writing, all of these visual elements are called *figures*.

A few simple rules for figures are as follows:

- Simplicity is crucial. Restrict the scope of the figure to show exactly your intended message.
- Craft a figure title that concisely and accurately describes the data in the figure. The guidelines given in Table 6.7 apply equally to figure titles.
- Number figures independently of tables and in the order they are discussed in the text.
- Whenever possible, use technical charting software (SigmaPlot, Kaleidagraph, Deltagraph) rather than business spreadsheet software such as Microsoft Excel.

TABLE 6.7 | Examples of faulty titles and how to correct them

Faulty Title	What's Wrong	Correction
Table 1.2 The Estimated Land Area And Subsurface Reservoir Volumes Needed For EGS Development.	Do not capitalize common nouns; do not start a title with the word "the"	Table 1.2 Estimated land area and subsurface reservoir volumes needed for EGS development
Table 1.2 Table of estimated land area and subsurface reservoir volumes needed for EGS development.	Do not start a title for a graphic with a description of the graphic	
Table 1.2 shows the estimated land area and subsurface reservoir volumes needed for EGS development.	Separate the table number and the title	
Table 1.2 Subsurface reservoir volumes	Do not write vague and undescriptive titles	
Table 1.2 Estimated land area and subsurface reservoir volumes needed for EGS development for 25, 50, 75, and 100 MW$_e$ plants.	Do not make titles excessively detailed; avoid repeating column headings	

When preparing graphs:

- Use two-dimensional charts as a rule; three-dimensional charts add clutter and make it difficult to compare data sets; "pseudo-3D" effects are meaningless and may obscure other data.
- Use simple black and white format; eliminate gray-scale shading, patterning, text marking of data series, and other effects; use only those items that are essential to make your point.
- Use a line for continuous data (theory); plot individual data points as a scatterplot, a histogram, or a column chart.
- Vary line type for different sets of continuous data; do not rely on color differences.
- Plot no more than six data sets on a single graph (unless there is a good reason).
- Label each axis clearly and include the units in parentheses (or alternatively in square brackets).
- Include a legend (data set identification) if there are multiple data sets.
- Position the legend preferentially (1) on the plot area, (2) immediately beneath the graph, (3) as part of the caption, or (4) immediately above the graph.
- Give each graph a caption (positioned *below* the figure in a report) or a title (positioned *above* the figure in a presentation).
- Do not display gridlines, chart shading, and text callouts of data unless there is a good reason for doing so.
- Do not simply accept Excel defaults—the default format is not standard for technical reports.

We frequently refer to MS Excel in the following sections when we provide examples of different types of graphs. While Excel is not the only or the best software available for making graphs, it is a good plotting program for novices for the following reasons:

- Data input and subsequent plotting of these data is relatively straightforward in Excel.
- Excel is readily available and is included in the Microsoft Office suite of computer software.
- If your organization has Excel on its computers, you can probably get personal assistance from an information technology or computer support staff member.

You may eventually switch to a higher-powered plotting program such as SigmaPlot, Kaleidagraph, or Deltagraph, but the experience gained by working with Excel should make this transition easier.

x–y Graphs

An *x–y* graph displays the relationship between *two or more quantitative parameters or variables*. The **independent variable** is the one that the engineer changes or manipulates. The **dependent variable** changes in response to changes in the independent variable. By convention, the independent variable is plotted on the horizontal or *x*-axis. The dependent variable is plotted on the vertical or *y*-axis. If there is not a causative relationship between the two parameters, then it does not matter which parameter is plotted on which axis.

Coordinate Axes
The coordinate axes for *x–y* graphs are selected based on the patterns shown by the numerical data. Three common axes are

- Linear (also called rectangular or Cartesian)
- Logarithmic (also called log-log)
- Semi-logarithmic

Linear coordinates are the most frequently used grid (see Figure 6.3a). The *x*- and *y*-axes are both linear, and the intervals are uniform. In other words, the distance between major divisions on each axis is constant. For ease of reading, the increments of the major divisions are multiples of 1, 2, or 5.

Graphs with logarithmic coordinates have a logarithmic scale on both axes (see Figure 6.3b); semi-log plots have one logarithmic and one linear axis (see Figure 6.3c). On a log scale, the distance plotted between successively higher numbers becomes successively smaller. The advantage of using a logarithmic scale is that a wide range of values, whose smallest and largest values differ by several orders of magnitude (powers of 10), can be plotted without loss of resolution. Furthermore, data that appears as a power law ($y = ax^n$)

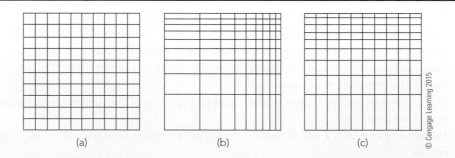

(a)   (b)   (c)

© Cengage Learning 2015

FIGURE 6.3 | Coordinate grids found in engineering graphs: **(a)** Cartesian (linear), **(b)** log-log, **(c)** semi-log

when plotted on linear coordinates can be transformed to a linear form on log-log coordinates (see the section "Applying Graphical Analysis"). Similarly, exponential functions appear as straight lines on semi-logarithmic coordinates.

Data Points—Fit a Line to Them or Not?

After having plotted data on an *x*–*y* graph with the appropriate coordinate grid, how do you decide whether to leave the points unconnected, add a line of best fit, or connect the points with straight or smoothed lines? The answer to this question depends on the purpose of your study and the nature of the data. Several engineering studies provide examples.

Points Left Unconnected   An *x*–*y* graph in which the points are not connected is called a **scatterplot**. Quite often a scatterplot represents a preliminary way to display measured data. If the purpose of the study was to determine the relationship between the parameters, the engineer would be more likely to recognize a trend from the scatterplot than from just the numbers in a table.

Figure 6.4 shows the characteristic head for a centrifugal pump as a function of the flow rate. Although the trend can be easily recognized in the scatterplot, adding the appropriate trendlines, as in Figure 6.5, makes the head-flow rate characteristic exceedingly clear. The second-order curve fit for pump head (proportional to pressure) as a function of mass flow rate (proportional to velocity) conforms to the theoretical expectation that pressure varies with the square of velocity.

Two points should be noted in Figures 6.4 and 6.5. The legend is included in the caption as an example of doing so, in case an organization or publisher requires no annotation on the plot surface. The use of scientific notation (for example, 1.0×10^{-4}) along the axes is usually recommended when the number of decimals exceeds the length of the scientific notation text. Council of Science Editors (2006) recommends using scientific notation for values of 10^4 and greater and for values of 10^{-4} and smaller.

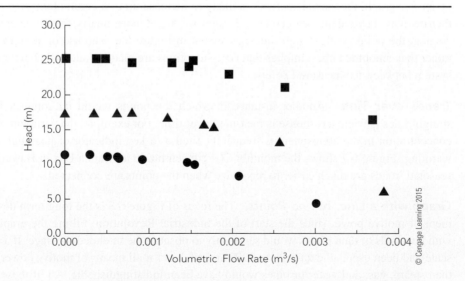

FIGURE 6.4 | Head characteristics of a centrifugal pump. Circles are for an operating speed of 2,000 RPM, triangles for 2,500 RPM, and squares for 3,000 RPM.

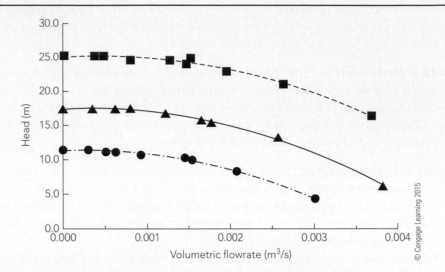

FIGURE 6.5 | Head characteristics of a centrifugal pump with second-order polynomial trendlines. Circles are for an operating speed of 2,000 RPM, triangles for 2,500 RPM, and squares for 3,000 RPM.

Points Connected by Straight or Smoothed Lines In experimental studies, the independent variable is usually changed at discrete intervals that are convenient to measure. For example, a biochemical engineer who is designing a novel heat-resistant enzyme might measure the activity of this enzyme at a few selected temperatures over the range of interest, not at every degree or even half-degree. When the sampled data points are plotted on an *x–y* graph, the enzyme activity is likely to increase up to a certain temperature and then decrease beyond that temperature. To show this trend more clearly, the engineer would connect the points with straight lines, as shown in Figure 6.6. The use of straight lines, rather than smoothed lines, implies that no assumptions are made about the behavior of the system between the measured points.

Trends over Time Another instance in which the points would be connected with straight lines is when a response is measured over time. For example, the variation of CO_2 concentration in the atmosphere is frequently cited as a key indicator of potential global warming. Figure 6.7 shows the monthly CO_2 concentration at Mauna Loa in Hawaii. The seasonal trends are much easier to recognize when the points are connected.

Graph with a Line, but no Points The focus of Figure 6.8 is the long-term development of motive power since the start of the Industrial Revolution. Filling the graph area with hundreds of data points would serve only to obscure the intended message. If a linear scale had been used, all data prior to 1900 and all data for all means of motive power other than steam, gas, and water turbines would have been indistinguishable. All of these other data are several orders of magnitude smaller than steam, gas, and water turbine values.

Linear Best-fit Lines If the data points lie close to a straight line and the *x* and *y* parameters are *expected* to have a linear relationship, a best-fit line (least-squares regression line)

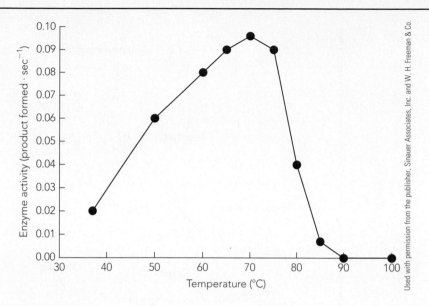

Used with permission from the publisher, Sinauer Associates, Inc. and W. H. Freeman & Co.

FIGURE 6.6 | Effect of temperature on the activity of a fictitious heat-resistant enzyme

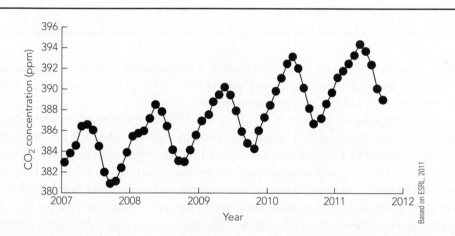

Based on ESRL, 2011

FIGURE 6.7 | Monthly average CO_2 concentration at Mauna Loa, Hawaii

can be added to the graph. Most graphing calculators and software programs with statistics capability will calculate the equation of the line for you.

A calibration curve is a common example in engineering when a **least-squares regression line** and its equation are used to predict one parameter when the other is known. Recall that the equation of a straight line is

$$y = mx + b \tag{6.4}$$

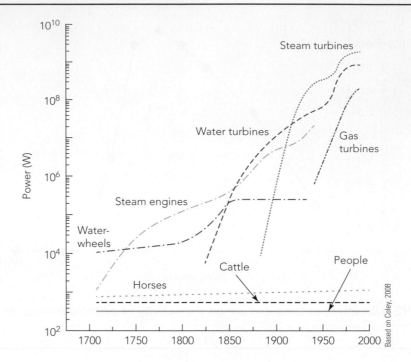

FIGURE 6.8 | Historical development of motive power from 1700 to 2000 showing an effective use of a logarithmic scale on the y-axis of a line graph

where m is the slope and b is the y-intercept. If x is known, for example, a rough estimate for y could be made simply by reading up and over on the graph. Substituting x into the equation would result in a more precise value for y.

Figure 6.9 shows a calibration curve for an orifice flow meter in flow with relatively large-amplitude low-frequency pulsations. The flow rate was obtained by capturing the system discharge in a bucket over a measured time and weighing the bucket before and after filling. Practical considerations did not permit capturing the discharge over a sufficient time period to obtain a representative average. The flow rate is expected to vary with the square root of the pressure drop across the orifice plate. By plotting the square of the measured flow rate as a function of the measured pressure drop (as a percent of full-scale meter deflection), a calibration equation was generated from the linear fit to the data.

On other occasions, a calibration equation might be sought for devices or systems that are known to have nonlinear characteristics. If one suspects an **exponential relationship** between dependent and independent variable, a semi-log plot is ideal for confirming this exponential dependence. Figure 6.10 shows a scatterplot with an exponential trendline fit for the loss coefficient for a butterfly valve as a function of the opening angle of the valve. The R^2 value indicates that the data are well represented by the straight line on the semi-log coordinates. The advantage of using a semi-log plot is the linearity of the resultant curve fit, permitting a very quick visual check of the appropriateness of the chosen trendline.

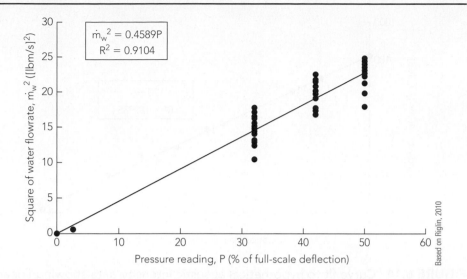

FIGURE 6.9 | Calibration of an orifice flow meter in a flow with low-frequency pulsations, yielding a calibration curve with a better than 90% regression coefficient

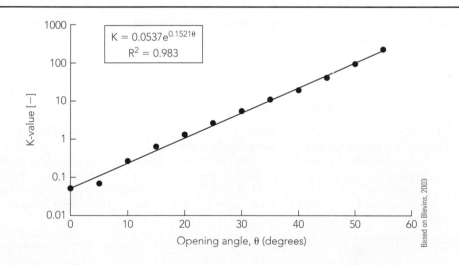

FIGURE 6.10 | Loss coefficient value for flow through a butterfly valve as a function of the angle of the valve

Other devices, systems, and phenomena may have **power law relationships** between input and output parameters. A **power law** is expressed as the following relationship between x and y:

$$y = Ax^n \tag{6.5}$$

(called *trendlines* in Excel) might fit the data reasonably well. However, the curve fit selected should agree with theory.

Knowing standard functions helps you recognize how two parameters are related. In the following sections, we will describe the equation and form of several common functions. By understanding how these functions behave on different coordinates, it may be possible to identify the correct equation for the relationship between any two parameters of interest.

Rectangular Coordinates (or Cartesian Coordinates)

Rectangular coordinates are the most frequently used type of axes. The axes are marked with uniform scales, in which the distance between major divisions is constant. Recommended increments of major divisions are 1, 2, or 5 to an appropriate power for ease of reading.

TABLE 6.9 | Hypothetical data for plotting exercise

Voltage (V)	Current (A)
0.00	0.0005
5.03	0.006
9.92	0.0111
14.99	0.016
19.95	0.0211
25.04	0.026
29.9	0.0311
34.98	0.0361
40.07	0.041
44.95	0.0459

For example, consider the data in Table 6.9, which could be obtained from the measurement of current (in amps) through an electrical resistor as a function of voltage (in volts). If, in our experiment, we varied the voltage (which is the independent parameter) and measured the current (which depends upon the value of the voltage), then we should plot the voltage along the *x*-axis as the independent parameter and the current along the *y*-axis as the dependent parameter.

When we plot the pairs of data from Table 6.9, we see that the points come very close to falling on a straight line, as shown in Figure 6.22. Such a relationship is said to be linear. Equation 6.4 (repeated here as Equation 6.6) provides the mathematical expression for any straight line as

$$y = mx + b \tag{6.6}$$

where *m* is the slope and *b* is the *y*-intercept. The slope, *m*, can be found by taking the change in the *y*-value from one point to the next and dividing it by the change in the *x*-value. Mathematically, we can write an expression for the slope as follows:

$$m = \frac{\Delta y}{\Delta x} = \frac{y_2 - y_1}{x_2 - x_1} \tag{6.7}$$

where (x_1, y_1) and (x_2, y_2) are any two data pairs. Further, whenever the data points fall on a straight line, the value of *m* is independent of the pairs of data points we select to use in the calculation. Try it! You will find that regardless of the two sets of points you choose on Figure 6.22, the value of *m* will always turn out to be 0.001 (within the accuracy of the data). For any data set that forms a straight line, any pairs of data points can be used in Equation 6.7 to calculate the slope *m*. For data sets that do not form a straight line, however, the slope *m* depends on the points you choose for its calculation.

To generate a "best-fit" line to the data, the linear regression algorithm in Excel can be used. Knowing that there can be no current flow without an imposed voltage, the first value of current appears to contain a small (and perhaps systemic) error. The

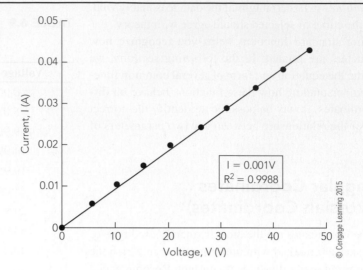

FIGURE 6.22 | Linear relationship between input voltage and measured current through a resistor. The figure is formatted appropriately: Tick marks extend inward from the axes, grid lines are absent, the background color is white, the equation of the regression line is displayed, and there is no border around the figure.

source of this error should be carefully investigated, but we can instruct Excel to force the regression line to include the (0, 0) point. This means the value of b in Equation 6.6 will be zero.

One of the fundamental reasons engineers plot their experimental data is to see if the data form a straight line. If this is the case, then you can find either the x-value or the y-value when the other value is known by using the best-fit equation.

Logarithmic Coordinates

Many relationships in engineering are not linear. For example, consider a power law equation, one in the form of Equation 6.5, here rewritten with different constants as

$$y = bx^c \tag{6.8}$$

For an assumed value of $b = 1.5$ and for values of c ranging from -5 to 5, Equation 6.8 can be plotted on linear coordinates, as in Figure 6.23. Note that the slopes of the curves are functions of the x-coordinate, not constant as it was for the straight line examined previously. When this same equation is plotted on coordinates with logarithmic scales on both axes (called a log-log plot), however, the curves are transformed into straight lines (Figure 6.24). Notice that on a log-log plot, the distance between any two numbers along the axis is proportional to the difference in their logarithms. This means that the distance from 1 to 2 is proportional to the log (2) − log (1), and this is the same as the distance between 10 and 20, log (20) − log (10). The distance between successively higher numbers becomes successively smaller.

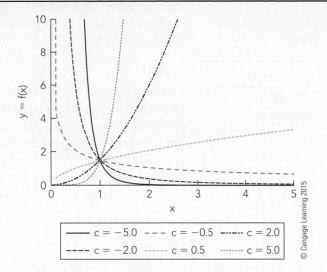

FIGURE 6.23 | An example of the function $y = bx^c$ where $b = 1.5$ and $-5.0 \leq c \leq 5.0$ plotted on Cartesian (linear) coordinates

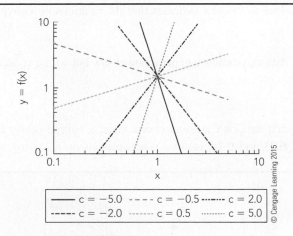

FIGURE 6.24 | Plotting power law functions, $y = bx^c$ with $b = 1.5$ on log-log coordinates yields straight lines for all values of c. Note the relative expansion of the distance from 0.1 to 1.0 and the compression of the distance from 1 to 5 when compared with the same data in linear coordinates in Figure 6.23.

Graphing Equation 6.8, for a range of c values, on log-log coordinates produces straight lines. Straight lines would also result from taking the logarithm of both sides of the equation and plotting these log values in rectangular coordinates. (Try this as an exercise with the data in the next example.) An added advantage of the log-log plot is that data over

TABLE 6.10 | Specific volume of saturated steam varies with the absolute pressure.

v (ft^3/lbm)	p (psia)
26.43	14.70
22.40	17.53
19.08	20.80
16.32	24.54
14.04	28.83
12.12	33.71
10.51	39.25
9.147	45.49
7.995	52.52

© Cengage Learning 2015

multiple orders of magnitude, that is, data ranging from very small to very large numbers, can be clearly displayed.

EXAMPLE

Consider the data in Table 6.10 taken from experimental measurements on saturated steam (i.e., steam just on the verge of beginning to condense). Let v = specific volume of the saturated steam and p = pressure. Figure 6.25 shows these data plotted in rectangular coordinates; note that the general shape of the curve appears to be either parabolic

$$y = a/x^2 = ax^{-2} \qquad (6.9)$$

or hyperbolic

$$y = a/x = ax^{-1} \qquad (6.10)$$

By taking the logarithm of either equation and adding a constant, the same data form a straight line on log-log coordinates, as shown in Figure 6.26. The R^2 value of 1 indicates that the power law regression curve fits the data perfectly. The linear characteristic of the log-log plot of the data confirms this power law relationship. In this example, the exponent in the curve fit is very close to -1 and indicates that the relationship is hyperbolic rather than parabolic.

On occasion, data may follow a general power law but with a vertical offset. Such data are described by

$$y = a + bx^n \qquad (6.11)$$

If one plots such data on Cartesian coordinates, a curved power law line, similar to those in Figures 6.23 and 6.25, results. When plotted on a log-log scale, the data will

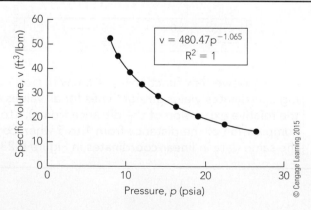

$$v = 480.47p^{-1.065}$$
$$R^2 = 1$$

© Cengage Learning 2015

FIGURE 6.25 | Specific volume of saturated steam has a nonlinear pressure dependence.

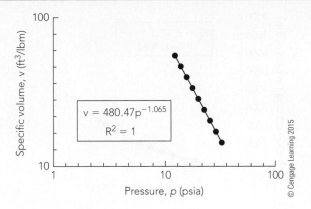

© Cengage Learning 2015

FIGURE 6.26 | Specific volume of saturated steam displays a linear characteristic with pressure when plotted on log-log coordinates.

approach a straight line, but the constant a prevents the data from forming a perfectly straight line. To get a straight line on log-log coordinates, the value of a must be determined and subtracted from each y value before plotting, making the right-hand side of Equation 6.11 look like that of Equation 6.8.

To determine the value of a, one selects two data points, (x_1, y_1) and (x_2, y_2). From either the raw data or from a rough plot of the raw data, one picks a third point (x_3, y_3) such that $x_3 = (x_1 \cdot x_2)^{1/2}$. The value of a can then be estimated by

$$a = \frac{y_1 y_2 - y_3^2}{y_1 + y_2 - 2y_3} \qquad (6.12)$$

Once the value of a is known, the quantity $(y - a)$ can be plotted versus x on log-log coordinates, and the resulting plot will approximate a straight line (as in Figure 6.26).

EXAMPLE

As an example of a power law with vertical offset, consider the following experimental data concerning the horsepower required to drive a ship through the water at a selected velocity (Table 6.11). Velocity, as the independent variable, is plotted on the x-axis, and power, as the dependent variable, is plotted on the y-axis. When these data are plotted in rectangular coordinates, as shown in Figure 6.27, the required horsepower is seen to increase with some power of velocity, but this curve does not appear to intersect the origin if extrapolated. A power law equation for the data gives a good fit for low to mid-range velocities, but deviates at the high end of the speed range. While the overall correlation

TABLE 6.11 | Engine power output required to produce a given ship velocity

Velocity (knots)	Power (hp)	hp-129 (hp)
4.	210.	81.
5.	290.	161.
6.	400.	271.
7.	560.	431.
8.	780.	651.
9.	1084.	955.
11.	1810.	1681.
12.	2300.	2171.
16.	5380.	5251.

© Cengage Learning 2015

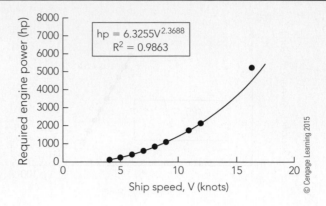

$$hp = 6.3255V^{2.3688}$$
$$R^2 = 0.9863$$

© Cengage Learning 2015

FIGURE 6.27 | Required engine power to propel a ship through the water at a specified speed

appears to be acceptable with an R^2 of 0.9846, there is room for improvement in the high-velocity range.

To improve the curve fit, the data are plotted on log-log coordinates, as shown by the open circles in Figure 6.28. This curve is not a straight line but tends toward a straight line at the larger values of velocity. Such a trend indicates that a constant, a, is needed in the equation, as shown in Equation 6.11. To find the value of this constant, we will choose the velocity values $x_1 = 4$ and $x_2 = 16$ judiciously so that x_3 is an integer value ($x_3 = (x_1 x_2)^{1/2} = 8$). Substituting the corresponding y values (power in hp from Table 6.11) into Equation 6.12

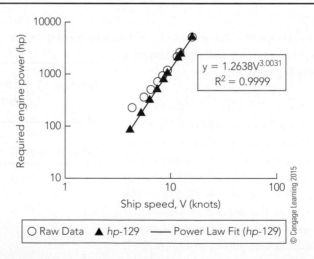

$$y = 1.2638V^{3.0031}$$
$$R^2 = 0.9999$$

O Raw Data   ▲ *hp*-129   —— Power Law Fit (*hp*-129)

© Cengage Learning 2015

FIGURE 6.28 | Ship power requirements for a specified speed showing evidence of an additive constant (circles) and with the constant subtracted off (triangles). The *y*-value in the solid line power law correlation is *hp* − 129.

yields $a = 129.4$. The third column of Table 6.11 shows the result when 129 was subtracted from each of the original power values.

When the $hp - 129$ values are plotted (as triangles in Figure 6.28), the data fall on a straight line and the correlation coefficient R^2 improves to 0.9999. The regression line equation can be rewritten by replacing y with $hp - 129$ and rounding for significant figures. Note that the resulting equation that describes the data in Table 6.11 has the same form as Equation 6.11:

$$hp = 129 + 1.26\,v^3 \tag{6.13}$$

Equation 6.13 allows engineers to calculate the power required for any velocity. To prove to yourself that this equation is valid, program your calculator with this equation to check that it accurately reproduces (to within a few percent) the data in Table 6.11.

Semi-Logarithmic Coordinates

There is a third characteristic type of data that cannot be plotted as a straight line in either rectangular or log-log coordinates. This type of data is exponential data, typically described by an equation of the form

$$y = ae^{bx} \tag{6.14}$$

where a and b are both constants. To generate a straight line for this type of data, it is necessary to plot the independent parameter, x, on a logarithmic axis while plotting the dependent parameter on a linear (rectangular) axis. When one axis is logarithmic and the other is linear, the coordinates are called **semi-logarithmic**. If we let $w = \ln(y)$, the inverse function of the exp (bx) function, then we can rewrite Equation 6.14 as

$$w = \ln(y) = \ln(a) + \ln(e^{bx}) = \ln(a) + bx \tag{6.15}$$

Equation 6.15 shows that w is a linear function of x, indicating that if we plot $\ln(y)$ versus x, we should expect a straight line.

EXAMPLE

Consider how atmospheric pressure varies with elevation. Table 6.12 provides typical barometric pressure variation with altitude when z is in ft and p is in inches of mercury. A plot of the data in Table 6.12 will not form a straight line in either rectangular or log-log coordinates. When the pressure is plotted on a logarithmic vertical axis and the elevation on a linear horizontal axis, as in Figure 6.29, a straight line is obtained. The slope of the line is equal to the constant b, and the y-intercept is equal to the constant a in Equation 6.15.

If there is an additive constant c on the right-hand side of Equation 6.14, that is,

$$y = ae^{bx} + c \tag{6.16}$$

TABLE 6.12 | Variation of atmospheric pressure with elevation above sea level

Elevation, z (ft)	Standard Pressure, p (in Hg)
0.	30.
886.	29.
2753.	27.
4763.	25.
6942.	23.
10,593.	20.

© Cengage Learning 2015

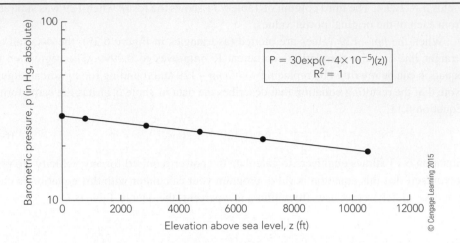

FIGURE 6.29 Exponential variation of barometric pressure with altitude plotted on semi-log coordinates

its value can be estimated by picking two arbitrary points (x_1, y_1) and (x_2, y_2), and from them determining a third point (x_3, y_3) where $x_3 = 1/2(x_1 + x_2)$. The value of c is then given by

$$c = \frac{y_1 y_2 - y_3^2}{y_1 + y_2 - 2y_3} \qquad (6.17)$$

A straight line is then obtained by plotting $(y - c)$ on the logarithmic ordinate and the value of x along the linear abscissa.

APPLYING GRAPHICAL ANALYSIS

As a practicing engineer, you are likely to work on different projects that involve the analysis of data. In an experimental study (Figure 6.30), you would first familiarize yourself with what is already known about the topic. After reading published papers (see Chapter 2), proprietary company reports, and other relevant literature and/or discussing the project with experts in the field, you would plan and carry out your experiments. After collecting data, you would select an appropriate visual element that helps you understand what the results mean. x–y graphs are one of the most frequently used graphs in engineering reports because they show the relationship between two parameters. When the relationship is linear, the equation of a straight line makes it easy to determine one parameter when the other is known. When the relationship is not linear, you would call upon your experience and training with the form of standard functions and, when possible, transform these functions into a straight line (see the section "A Little Math: Graphs of Standard Functions"). During the analysis process, it is important to correlate the data with theory in order to determine an appropriate final form of the graph.

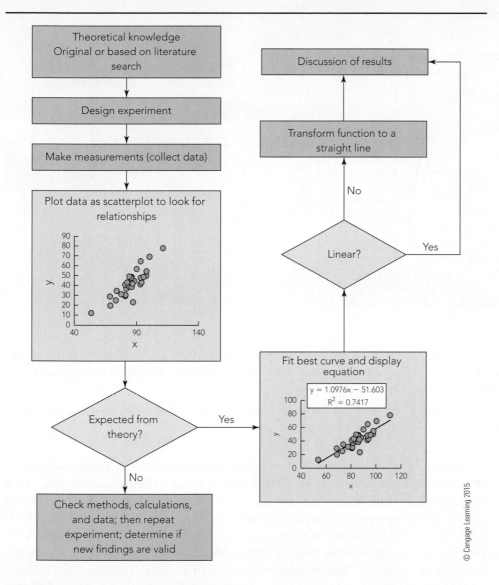

FIGURE 6.30 | Overview of the process from planning an experimental study to analyzing the data

DRAWING SCHEMATICS, DIAGRAMS, SKETCHES, AND ILLUSTRATIONS

Figures are not used just to display the results of a study; they can also enhance the reader's understanding of the procedures section of a technical report. For example, the test setup might be depicted by a photograph, an illustration, or a schematic. Diagrams are useful for illustrating processes and conceptual relationships. Many tools are available for preparing visual elements, as most students already know.

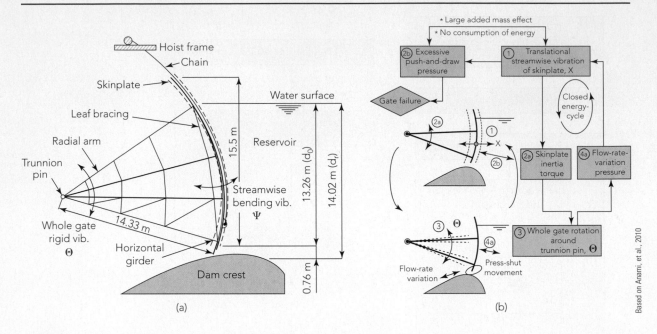

FIGURE 6.31 | Examples of drawings of structures and processes used to illustrate technical reports. (**a**) A side view of a large radial gate and (**b**) the mechanism by which such gates can undergo self-excited vibration

Microsoft Office Word allows you to draw a limited number of shapes and objects and combine them to create moderately complex visual elements (see Appendix I for examples). Microsoft Office Visio provides a broader range of menu-driven flowcharting and schematics options. The two images of moderate complexity shown in Figure 6.31, for example, could be created with the drawing tools in the MS Office Suite. OmniGraffle provides menus with a similar range of shapes, objects, and templates for Mac users and for iPAD applications. Other drawing packages such as Adobe Illustrator, Corel Draw, and Macromedia FreeHand provide even greater flexibility. The learning curve for using these drawing packages is steeper than that for MS Office software. Whether you decide to invest your time in learning how to use dedicated drawing software will be determined, in part, by how often you will need to use it in your work. In a corporate environment, there may be a graphics department to assist in the preparation of professional grade illustrations.

Engineering students who have access to CAD packages such as AutoCAD, SolidWorks, and ProE can produce drawings of components and assemblies. Exporting the drawings as JPEG, PDF, or PNG is usually straightforward; the format is selected in the **Save as** dialog box.

Processing Images

To produce publication-quality images, at least one of these postprocessing actions may be required:

- Adjustment of exposure, brightness, and contrast; the settings may be acceptable for screen viewing, but they may not be optimal for printing
- Removal/reduction of image distortion or noise
- Removal of proprietary information
- Removal of artifacts—grease-proof pencil marks, scratches, and reflections
- Insertion of appropriate annotation—markers and text
- Scaling and resizing—image too large or too small
- Sharpening for publication

Leave any resizing and sharpening effects to the very end, as these are destructive to image quality if done repeatedly.

Regardless of the software you use to prepare visual elements, there may be some reduction in the quality of the images when they are transferred to your report. To inform your use of various available file formats and their expected degradation, see Table 6.13.

TABLE 6.13 | Graphic file format characteristics (after Wang, 2008)

File Characteristics	Description	Examples
Lossless formats	Preserves the quality of the original. Can be "losslessly compressed" (typically, 2:1 compression) without degradation. TIFF and PSD formats permit layer information to be preserved and are universally supported by professional publishing software	Photoshop (PSD) Tagged Image File Format (TIFF) Portable Network Graphics (PNG)
Native capture formats	Preserves all the quality of the original but requires significant editing (e.g., adjustment of window and level, resolution, color mode, etc.) before being ready for print or screen publication	RAW (native capture) format specific to camera manufacturer DNG (Adobe digital negative), a "universal" RAW format DICOM original files
Lossy formats	Deliberately discards data to obtain much smaller file sizes to facilitate both storage and speed of distribution. GIF is very poor for continuous-tone images. High levels of JPEG compression cause marked loss of quality.	Joint Photographic Expert Group (JPEG) Graphics Interchange Format (GIF)

Based on Wang, 2011

SUMMARY

Following the guidelines presented in this chapter and with a little bit of practice, you will be able to prepare professional quality engineering graphs of any data set you might record in a lab. Certainly, professional looking graphs will do much to enhance your professional reputation in your postcollegiate job. Although it will become easier to make graphs with practice, you still need to use your knowledge of how systems and processes behave to decide on the appropriate way to plot and interpret the data.

After having read this chapter and the previous two on technical writing principles, you should have a better idea of the technical report format, the accepted structure of the component parts of the report, and the ways to display and analyze data.

EXERCISES

The following graphs contain formatting errors. Copy the data into a spreadsheet and plot the graphs according to the instructions provided in this chapter.

1. Conduction heat transfer through three tubes with different lengths is presented as a function of the ratio of outside tube diameter to inside tube diameter.

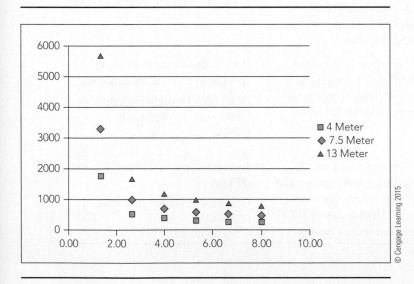

© Cengage Learning 2015

L (m)	r_1 (cm)	r_2 (cm)	r_2/r_1 [–]	$\dot{Q}$ (kW)
4	1.5	2	1.33	1752
4	1.5	4	2.67	514
4	1.5	6	4.00	363
4	1.5	8	5.33	301
4	1.5	10	6.67	266
4	1.5	12	8.00	242
7.5	1.5	2	1.33	3284
7.5	1.5	4	2.67	963
7.5	1.5	6	4.00	682
7.5	1.5	8	5.33	564
7.5	1.5	10	6.67	498
7.5	1.5	12	8.00	454
13	1.5	2	1.33	5693
13	1.5	4	2.67	1670
13	1.5	6	4.00	1181
13	1.5	8	5.33	978
13	1.5	10	6.67	863
13	1.5	12	8.00	788

© Cengage Learning 2015

2. Stress on the upper beam surface due to a 10-lb$_m$ weight on the tip of a 15-in. long, 0.5-in.-wide cantilever beam is plotted as a function of the beam height, h.

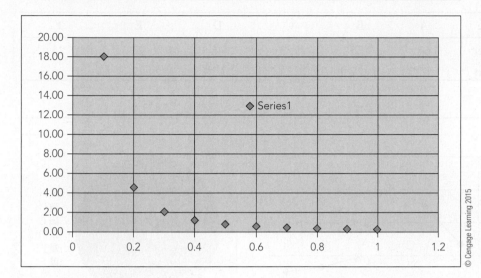

h (in)	Stress (ksi)
0.10	18.00
0.20	4.50
0.30	2.00
0.40	1.13
0.50	0.72
0.60	0.50
0.70	0.37
0.80	0.28
0.90	0.22
1.00	0.18

3. Approximate values for power consumption for various computer memory devices are compared in the table and in the figure. Correct the formatting errors in both the table and the figure to make them suitable for a technical paper.

(in nanowatts)	Memristor	Flash	Phase-Change	Spin-torque Transfer
Baseline Power	12	160	131	120
Read Power	34	60	54	56
Write Power	56	97	88	67

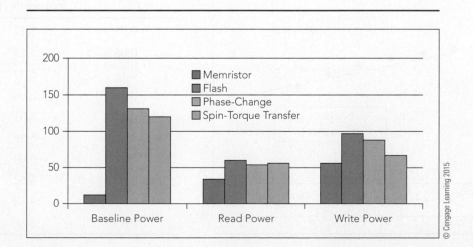

4. Average grade distributions from many years ago are plotted in the graphs below, and are intended for use in a written document. In addition to correcting formatting errors, explain under what circumstances, if any, each type of graph might be appropriate.

Grade	A	B	C	D	E	F
Number	67	125	153	46	32	25
Percent	15.0	27.9	34.2	10.3	7.1	5.6

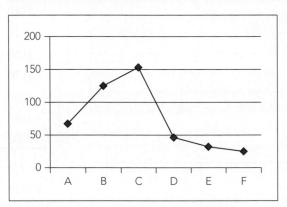 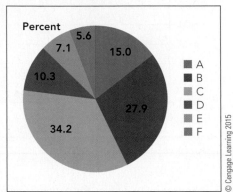

5. The measured damped natural frequency is plotted as a function of the damping ratio. An assumed linear trendline has been added to the data. Correct the formatting of the graph and determine if a better fit can be found using a different trendline.

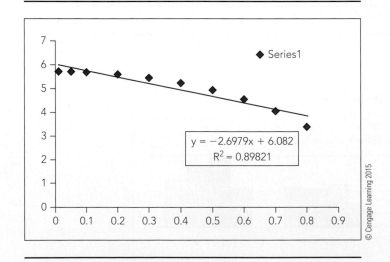

Damping ratio (%)	Damped natural frequency (Hz)
0.01	5.77
0.05	5.77
0.1	5.74
0.2	5.66
0.3	5.51
0.4	5.29
0.5	5.00
0.6	4.62
0.7	4.12
0.8	3.46

6. The cost of energy has changed substantially over the past decade. The following data were presented graphically in a newspaper (*Sunbury Daily Item*, 25 January 2013, p. A2). Correct the formatting errors in the graph and, if possible, update the data presented to include the most recent years.

Year	Retail Electricity, Price ($) per MBtu	Fuel Oil, Price ($) per MBtu	Natural Gas, Price ($) per MBtu
2000	27.9	9.3	8
2001	28.2	8.8	11
2002	28.5	8	9
2003	28	10	10.5
2004	28.3	11.5	12
2005	29.1	15	13.7
2006	30.5	17.5	16
2007	32.4	19.3	14.5
2008	33.5	24.5	15.5
2009	34.5	18	14.4
2010	37.5	21.5	12.6

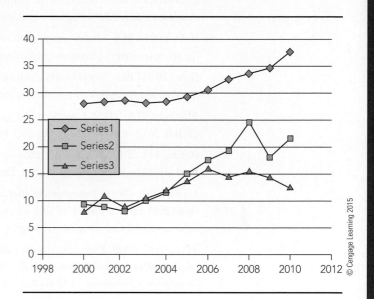

© Cengage Learning 2015

REFERENCES

Anami, K., Ishii, N., Tsuji, T., Oku, T., Goto, M., Yagi, K., and Knisely, C.W. 2010. Coupled-mode dynamic instability of tainter gates with parallel bending vibration of the skinplate, Montreal ASME Conference Summer 2010 FEDSM-ICNMM2010-30400.

Bergman, T.L., Lavine, A.S., DeWitt, D.P., and Incropera, F.P. 2011. Introduction to heat transfer, 6th ed. New York: John Wiley.

Blevins, R.D. 2003. Applied fluid dynamics handbook. Malabar, FL: Krieger.

BLS. 2010. U.S. Bureau of Labor Statistics, BLS spotlight on statistics: Back to college, September 2010. Accessed 26 November 2011 at <http://www.bls.gov/spotlight/2010/college/data.htm>.

BLS. 2011. U.S. Bureau of Labor Statistics, BLS Current Population Survey. Accessed 3 May 2013 at <http://www.bls.gov/emp/ep_chart_001.htm>.

Coley, D. 2008. Energy and climate change: Creating a sustainable future. New York: John Wiley.

Council of Science Editors. 2006. Scientific style and format: The CSE manual for authors, editors, and publishers, 7th ed. Reston, VA: Council of Science Editors.

EIA. 2011. Annual energy outlook 2011 with projections to 2035. U.S. Energy Information Agency, DOE/EIA-0383(2011)|April 2011. Accessed 8 August 2013 at <http://www.electricdrive.org/index.php?ht=a/GetDocumentAction/id/27843>.

Most online sources suggest using an email disclaimer on both internal and external messages to limit one's liability should the message reach an unintended recipient. Within the legal community, however, there appears to be less consensus on the functionality of disclaimers. Even if disclaimers do not limit your liability, they may prevent lawsuits if the claimants believe that the disclaimers have validity. Some sources suggest that a disclaimer can be a good marketing tool in that it gives a professional and trustworthy impression. On the other hand, Moss (2007) argues that since the courts have generally ruled that sending email is a form of public disclosure, the confidentiality disclaimer is worthless. Moss says confidentiality disclaimers are a waste of megabytes in file transfer and storage.

Further guidance on writing emails can be found in Flynn and Flynn (2003) and Baude (2006).

LETTERS

The writing of letters, especially handwritten personal letters, is a dying art. Business letters, while also declining in frequency, are nonetheless an essential part of doing business. Often such letters are considered to represent contractual obligations. Other times they will serve as legal documentation of intent and premeditation in case of a legal challenge. Without written documentation, it is very difficult to establish the sequence of actions and other legal commitments that too often form a part of litigation.

In any business capacity—as an engineer, a manager, an administrator, a partner or owner of a business, to name a few—an educated individual should be able to produce a well-written letter. The content of a business letter is not drastically different from that of a memo, but the business letter is formatted differently and is often the basis for communication between companies. After presenting an overview of business letter formats, we will describe specific types of letters in greater detail: letters of complaint, reference letters, transmittal letters, and cover letters for job applications.

Business Letters

Three predominant considerations in preparing a business letter are:

1. Audience:
   Who will receive the letter? How much background information will the recipient require? How do you expect the recipient to respond?
2. Purpose:
   What is your reason for writing the letter?
   - To persuade the recipient to take some course of action
   - To promote your business or product line
   - To inform the recipient
   - To establish a new business relationship with the recipient
   - To end a business relationship with the recipient
   - To satisfy a legal requirement by writing to the recipient
3. Format:
   Business letters generally have one of three common formats:
   - Block
   - Modified block
   - Semi-block format

An entry-level engineer should use the company-defined format when writing business letters. These three formats are described in more detail in the following sections.

Block Format

Figure 7.7 shows the elements of a business letter in block format, the most commonly used format in which all elements are left aligned, with at least a 1" margin on all sides. Some sources and organizations prefer wider margins, up to 1.5". In this letter format, a blank line separates paragraphs, and there is no indentation to mark the beginning

Company letterhead	**Nishimura Industries, Inc.** Custom Instruments, Mechanisms, and Machinery 25 Goldrush Drive, Aresquell, AK 99999
Date for future reference	(3 to 9 blank lines) Date (1 to 3 blank lines)
Inside address	Mr./Ms./Dr. Full Name Title/Position of Recipient Address Line 1 Address Line 2 Address Line 3 Country (if international) (Blank line)
Keywords introducing the topic	Subject: (if appropriate) (Blank line)
Salutation	Dear Mr./Ms./Dr. Last Name: (Blank line)
Define the situation	Paragraph 1 establishes why the letter is being sent. _____ . _____ last line of paragraph 1. (Blank line) Paragraph 2 will add details or further explanation of the topic of the letter. _____. _____. Add the details and explanations needed in this paragraph. _____ _____ last line of paragraph 2. (Blank line) Paragraph N – added topics or facets of the issues involved in the topic of the letter ____ last line of paragraph N. (Blank line)
Reiterate takeaway message	Final text paragraph. Reiterate the actions, outcomes, the main topics, or whatever was the purpose for the letter. Tell the reader what you want the reader to take away from the letter. (Blank line)
Courtesy of thanking reader	Closing. Acknowledge the assistance you are requesting, the consideration you have asked for, the time the reader has taken to read and, hopefully, respond to your letter. (Blank line)
	Sincerely, ⬅ (Note: The closing may be Sincerely, Sincerely (Blank line) yours, Yours truly, Very truly yours, Yours very (Blank line) truly. These expressions can be used (Blank line) interchangeably.)
Record of who you are	Printed Name Title of Writer
Writer's initials and typist's initials	Enclosure: Title (if any enclosure is included) XYX/bcb

FIGURE 7.7 | Elements of a business letter in block format

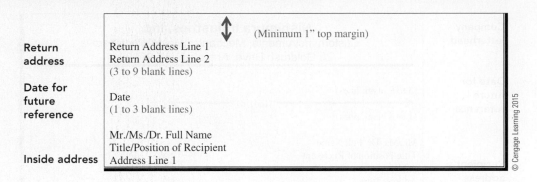

FIGURE 7.8 | Top portion of letter in block format without letterhead, showing left aligned return address

of a new paragraph. If the letter is not on company letterhead, as in Figure 7.8, the letter would begin with a left-aligned return address line. The rest of the formatting would be identical to that in Figure 7.7.

The variable spacing after the letterhead (or return address) and after the date line permits the vertical centering of the letter on the page for visual appeal. If the letter spills over to a second page that contains only the signature, the spacing can be reduced to keep the letter on a single page.

Modified Block Format   The modified block format is a variant of the block format. The major difference between the two is the indentation of the return address (if no letterhead is used), the date, and the signature. The tab setting for these three elements is approximately two-thirds of the way across the page; it should be consistent for all three elements. Figure 7.9 shows the layout for the top portion of a letter in modified block format without letterhead, and Figure 7.10 shows the same modified block format with company letterhead. As was the case in block format, paragraphs are separated by blank lines, and the first line is not indented.

A third business letter format, the semi-block format, is an older, rarely used format that is identical to the block format except that the first line of each paragraph is indented one half an inch. The semi-block format fell out of use in the mid- to late 1980s.

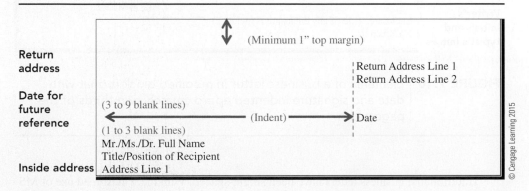

FIGURE 7.9 | Top portion of a business letter in modified block format without letterhead, showing position of return address

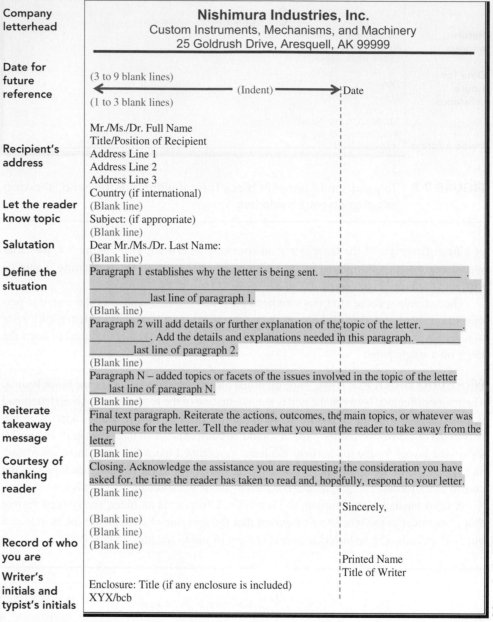

Company letterhead

Date for future reference

Recipient's address

Let the reader know topic

Salutation

Define the situation

Reiterate takeaway message

Courtesy of thanking reader

Record of who you are

Writer's initials and typist's initials

Nishimura Industries, Inc.
Custom Instruments, Mechanisms, and Machinery
25 Goldrush Drive, Aresquell, AK 99999

(3 to 9 blank lines)
←———————— (Indent) ————————→ Date
(1 to 3 blank lines)

Mr./Ms./Dr. Full Name
Title/Position of Recipient
Address Line 1
Address Line 2
Address Line 3
Country (if international)
(Blank line)
Subject: (if appropriate)
(Blank line)
Dear Mr./Ms./Dr. Last Name:
(Blank line)
Paragraph 1 establishes why the letter is being sent. _____ .
_____ last line of paragraph 1.
(Blank line)
Paragraph 2 will add details or further explanation of the topic of the letter. _____.
_____. Add the details and explanations needed in this paragraph. _____
_____ last line of paragraph 2.
(Blank line)
Paragraph N – added topics or facets of the issues involved in the topic of the letter
____ last line of paragraph N.
(Blank line)
Final text paragraph. Reiterate the actions, outcomes, the main topics, or whatever was
the purpose for the letter. Tell the reader what you want the reader to take away from the
letter.
(Blank line)
Closing. Acknowledge the assistance you are requesting, the consideration you have
asked for, the time the reader has taken to read and, hopefully, respond to your letter.
(Blank line)

Sincerely,

(Blank line)
(Blank line)
(Blank line)

Printed Name
Title of Writer

Enclosure: Title (if any enclosure is included)
XYX/bcb

FIGURE 7.10 | Elements of a business letter in modified block format with date and signature indented approximately two-thirds of the page width

Traditionally, business letters have been single-spaced. With the widespread use of MS Windows-based software, however, the spacing may change. In Word 2010, the default spacing is 1.15 lines, claimed by Microsoft to make the text more readable.

Types of Business Letters

Common types of business letters include

- Informational letters
- Letters of complaint
- Reference letters and
- Transmittal letters

Informational Letter   As the name implies, informational letters provide the recipient with information. Action by the recipient may or may not be required. Figure 7.11 shows a fictitious informational letter prepared in block format on company letterhead.

Nishimura Industries, Inc.
Custom Instruments, Mechanisms, and Machinery
25 Goldrush Drive, Aresquell, AK 99999

September 24, 20XX

Mr. Axel Eichner
Ostermann Environmental Group
245 Heimdorf Allee
D-99675 Unterbirkenwald
Germany

Subject: Your order dated August 23, 20XX

Dear Mr. Eichner:

I am pleased to confirm that we have received all sub-components needed for the construction of your custom multi-gas concentration sensor. We anticipate completing the assembly within the next 10 business days.

The unit, after testing and calibration, will be shipped to you via an international courier. Our shipping department will prepare the necessary customs forms and they will accompany the shipment to expedite processing of the shipment by the German customs authorities.

We anticipate shipping the unit on or before October 25, 20XX.

Thank you again for your order. We look forward to continuing to serve your needs for custom instrumentation in the future.

Sincerely,

Ernest Q. Trautman
Ernest Q. Trautman
Head, Customer Relations and Quality Control

FIGURE 7.11 | A sample informational letter in block format with company letterhead

Letter of Complaint In a letter of complaint, the writer expresses his or her dissatisfaction with a product, process, or service. The opening of the letter should provide details such as the purchase or service date and the precise nature of the problem. The writer should then politely suggest what action is desired. As much as possible, the writer should keep the letter succinct so that the recipient can quickly and accurately assess the situation. In addition, a letter that describes the complaint objectively and factually is more likely to get a positive response than one that is emotionally charged. A fictitious letter of complaint with annotation is given in Figure 7.12.

Return address	45832 Alameda Ave. Lewisburg, PA 17837
Date	December 9, 20XX
Recipient address	Mr. Evan J. Timakura IO-Universal Customer Service 12 Porta Luma Drive Ranchos Palos Rojos, CA 93245
Salutation	Dear Mr. Timakura:
Clear, concise statement of the problem	I am writing as a follow-up to our telephone conversation yesterday. I wish to reiterate in writing my dissatisfaction with your IOStuph-IT product. Your product has failed to meet your specifications for compressed storage capacity for archival records. We used your specifications in determining the required number of units to store five years' worth of archival data. Our estimated GB per year appears to be accurate, but we noticed last week that our IOStuph-IT storage is currently 87% full with just over two and a half years' worth of data. It appears your specified storage capacity per device is off by a factor of two.
Desired resolution	You suggested a plan in our telephone conversation to add to our storage capacity at no added cost on our side. We hope that you can gain your supervisor's approval of this plan. You said that IO-Universal would meet the specified storage capacity. We are pleased that you are willing to stand behind your product and make an effort to maintain our business relationship. As I told you on the telephone, failure to resolve the current situation with your IOStuph-IT units would force us to consider purchasing from other suppliers in the future. I earnestly hope that this decision will not be necessary.
Positive statement about relationship	Our previous experience with IO-Universal products has been very positive. We would like to continue doing business with your firm in the future. Supplying us with sufficient storage capacity for five years' worth of data would restore our confidence in your firm's ability to meet our archival needs.
Express gratitude	Thank you for your time and willingness to resolve the issues with your IOStuph-IT units. Sincerely, *Brice Savage* Brice Savage IT Manager E-Storage, LLC

FIGURE 7.12 | A letter of complaint

Responsible businesses, agencies, and organizations will always respond to letters of complaint because they know that customer satisfaction is important for their reputation and, consequently, to their success. Their initial response might be as simple as "We are looking into your complaint." If they agree that the problem was their fault, they will take steps to resolve it. If there are extenuating circumstances, they may explain why they are unable to take the action you suggest.

If you are the recipient of a letter describing a legitimate complaint, apologize for the misunderstanding or inconvenience, state how you will correct the error, and end your letter with the hope that the customer will be satisfied with the action taken and will continue doing business with you. On the other hand, if you determine that a claim is unfounded, you will want to provide ample evidence that you understand the issues raised in the original letter of complaint and that you have carefully considered how to proceed. The care with which you handle the situation and particularly the manner in which you convey the "bad news" will dictate whether you have lost a customer or client forever or whether you are able to save the relationship.

Letter of Reference

In academia, a reference letter is usually prepared to evaluate a student's or a colleague's character and abilities, most commonly to support an application for employment or an award. Frequently in industry, company policies regulate the disclosure of information concerning the performance of an employee. Check company policy before preparing any type of reference for a business colleague or subordinate.

The usual format of a letter of reference is as follows:

- Introduction
  » Establish the type and duration of the relationship you have had with the applicant.
  » Identify yourself, giving your job title to establish your credibility.
- Overview of the applicant's strengths and positive characteristics
  » Describe specific abilities or qualities
  » Provide examples that demonstrate those traits
- When asked to identify potential weaknesses, try to portray a weakness as a strength.
  » For example, an applicant tends to take longer than anticipated to complete tasks. *Weakness:* The applicant lacks focus. *Strength:* The applicant is thorough and detail-oriented.
  » For example, an applicant always has a quick response to almost any situation. *Weakness:* The applicant responds without adequate thought. *Strength:* The applicant is quick on his or her feet.
- Concluding paragraph
  » Reiterate that the applicant is a good prospect.
  » Explain why the applicant is a particularly good fit for the position or the award.
  » Indicate your willingness to be contacted for follow-up, if needed.
  » Include your contact information, if not already provided in the letterhead.

An annotated sample reference letter is given in Figure 7.13.

Transmittal Letter

When a technical document (such as a report or proposal) is distributed, it is customary to send a short letter along with the report explaining the nature

ARESQUELL
UNIVERSITY
One University Avenue
Aresquell, AK 99999

Mechanical Engineering
Department

XXX-YYY-ZZZZ
mecheng@aresquell.edu

Date   November 12, 20XX

Recipient address

NSF GRFP
Washington, DC

Subject: Letter of Reference for <u>Emily Ariel Devine</u>

Salutation   Ladies and Gentlemen:

Explain relationship with the applicant

As a member of the Mechanical Engineering Department at Aresquell University, I am pleased to write to you about the qualifications of our student Emily Ariel Devine. I have known Emily Ariel Devine (who goes by Ariel) since the beginning of fall semester of 20YY when she was enrolled in my Thermodynamics class. In her junior year (the 20YY-YY academic year), she was a student in my Fluid Dynamics, Heat Transfer, and Aerodynamics classes. Also during her junior year, she worked on an externally funded research and development project with me. This year, her senior year, she is a member of the senior design project team that I am advising. I know Ariel better than the average student.

Focus on classroom

After a few lectures in Thermodynamics, I knew that Ariel was an exceptional student. Ariel always came prepared for class, always paid attention, could always answer any question I asked in class, even when most of her classmates could not. She frequently asked insightful questions after class. She continued to produce high quality work in Fluid Dynamics and Heat Transfer in her junior year. As a junior, she enrolled in a senior/graduate elective course in Aerodynamics. Ariel finished tied for the highest grade in the class (with another junior), ahead of all the seniors and all the masters' level graduate students in the course.

Activity outside of classroom

Ariel worked for me during her junior year on an externally funded project to develop a quasi-isothermal compressor. She is part of the team that is further refining the machine this year. Ariel has been a teaching assistant in Thermodynamics since her junior year. She holds problem-solving sessions for the students for several hours every week. Her methods of instruction are pedagogically sound. She asks leading questions and provides guidance to the students instead of just handing them the solutions.

Compare with previous successful applicant

In many ways, Ariel followed a path very similar to that of Jonas McKnight, another Aresquell graduate, who is now supported by an NSF Graduate Fellowship. What sets Ariel apart from Jonas is her exceptional ability with machine shop tools and her practical hands-on assembly of complex components. She says she owes her mechanical expertise to growing up around her father's machine shop business.

FIGURE 7.13 | A two-page sample recommendation letter for a fellowship application. The student, the professor, and the university are fictitious.

Header for second page	Recommendation Letter for Emily Ariel Devine Page 2 November 12, 20XX
	Without a doubt, Ariel is the single best student I have had in my courses for the past twenty years when I consider both her theoretical knowledge and her hands-on ability.
Outcome of research effort	Ariel worked on the development of an isothermal compressor with me for two years. In that time, she developed an alternative cooling scheme that reduced the exit temperature of the compressed gas by an additional 20°F. She also reduced the weight of the machine by more than 50%. Further, she developed a predictive computer code that correlated with the measured gas temperatures to within about 4°F everywhere in the machine.
Leadership abilities	Ariel is and will be a leader. She is president of our Tau Beta Pi and our Pi Tau Sigma chapters, as well as our local ASME chapter. During her presidency, membership in our ASME student chapter doubled. She is respected and held in esteem by her classmates. She will become a leader in her academic field. She also understands that her unique gifts also carry with them the obligation to assist others. She mentors younger students very effectively here at Aresquell. She works with a group on K-12 outreach. She has already learned the satisfaction of seeing students grasp concepts with which they have struggled. She will be an exceptional teacher, sharing her knowledge in a compassionate manner with anyone wanting to learn.
Overall assessment	My assessment of Ariel's capabilities is that she can accomplish any objective she sets for herself. She will excel as a graduate student; she has proven capabilities in conducting both experimental and analytical research projects concurrently and seeing them through to completion. Ariel has excellent communications skills in both oral and written formats. She works well in groups, as she is doing now on her senior project. She has exceptionally good relationships with the machinists and technicians here at Aresquell. Ariel is a do-er. She will accomplish her goals. She will have a substantial impact on our future energy supply when she completes her proposed project implementing a practical solar-powered Ericsson cycle. Based on my knowledge of Ariel, I expect that she will deliver the research program she has outlined on schedule. Ariel sees projects through to completion.
Recommend-ation	I recommend Emily Ariel Devine for an NSF Graduate Research Fellowship with the utmost enthusiasm and with the certainty that she will be a most productive Fellow. Please feel free to contact me via telephone or email if I can be of further assistance.
	Sincerely, *Kurt Korenthal* Kurt Korenthal, PhD, P.E. Professor

FIGURE 7.13 | A recommendation letter, continued

of the report and why it has been sent to the recipient. The transmittal letter is a formal business letter and should follow the usual format: date, return address (or letterhead), recipient's address, salutation, the body, and a closing. The body typically explains the reason for the letter, a statement of facts related to the attached document, any expected follow-up by the recipient with a time frame for response, and the writer's contact information. A sample transmittal letter is given in Figure 7.14.

Date	August 28, 20XX
Return address	Mechanical Engineering Department Aresquell University One University Drive Aresquell, AK 99999
Recipient address	Mr. Robert Gostallier Isenaitch Compressors, LLC 467 Juneau Highway Aresquell, AK 99999
Salutation	Dear Mr. Gostallier:
Concise explanation of why the report is being sent	Enclosed is the concluding report for the project "Preliminary Development of a Quasi-Isothermal Compressor," as specified in the contract we have with you. The purpose of this report is to provide you with our latest design for the quasi-isothermal compressor and the performance data we obtained with this design. The report also suggests the possible steps in moving toward commercialization of the compressor.
Provide contact information	We would be pleased to continue working with you to bring this compressor to the market. If you should have any questions concerning the report, or wish to discuss continued collaboration, please feel free to contact me in my office at YYY-ZZZZ or by email. Sincerely, *Kurt Korenthal* Kurt Korenthal, PhD, P.E. Professor, Mechanical Engineering Department Telephone: XXX-YYY-ZZZZ Email: kkthal@aresquell.edu

© Cengage Learning 2015

FIGURE 7.14 | A sample transmittal letter for a hypothetical project report shown on a page without letterhead (usually letterhead would be used for such a letter)

Cover Letter   When submitting your resume for consideration for a job, it is customary to include a one-page cover letter, which is similar to the transmittal letter above. The cover letter accompanying your resume explains how you learned of the job opening, why you are interested in the job, and how your qualifications match the job requirements. The body of the cover letter addresses the following points:

1. Specify the job for which you wish to be considered. Then briefly explain how you became aware of the job opening and why you are interested in the job.
2. Explain how your expertise matches the job description and why you are an excellent candidate for the job.

Letterhead	**Parker McQueen** 57 McKenzie Hall ◆ Aresquell University ◆ Aresquell, AK 99999 pmcq027@aresquell.edu ◆ (999) 897-9678
Date	October 18, 20XX
Recipient address	Mr. Lucas Donnelly Recruiting coordinator East Barrow Contracting Company 300 East Yukon Road Barrow, Alaska 99789
Salutation	Dear Mr. Donnelly:
Note which job is of interest	I am writing this letter to inquire about your advertised summer internship in enterprise energy management (EEM) in the Barrow, AK area. I am currently a junior in electrical engineering at Aresquell University. I would like very much to make my career in Alaska after graduation and hope through a summer internship to become more familiar with the role electrical engineers play in the continuing development of Alaska.
Explain why the job is of interest	Working for you would give me experience in integrating computer systems into building energy management. I would have the opportunity to ask questions while assisting other engineers in your company. Such interactions would be a great resource to me and provide me with the next level of experience I seek.
Elaborate on your qualifications	As mentioned above, I am currently an electrical engineering major at Aresquell University. I am also minoring in economics and have taken three courses for this minor. I believe that my economics background coupled with my engineering knowledge will permit me to be a strong contributor to East Barrow Contracting. For the past three summers, I worked for the GSA in Juneau, creating emergency action plans, organizing documents for an audit, and job-shadowing engineers. During the summer of 20YY, I developed and implemented a computerized code for enterprise energy management (EEM). My coursework and work experience coupled with my desire to learn would permit me to make an immediate contribution to your company's projects.
State your availability for an interview	I look forward to hearing from you. I am available to travel to Barrow for an interview at your convenience. Sincerely, *Parker McQueen* Parker McQueen

© Cengage Learning 2015

FIGURE 7.15 | An annotated sample cover letter to accompany a resume submitted for an internship

3. State your availability for an interview at the convenience of the prospective employer.
4. State action desired: You look forward to hearing from the employer.

A sample cover letter is provided in Figure 7.15. If you are sending your cover letter by email, omit the return address, date, and recipient address.

Other Types of Business Letters   Other types of business letters include letters of apology, letters of resignation, and job interview thank you letters. Samples of these types of letters and others are readily available online, for example, at Letter Writing Guide (2012).

Personal Letters

Many university students have never written a personal letter, not counting letters to Santa Claus. The telephone, email, texting, and skyping provide immediate gratification, whereas thoughtful letters take much longer to write. For those who do take the time, writing a personal letter can be an opportunity for self-examination and reflection. The act of writing a letter provides a moment of tranquility in a hectic life. Letters afford a venue for baring your soul to a friend or someone with whom you would like to form an enduring relationship.

Much of what is known about the private lives of some of the great people in history comes to us from letters they sent and received. What will history have to say about the personal trials and accomplishments of today's great individuals when the pace of life is antithetical to introspection and electronic messages have largely replaced paper? Personal letter writing may add value to the life of an engineer, much like time capsules that are buried for future generations.

Personal letters usually begin with the date followed directly by the salutation. Seldom is the return address of the writer or the address of the recipient (the inside address) included in a personal letter. The writer might start with a general inquiry of the recipient's health, the weather, or other situations affecting the day-to-day life of the recipient. The writer then provides corresponding information about his or her own daily life and matters of a general nature.

After the introduction, the writer then moves to the essence of the letter. What prompted the writer to contact the recipient? The motive may be a longing for home, concern for the recipient, a memory of past activities, a comment on a topic of mutual interest, or a response to the recipient's previous query.

Personal letters are often not as polished as business letters, but should follow the same principles when written to a business associate: Provide context, state the purpose of your letter, and tell the recipient what action you desire on his or her part.

Because so many students lack experience writing personal letters, we suggest that students learn to write business letters first, and then explore the genre of personal letter writing.

RESUMES

The word *résumé* is a French word that is used in English without the accents. A **resume** summarizes a job applicant's education, previous employment experience, skills, and abilities. A resume represents your professional identity. Many engineering students prepare their first resume in their sophomore or junior year when they apply for summer internships or co-op programs. Resumes for full-time, entry-level positions are typically one page long. Resumes for engineers with ten years or more of experience may extend to two pages.

Most job advertisements in academia request a curriculum vitae (CV), which is Latin for "course of life." A CV is more comprehensive than a resume. CVs typically contain your advisers' names, the titles of your theses and dissertations, publications, postdoctoral appointments, publications, and specific tasks carried out in the course of employment. Most entry-level engineers entering the workforce in the United States and Canada will not be asked for a CV. For this reason, we will focus exclusively on preparing a resume.

What Does Not Belong in a Resume?

In the United States, to avoid any issues related to potential discrimination, the following personal information should *not* appear on your resume:

- Date of birth
- Marital status
- Race or ethnicity
- A photograph

In addition, some authorities now recommend not stating a career objective on your resume. The thinking is that you can explain your objectives or reasons for applying for a particular position in your cover letter. Other authorities suggest that listing career objectives on a resume is still appropriate and that each resume should be customized for each potential employer. Ask your adviser or a career services professional at your university for their opinion.

There are also differing opinions on whether the traditional phrase "References available on request" should be included on a resume. Those who are opposed to including it argue that a prospective employer who desires references will ask for them, and it is understood that a job applicant will have a list of references ready to go. Applicants should ask individuals whom they would like to act as references for permission before giving prospective employers their contact information. When the references along with their contact information are listed on a resume, it saves prospective employers time. On the other hand, if you have a lot of information to put on your resume, omit the references.

What Does Belong in a Resume?

A resume for a college student or recent graduate *should* include the following information in this order:

- Name and contact information
- Education
  - » GPA
  - » Major (and minor, if applicable) and list of relevant courses. Do not include high school courses if you are in college.
  - » Graduation date (anticipated)
- Previous employment and skills
- Honors, awards, and accomplishments
- Relevant extracurricular activities, if they required a skill set pertinent to the advertised job

Chronological Format versus Skills Format   For job applicants who follow the traditional path of graduating from college and working continuously in their field, it is customary to list the items in each category (education, honors, and relevant activities) in chronological order and employment in reverse chronological order. For applicants who have gaps in their employment history or who are looking to make a career change, however, the skills format resume may be advised. In this format, relevant and transferable skills are listed as the first item on the resume. If a prospective employer needs the given skill set, the applicant's gaps in employment history may be viewed less critically. If the applicant is invited to an interview, he or she can explain why he or she did not work (outside the home) for an extended period. When skills come first on a resume, the emphasis is on the skills and not the continuity (or lack thereof) in employment history. Samples of both resume formats are provided in Figures 7.16 and 7.17.

Parker McQueen

Aresquell University, 57 McKenzie Hall
Aresquell, AK 99999

(999) 897-9678   pmcq027@aresquell.edu

EDUCATION

Aresquell University, 20XX-20YY
- Major in Electrical Engineering
- Minor in Economics
- GPA 3.67/4.0

Relevant Courses:

Intro. to Electrical Engineering	Intro. to Economics
Circuit Theory	Macroeconomics
Digital Design	Microeconomics

WORK EXPERIENCE

United States General Service Administration   Summer 20XX and Summer 20YY
Federal Building, Juneau, AK 99801-1807
Engineering Intern
- Reviewed building plans to find errors and oversights
- Created emergency action plans for fire department officials
- Organized financial documents for auditors
- Shadowed engineers during their daily professional activities (attending meetings, design reviews, and site visits)

United States General Service Administration   Summer 20ZZ
Federal Building, Juneau, AK 99801-1807
Environmental Engineering Intern
- Implemented ION EEM data acquisition routine to monitor utility use
- Implemented ION EEM data acquisition routine to monitor remote wind and solar power installations
- Created user-friendly displays of data from wind and solar power installations

SKILLS

Platforms: AIX, HP-UX, Linux (SUSE and Red Hat), Solaris (UNIX), DOS, Windows 10/9/7/NT/XP

Languages: Assembler, Basic, C/C++

Software: BitDefender, CedeCheck, Exceed, MS-Office Suite, MS Visio, HP OpenView, GNU, McAfee Antivirus, MATLAB, ION EEM

PROFESSIONAL AFFILIATIONS

IEEE, Student Member, Aresquell University Section

AWARDS

Dean's List (4 semesters)
Alpha Lambda Delta (First year Honor Society)

FIGURE 7.16 | A sample resume in chronological format. The person and details are fictitious.

Stephanie J. Mueller

4567 Eastern Park Drive • Darytown, PA 17567 • 987-654-3210 • sjm034@somenet.com

PROFILE
Energetic professional with excellent problem-solving and professional communication skills, extensive knowledge of fire safety engineering practice, eight years of professional experience as Fire Safety Engineer with progressively greater management and leadership responsibilities.

SKILLS SUMMARY
- Excellent written and oral communication skills
- Excellent interpersonal skills
- Well-honed, hands-on knowledge of fire detection systems and alarm system design
- Successful design experience of fire suppression systems
- Extensive familiarity with national standards, as well as several local codes
- Trained as a first responder; able to respond efficiently in emergencies
- Exceptional recognition of potential fire hazards
- Training as an insurance site inspector

EXPERIENCE
Manfred Schroeder Fire Consultants
245 West Main St., Derrsport, PA 17664

1996-2000 Senior Fire Design Engineer
Duties entailed interfacing with clients, design of fire safety systems, overseeing and verifying code compliance, report preparation for clients, site inspection of installed fire safety systems in industrial buildings.

1992-1996 Fire Design Engineer
Duties entailed design of fire safety systems; assisting senior engineer in site inspections

Major Commercial Insurance
345 West Dayton Highway, Cincinnati, OH 45247

1983-1985 Fire Risk Insurance Underwriter
Duties entailed determining clients' insurance premium after site inspection to assess fire risk

PROFESSIONAL AFFILIATIONS
Society of Fire Protection Engineers, member
American Society of Civil Engineers, member

EDUCATION
University of Cincinnati, Associate Degree in Fire and Safety Technology, 1983
Carleton University (Canada), B.S. in Civil Engineering with concentration in Fire Safety Engineering, 1988
University of Canterbury (New Zealand), Masters of Engineering in Fire Engineering (MEFE), 1991

ADDITIONAL INFORMATION
2000-2012 Occasional Fire Safety Engineering Consulting as time permitted while I was a stay-at-home mother with 2 children
1992 to present Attendance at Annual National Meeting of Society of Fire Protection Engineers

© Cengage Learning 2015

FIGURE 7.17 A sample resume in the skills format, useful when there are gaps in employment history. The person and details are fictitious.

TABLE 7.1 | Checklist for a resume in chronological format

Name and contact information
- Include home address; do not list work address
- Include email address
- Include home and mobile phone numbers

Education
- List degree, institution, location, and date of completion
- Include major (and any minors), certificates, certifications, skill-based training, etc.
- Do not include high school courses

Related Work Experience
- Cite position title, organization, location, and dates
- Give a brief description of primary responsibilities
- Use action verbs* in your description; quantify where possible
- Stress accomplishments and related skills
- For professional experience more than 15 years old, just list title, organization, location, and dates; do not include job description

Professional Activities/Associations/Affiliations

College or community-related activities; include positions held

Accomplishments and honors

Presentations, publications, and papers

© Cengage Learning 2015

* Examples of action verbs for use in resumes are given on the Boston College website (Resume Action Verbs, 2012), among others.

Many resume templates can be found online and at college career services centers. Do not just copy a template; instead, tailor your resume to the position for which you are applying. Regardless of whether you choose a chronological or a skills format, make your resume a masterpiece, a reflection of the type of person you are. Seek advice from as many sources as possible. Make sure your resume looks professional and visually pleasing. Proofread (and have someone else proofread) it carefully for proper grammar, correct spelling, and correct punctuation. Make sure all information such as dates, addresses, and names of universities and employers is accurate. Include only essential information and provide substantial white space. See Table 7.1 for a checklist of resume components.

Resume Not Required   Recently, companies with a strong Internet presence have asked job applicants *not* to send a resume (Silverman, 2012). Instead, these companies request links to pages that show the applicant's presence on the web (Twitter, Tumblr blog, and LinkedIn, among others). The applicants frequently are asked to submit a video presentation of their qualifications demonstrating their interest in the position. Clearly, the

---

Recruiting Tip—Send a Thank You Note or Email

After an interview, send your host a thank you note. This courtesy portrays you as a considerate and thoughtful person. A thank you note may include one or more of the following components:

1. Thank your host for inviting you for an interview for (state title of position) on (date). Thank your host for his or her hospitality and time.

2. Express your heightened interest in the position as a result of the information gained from the interview. Reemphasize briefly how your qualifications make you a good fit for the position. Highlight specific capabilities that would help you execute the most important tasks discussed during your interview.

3. Let your host know that you look forward to hearing back from the company in the near future. Express your willingness to visit the company again for further discussions about your application and the job.

Close by providing your contact information and the best times to reach you. Thank your host again for the opportunity to meet with him or her.

---

applicant will need to be quite clever to display the expected skill set without supplying information that violates antidiscrimination laws. Time will tell to what extent multimedia job applications become commonplace.

International Employment Applications

Engineers contemplating working abroad must be aware that the documentation format used in other countries may be different from that used in the United States. Several websites describe differences between the European CV and the North American resume (see, for example, American Resume vs European CV, 2012; and CVs and Resumes Compared, 2012). The website CVs in the European Job Market (2012) provides a template for a common EU CV format, but this site cautions that this format has not yet been adopted everywhere, especially not in Germany, France, and Switzerland.

In the English-speaking southern hemisphere in the countries of Australia and New Zealand, the CV is still the preferred job application document. Authorities on the Australian and New Zealand CV format note that there are slight differences between them and also that they differ from the British version. Australian resumes are expected to have a fair amount of white space and run two to four pages in length. Australian spelling is required. For details on the Australian CV, see Howard (2012). The New Zealand CV seems to be a hybrid—somewhat longer than an American resume, but perhaps a bit shorter than the British CV. The New Zealand CV is usually only one to two pages in length. Further details can be found at New Zealand CV Writing (2012), Working in New Zealand (2012), and the University of Canterbury (2012) website.

SUMMARY

In any form of professional correspondence, the writer must clearly state the purpose of the correspondence and specify the desired response from the recipient. The recipient is likely to respond more quickly when the content is tailored to his or her needs and a standard, familiar format is used for memos, emails, and business letters.

When preparing your resume, make sure that it represents the professional image you wish to project. Include your education, your previous employment, and your transferrable skills and abilities. If you wish to work outside of the United States, prepare your employment documents according to the standards of your anticipated host country.

EXERCISES

1. Identify a company in a field in which you might like to work. Visit the company's website and identify the office to which job inquiries are to be directed. Prepare an inquiry about a job you are interested in and email your inquiry to your instructor for feedback on your email style.

2. Find an advertisement for a position in your field in a professional publication. It can be a permanent position or an internship. Respond to the advertisement by sending the requested information to your instructor for feedback.

3. Prepare a letter of complaint about a purchase, a contract, or another business transaction. Specify the action you want the recipient to take to remedy the problem.

4. Interview a classmate to determine the person's strengths and weaknesses. Prepare a letter of reference that portrays the person's strengths and weaknesses in a positive light. Try to be honest about the strengths and weaknesses, but still give the person a favorable reference.

5. Prepare a transmittal letter for an industrial report you have written. Assume that you are transmitting the report via your direct supervisor to the Director of Research and Development in the organization.

6. Prepare your resume in chronological format.

7. Prepare your resume in skills format.

8. Write a short, friendly, personal letter to a business associate with whom you have shared a memorable moment.

REFERENCES

American Resume vs European CV. 2012. The American resume vs. the European CV. Accessed 23 February 2013 at <http://biotechnologycareers.blogspot.com/2009/07/american-resume-vs-european-cv.html>.

Baude, D.M. 2006. The executive guide to email correspondence: Including model letters for every situation. Pompton Plains, NJ: Career Press.

Blodget, H. 2011. Bombshell: Huge company bans internal email, switches totally to facebook-type-stuff and instant messaging. Accessed 23 February 2013 at <http://articles.businessinsider.com/2011-12-04/tech/30473966_1_abc-news-internal-emails-messages-employees>.

Business Email Etiquette. 2012. Business email etiquette expert. Accessed 23 February 2013 at <http://www.businessemailetiquette.com/>.

CVs and Resumes Compared. 2012. CVs and resumes compared. Accessed 23 February 2013 at <http://biotechnologycareers.blogspot.com/2009/07/cvs-and-resumes-compared.html>.

CVs in the European Jobmarket. 2012. CVs in the European jobmarket. Accessed 23 February 2013 at <http://www.eurobrussels.com/cv_euro.php>.

Flynn, N., and Flynn, T. 2003. Crisp: Writing Effective Email, Revised Edition: Improving Your Electronic Communication, revised edition. Rochester (NY): Axzo Press.

Howard, G. 2012. Writing a resume "Aussie" style. Accessed 23 February 2013 at <http://www.topmargin.com/documents/australian_resumes.pdf>.

Kehrer, D. 2012 Business Email Etiquette—How to avoid pitfalls and make your business email look and read professional. Accessed 23 February 2013 at <http://www.business.com/guides/business-email-etiquette-1637/>.

Kim, S. 2011. Tech firm implements employee 'zero email' policy, ABC News Blog, 29 November 2011. Accessed 23 February 2013 at <http://abcnews.go.com/blogs/business/2011/11/tech-company-implements-employee-zero-email-policy/>.

Letter Writing Guide. 2012. Letter writing guide. Accessed 23 February 2013 at <http://www.letterwritingguide.com/>.

Moss, J. 2007. Confidential conundrum. Accessed 23 February 2013 at <http://www.snewsnet.com/snews/gt_upload/Fitness07_lawreview.pdf>.

New Zealand CV Writing. 2012. NZS.com New Zealand CV writing. Accessed 23 February 2013 at <http://www.nzs.com/new-zealand-articles/business/cv-writing.html>.

Purdue OWL. 2012. Email Etiquette. Accessed 23 February 2013 at <http://owl.english.purdue.edu/owl/resource/636/01/>.

Resume, 2012. Resume, resumé, or résumé? Accessed 23 February 2013 at <http://painintheenglish.com/case/193>.

Resume Action Verbs. 2012. Boston College action verb list. Accessed 23 February 2013 at <http://www.bc.edu/offices/careers/skills/resumes/verbs.html>.

Saalfield, P. 2005. Internet misuse costs businesses $178 billion annually. Accessed 23 February 2013 at <http://www.infoworld.com/t/applications/internet-misuse-costs-businesses-178-billion-annually-996>.

Silverman, R.E. 2012. No more resumes, say some firms, The Wall Street Journal (wsj online, 24 January 2012 edition). Accessed 23 February 2013 at <http://online.wsj.com/article/SB10001424052970203750404577173031991814896.html>.

University of Canterbury. 2012. Preparing a curriculum vitae (CV). Accessed 23 February 2013 at <http://www.canterbury.ac.nz/careers/career_resources/GTJH/job_hunting.shtml>.

Working in New Zealand. 2012. Top tips for creating a CV for the NZ job market, Accessed 23 February 2013 at <http://www.workingin-newzealand.com/jobs/job-tools/cv>.

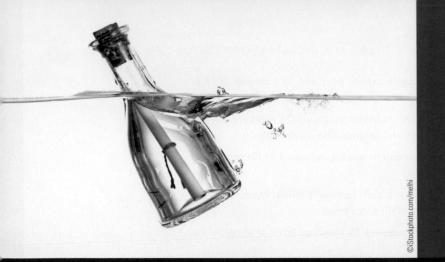

Chapter
8

PROPOSALS, COLLABORATIVE WRITING, AND PROGRESS REPORTS

In industrial settings, as well as in academia, good ideas often abound, but if no one knows of the ideas, they shrivel and die just like unpicked fruit. Unlike fruit, however, the next season may not see a regeneration of the same idea. As discussed in Chapter 1, the ability to explain innovative concepts to other people is essential for helping your ideas become reality. The discussion of proposal preparation in this chapter will provide a framework for communicating your ideas.

Discussions of ideas among a team of people can stimulate further innovative thought. When working as a team, the members will need to agree on the language they will use to communicate their ideas to managers or sponsors. The team will also need a plan for working on different parts of the presentation document, often concurrently. The discussion of collaborative writing will offer a range of options on how team members might structure their collaboration.

Once you have successfully found funding for your project, or even if you are working on a project funded in some other way, you will be obligated to provide periodic reports. Your progress reports should ideally follow the plan put forth in the funded proposal. In research, however, one or more new developments may emerge in the course of the work. If a new development is germane to the proposal, you will need to justify any deviations from the proposed plan of action.

The writing style of both the initial proposal and the progress reports needs to be persuasive. The writer must first demonstrate that the actions are necessary and then convince the audience both that the action taken was appropriate and that the outcomes of the action are significant. This chapter provides an overview of proposal and progress report writing, and underscores the need for preparation in any collaborative writing project.

PROPOSALS

To move your idea forward, it is often necessary to look for internal and external funding sources. To find funding sources, build and maintain relationships with funding agency program officers and monitor the websites of organizations that typically issue Request for Proposals (RFPs) in your field. Most likely, no single RFP will be looking precisely for the idea you have formulated. Your task, therefore, is to find a way to link your idea with the RFP language. If you are able to connect all parts of your idea with the details of the RFP, and if other applicants cannot, your proposal has a good chance of being funded. Often, however, the funding source may have more proposals than available funds. In this case, you must make your proposal more than persuasive; you must make it the best proposal submitted.

Often in an innovative research proposal, the narrative portion of the proposal may be 15 to 30 or more pages long. In some RFPs, the maximum text length may be specified. Most writers discover that writing a succinct comprehensive proposal is more challenging than writing a long and rambling one. Invariably, the shorter proposal is a more comprehensible piece of writing because the writer has distilled the concepts down to their essence.

We will distinguish between two types of proposals:

1. A **full proposal** of 15 to 30 (or more) pages, which might describe multiple years of effort in multiple phases of a project, usually submitted to an external funding agency.
2. A **short proposal**, limited to perhaps 2 to 10 pages, which spans a correspondingly shorter period and has a more limited scope. A short proposal may serve to support a preliminary letter of intent. Upon approval of the letter of intent, a full proposal will then be submitted. Theses and dissertation proposals are included in the short proposal format because the expected content more closely matches that of the short proposal.

Purpose of a Proposal

In any proposal a writer needs to convey convincingly three fundamental points:

1. The proposed work is significant, is consistent with the current state of the art, and fills a clear need.
2. The person proposing the work is the right person to undertake the project.
3. The proposed project is feasible, and the proposed methodology is appropriate.

Style of a Proposal

Regardless of length, Glisson (2004) notes several stylistic features common to all proposals:

- The purpose is stated concisely right at the beginning. Strong, concise, and precise language is used in the first paragraph to convince the reader that the ideas have merit.
- Relevant background information is summarized.
- The language is readily understandable (avoid jargon and unexplained acronyms).
- The sentences are short to moderate in length to enhance readability.
- The proposed ideas are stated clearly and are not hidden between unnecessary words.
- All the information provided convinces the reader that the project is worth funding.

Preparing to Write

Before beginning to write a proposal, clarify in your own mind exactly what you will be proposing. Without a clear and well-formulated idea, your proposal will be much like trying to provide driving directions without knowing your destination. You may eventually reach a destination somewhere, but is this the same destination desired by funding agencies?

There is no doubt that the writing process helps you clarify your intentions. Indeed, writing down preliminary ideas in brainstorming sessions with others, often with multiple iterations of insertions, deletions, revisions, changes of directions, and even new discoveries is necessary to fine-tune the vision. Especially in group projects that involve engineers, managers, salespeople, and other stakeholders, these preliminary discussions may change the original idea significantly but ultimately strengthen the final proposal.

Once the vision has been formulated, you, as the author of the proposal, must

- State the purpose clearly
- State explicitly the need the proposal fills
- Ensure that the proposed work is consistent with the philosophy and mission of your institution (or course)
- Ensure that the proposed work is compatible with the mission of the funding agency (if applying to one) and addresses the requirements of the RFP

After the problem is identified and a potential funding agency is chosen, the next part of an outline for a proposal should answer the question, "How will the project be done?" The outline should address:

- Method(s) and approach(es)
- Timeline
- Resources (includes both physical resources and personnel)
- Projected outcomes
- Assessment plan (how you will measure the project's success)
- Dissemination plan (sometimes required—how you will inform others of the outcomes)

A critical part of the "How?" is the budget, which is often presented in a table or in a separate spreadsheet (depending on the requirements of the RFP or the funding agency). Initially, in the planning stage of the project, a crude budget must be formulated to determine whether the project fits within the defined funding limits. If the preliminary budget is prohibitive, the scope of the project may need to be scaled back by removing the least cost-effective outcome, for example. If the project cannot be scaled back, it should be abandoned or put on hold until additional funding opportunities are identified. If the preliminary costs are reasonable, the details of the project are fleshed out and a more detailed budget is created.

Components of a Full Proposal

Many potential funding sources, especially government agencies, have well-defined formats that are required for all proposals. Follow the format precisely because failure to do so may immediately disqualify a proposal before anyone even reads it. In the absence of guidelines from the funding source, the following structure may be suitable for your full proposal:

1. The cover page
2. The executive summary

3. The narrative
4. The budget
5. References or bibliography
6. Supporting documentation
7. The authorization page signed by a representative of the principal investigator's organization

We will describe each section in detail below. Depending on your specific proposal, the order and names of these sections may be different.

Cover Page/Title Page

If the format is specified by the funding agency, follow the instructions exactly. Otherwise, it is customary to supply the following information:

• Title of project—make it concise and precise with no ambiguity
• Submitted to—write the name of the funding agency
• Date
• Names of the Principal Investigators (PIs)
• Institution(s) to which the PI(s) belong
• Signature of the authorized institutional representative

Project Overview/Executive Summary/Abstract

As with the abstract of a technical paper or report, the project overview/executive summary/abstract should be written after the rest of the proposal is prepared. In the summary, answer the following questions:

• What do you want to do?
• Why do you want to do it?
• How will you do it?
• Who will work with you (if collaborative)?
• How does the proposal suit the mission of the funding agency/RFP?
• What are the anticipated outcomes?

After reading the overview, any reviewer will *unambiguously* understand the originality of the proposed project, the rationale for the proposed work, the expected outcomes, the expertise of the PI or PIs, the scale of the project, the need for the proposed work, and the significance of the expected outcomes.

Writing Persuasively

Directly answer the question, "Why is your proposed work important?"

"The results of the proposed experiments will be important because . . ."
"It is important to know the answer to this question because . . ."

Focus on the impact and relevance of the proposed work.
Make the case that the topic is significant.

Cite your own previous work throughout the proposal.
Make the case that you are the right person to do the work.

(after University of New Brunswick, 2004)

Narrative   The main text of the proposal is sometimes called the narrative, but in other cases it may be subdivided into the sections found in a technical paper.

Introduction/Background/Need and Fit   Here it is essential that you make a strong case for what you will do, how it fits with the state of the art, why it is significant, and how it aligns with the mission of the funding agency. This section should include:

- Current state of the art; what is known; what is lacking
- Background providing perspective for the reader
- Statement of need in the context of existing information
- Its fit with the host organization and funding agency
- Purpose, goals, and measurable objectives
- Compelling and logical reason for support

Closing the Loop

The description of the proposed project must be crafted to be compatible with the interest of a particular funder. Make the appropriate connections, so that any reviewer can readily see that the project aligns well with the purpose and goals of the funding source. This loop closure is the single most critical aspect of the proposal's narrative because it affects the reviewer's assessment of how compelling your proposal is. Make your proposal fit the RFP. If your proposal does not, and cannot, fit the RFP, seek a different source of funding.

Method   Describe the approach. How will the project be accomplished, and why did you choose the proposed approach:

- Experimental
- Analytical
- Computational
- Combined

Describe the methods and processes. What specifically will you do? Provide details of instruments and processing, as well as any other forms of acquiring data. Explain why the proposed methods are appropriate for your study. State how you will validate your methods. What is appropriate may vary tremendously across disciplines.

Describe the outcomes. What are the anticipated results, and what will you do with them? Will you publish a book, technical report, journal paper(s), or conference paper(s)? How will society benefit from your research?

Describe plans for completing all required documentation and how you will handle contingencies.

- Progress report plan
- Final presentation of results, for example, report, presentation, journal article
- Assessment plan (may be a separate section)
- Dissemination plan—if required
- What you will do if the focus of your research project deviates substantially from the proposal

Describe ownership of intellectual property (IP).

- Who owns what portion of any IP generated, and under what circumstances?
- Who approves publications?

Describe the role of personnel and key staff. Who will do what?

Project Timeline   The project timeline provides mileposts for completing different stages of a project. The timeline

- Shows project flow
- Includes start and end dates
- Lists a schedule of activities, including anticipated progress reports
- Describes projected outcomes

Personnel and Available Expertise   What are the credentials of the PIs? Provide information that certifies the PI's ability to successfully undertake the proposed work. Also include institutional or individual track records and resumes. Finally, describe any additional expertise to be provided by staff hired for the project.

Resources   Include both existing resources and resources that are needed to complete the project. Existing resources include

- In-kind matching
- Space, equipment, and utilities

Needed resources may include

- Equipment
- Supplies
- Additional personnel

Method of Evaluation/Assessment (if required as a separate component)   Some funding sources require a separate assessment plan. Inquire about the funder's expectations before submitting your proposal.

Budget   Many funders provide mandatory budget forms, which may require that you list

- In-kind support
- Matching revenue (where appropriate)

Be prepared to be flexible about your budget in case the funder chooses to negotiate costs.

List of References/Bibliography   The list of references shows the funding agency that you are familiar with the literature. Citing sources enhances your own credibility and provides substantiation for your claims.

Supporting Materials/Appendices   The following materials may be included in an appendix to provide support for your proposal:

- Letters from collaborators
- Resumes of key personnel

TABLE 8.1 | Proposal submission checklist

☐ Narrative is connected convincingly to RFP or mission of funding agency.

☐ Need for the project is well documented and consistent with the literature.

☐ Proposed methods are feasible and well thought out.

☐ Proposal is submitted on time.

☐ Proposal is in the requested format.

☐ Requested number of copies has been submitted.

☐ Proposal has original authorized signatures.

© Cengage Learning 2015

- Letters of institutional support for release time, equipment usage, and so on
- Results from previous related grants

Authorization by Host Organization/Supervisor   A signed authorization certifies that the proposal has been approved by the PI's organization.

Submission

Besides making sure the format and content of your proposal meet the funding agency's requirements, check for instructions on submission. Is the proposal to be submitted electronically or in printed form? If printed, are there instructions for binding, such as staple top left, three-hole punch, or hard bound? A neat, orderly document on high-quality paper makes a good impression, but you should limit expenditures on binding and frills beyond what is required. You do not want to give the impression that the proposing organization has plenty of resources or that it needlessly spends money to curry favor. Read carefully the RFP specifications of what is expected in the proposal and comply precisely with the instructions. Table 8.1 provides a checklist for proposal submission.

Short Proposal Formats

Short proposals are often limited to just a few pages. This is especially true of proposals that accompany a preliminary letter of intent, in which the author gives a short outline of the concept that, if approved, will be included in a subsequent full proposal. A short proposal to an external funding agency contains many of the same components as the full proposal, just in a very condensed format. Table 8.2 presents an outline for short proposals.

In the academic setting, undergraduate students frequently write semester project proposals for their courses, seniors may write a senior thesis proposal, and graduate students will routinely write their thesis or dissertation proposals under the guidance of their faculty advisers.

As stated earlier, writing short proposals may be even more challenging than writing full proposals because the writing style must be very concise to stay within the page limit. Writing concisely is a daunting task. The writer must weigh the significance of each word

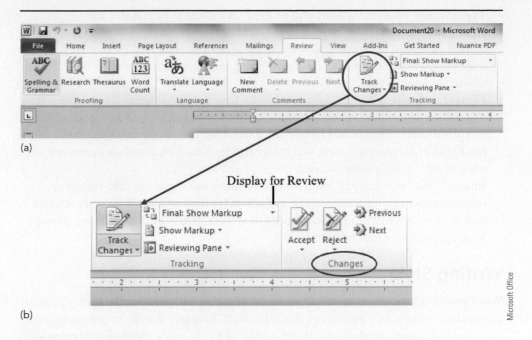

FIGURE 8.1 | Location of the track changes feature on the MS Word Review tab. **(a)** Track changes is turned off. **(b)** Track changes is turned on and commands in the adjacent Changes group are shown

details. Here, we will briefly show the location of the track changes feature and explain its utility. The **Track Changes** button is located on the **Review** tab in MS Word 2007 and 2010 (Figure 8.1a). To activate the command, click on the button. When **Track Changes** is turned on, the button is orange, as shown in Figure 8.1b. Any changes to the manuscript will be recorded until you turn off the feature. Changes made by each person in the manuscript will be recorded in a distinct color.

The user can choose from one of four views of the marked-up manuscript by clicking the down arrow on the right side of the **Display for Review** box. The **Final: Show Markup** option displays the edits as strike-through text, with the corrected text in the original (default black) color. Place the cursor over the edited text and click the **Accept** or **Reject** button (see Figure 8.1b) to carry out the corresponding action. Other viewing options are **Final, Original: Show Markup,** and **Original**. The **Final** view is handy for checking document formatting without the distractions of all of the markup. Regardless of the view, collaborative writers can share their edits with their colleagues in a way that allows everyone to see both the original and edited versions of the manuscript.

Document Format

Agreeing on a uniform document format before beginning to write will avoid tedious formatting revisions later on. Some considerations relative to format include:

- **The sections and section headings.** What sections will be included and how will they be named?

- **Typography.** What line spacing will be used? What font style and size are appropriate? What will be the style of the different level headings? Will the first line of each paragraph be indented, or will the paragraphs be separated by an empty line? How will vertical lists be formatted: using bullets, numbers, tabs, or borderless tables?
- **Figures and tables.** Will figures and tables be inserted in the text as close to the sentence in which they are first referenced, or will visual elements be attached at the end of the manuscript? If different people will be working on different sections, what kind of numbering system will ensure that the visual elements are numbered sequentially in the final manuscript?
- **Images.** Make sure that any images imported into a document file have sufficient resolution and a uniform appearance. Acknowledge images taken from other sources and, if the image will be included in published work, obtain prior written permission from the original author or publisher.

Writing Style

When people with different writing styles work together on a manuscript, it may be challenging to produce a final manuscript that reads well. To ensure that the overall style is uniform and consistent, collaborators should decide on a style guide for their writing project, which may include the following topics:

- **Vocabulary.** Using consistent vocabulary enhances reader comprehension. When there are multiple expressions for the same thing, choose one term and use it consistently throughout the manuscript. For example, when discussing design possibilities, agree on one word to describe the possibilities. Will you refer to them as configurations, arrangements, formations, variants, alternatives, or options? Define jargon and abbreviations that may not be familiar to your audience.
- **Notation and numbers.** What will be the symbol for diameter, for angular deflection, for velocity? By convention, numerals are used when they are followed by units, but in what situations might numbers be spelled out?
- **Person and voice.** Use first person (I or we) to avoid ambiguity about who is doing what. Use passive voice to describe procedures, where the action is more important than who is doing it. Manuscripts are easier to read and comprehend when a mixture of active and passive voice is used appropriately. For a more detailed discussion of the use of passive and active voice, see Chapter 5.
- **Tense.** In technical reports, past tense is used to describe your own results and when citing another author's work directly. Present tense has the connotation that the statement is generally accepted by the audience. For a more detailed discussion of tense, see Chapter 5.
- **Tone.** Technical reports are generally factual and objective. It may be appropriate, however, to adopt a persuasive tone when describing, for example, the significance of the problem and the proposed work as well as the expertise of the PIs.
- **Punctuation, spelling, and grammar.** In the United States, the period is placed inside the end quotation mark. In other English-speaking countries, however, the period is placed outside. Will there be one or two spaces after periods and colons? Will a comma be used before the "and" when listing items in series? For words with alternate spellings, which spelling will be used? Similarly, will compound words be

hyphenated, written as one word, or left as two? With multiple correct answers to each of these questions, consistent usage among collaborators is important.

Schedule and Deadlines

Schedules are great motivational tools when the deadlines are realistic and everyone in the group is committed to meeting those deadlines. Some mileposts to include in a schedule for a writing project include

- The manuscript outline
- Research and budget
- First draft
- Editing
- Revision
- Final check and assembly

Revision

Revision is a critical part of any writing project and requires attention to

- Technical accuracy
- Style, consistency, and readability
- Mechanical errors, such as spelling, grammar, and typos, and formatting

Depending on the strengths of the individual collaborators, different aspects of the editing and revision process may be allocated to different group members. See Chapter 5 for suggestions for the revision process.

Individual Responsibilities

Perhaps the most difficult aspects of collaborative work are dividing the labor equitably and keeping everyone on track. Successful collaboration requires good planning, good communication, and a willingness to compromise for the good of the whole. When it comes to group dynamics, the following points should be resolved at a preliminary meeting:

- **Select a group leader.** This person is the key to moving forward. The group leader must understand the details of the project about which the group is writing and must be a proficient writer. The group leader must also have a good sense of the end objective and keep the group moving toward that objective without being overbearing. The leader needs to listen to input from individual group members and find a way to synthesize diverse opinions into a coherent plan of action. For international projects, the leader must have a strong awareness of the cultural differences among members of the group. At the conclusion of the writing process, the group leader should draft the abstract or the executive summary of the project, since the leader is the only member who has dealt in detail with all of the constituent sections.
- **Identify the strengths of group members.** Who has expertise in searching engineering databases and gathering information? Who has experience with technical and computer-related problems? Who is willing to get pricing on materials needed to construct the apparatus? Who will be responsible for contacting the professor when

assistance or clarification is needed? Who is a detail-oriented individual who would make an excellent proofreader?

- **Assign responsibilities based on strengths.** Individual collaborators may be assigned multiple tasks, but the division of labor should be fair and equitable. For example, in addition to writing a section, an individual may also be the group's go-to person when it comes to technical difficulties. The group leader might not be assigned to write any particular section, but is responsible for ensuring that the sections produced by multiple writers are consistent. Other individuals may be involved primarily with materials acquisition and budget; others may only be responsible for editing and revision, with less input at the actual writing stage.
- **Make sure that everyone** who has been given a writing assignment **understands the task**, adheres to the style guide, and works hard to stick to the deadlines.
- **Come up with contingency plans** when things go wrong. If a group member is unable to meet a deadline (possibly due to illness or another unforeseeable situation), how will the unfinished activity be completed? How will disagreements be resolved?
- **Give all group members the opportunity to proofread** the final manuscript before submission. Keep in mind that when your name goes on a manuscript, you are responsible for the entire content and format, even if you personally did not write every section.

PROGRESS REPORTS

Another piece of writing that is often collaborative in nature is the progress report. Most frequently, a progress report is used to update an oversight entity about the progress of an engineering design project or a research project, funded either internally or externally. Funding agencies may have a particular prescribed format for progress reports. If your proposal received funding, the details for the reporting process should have been provided at the time of approval. The National Science Foundation website (National Science Foundation, 2012a; 2012b) documents a new common reporting format, called the Research Performance Progress Report (RPPR). The details of the report format will most likely evolve over time, but the focus is on creating a standard format for reporting progress on all federally funded projects. See the cited websites for further details.

In the absence of other institutional guidelines, the template for progress reports in Table 8.7 can be used. As shown, progress reports are typically in memo format. Chapter 7 contains details about memo writing, with instructions for formatting memos. The progress report itself is relatively succinct, usually only one to three pages in length. Progress reports often contain multiple attachments that document the written summary of progress made on the project and are often components in the final project outcomes. The attachments may include plots of significant test results, drawings of components or test equipment setups, copies of submitted journal papers or conference reports, and any necessary detailed financial budget sheets. Write progress reports in such a way that the information can be incorporated directly into the final project report.

When you prepare your progress report, use the template presented in Table 8.7 as a checklist to ensure that each component is included.

TABLE 8.7 | Template for progress report

DATE: _____

TO: _____

FROM: _____

RE: Progress Report on _____

© Cengage Learning 2015

Executive Summary Paragraph (often not labeled as such) _____
_____

Accomplishments during the reporting period relative to planned activities

- Introduction to goals of project for the reporting period
- Accomplishments for the reporting period compared with proposed work plan
- Assessment of success or failure of planned activities and processes
- List of milestones attained
- List of any unattained milestones and new projected date for attainment
- List of changes from original work plan and any substitute milestones

Interpretive discussion of

- Successes and what they mean for continued progress
- Issues that arose causing delays, prohibiting attainment of milestones
- Remedial steps taken to ensure attainment of milestones in the future
- Documentation of any needed change in project due to unforeseen complications or developments

Projections for next reporting period

- Timeline of planned activities
- Anticipated accomplishment of milestones
- Major expenditures for next reporting period

Financial update

- Planned expenditures completed
- Unanticipated expenses
- Delayed expenditure
- Project balance

List of attachments documenting narrative discussion
Attachments on subsequent pages

Standard memo format

Provide executive summary of report

Part 1 is a synopsis of activities during the reporting period

Part 2 is a discussion of successes, delays, failures, and changes in plans

Part 3 is an explanation of anticipated activities in the next reporting period

Part 4 is a review of finances

Attachments

Annotated Senior Design Project Progress Report

The following progress report for a senior design project is based on an actual student report, but the authors' names were changed and the report was entirely re-written for presentation in this book. The page format was modified to make room for the annotations.

Memo format

DATE: 13 October 20XX

TO: Professor Charles Knisely, Project Adviser

FROM: Jan Meyer, Jutta Kastner, and Erik Borque

RE: Biweekly progress report on Mash and Wort Design Project

Summary paragraph

The Mash and Wort Design team has explored various design configurations over the past two weeks, breaking the overall design challenge into various sub-systems. Brainstorming for each sub-system was undertaken, initially individually by each team member and then subsequently as a group. A key concept that recurs and affects all subsystems is the control of temperature. Another constant concern is keeping the overall system cost within the allocated budget.

Synopsis of accomplishments

The heating of the mash and subsequent preparation of the wort require the following sub-system processes:

- Power supply
- Mash heating
- Mash mixing
- Sparge water heating
- Sparger design
- Fluid pumping
- Wort heating
- Wort mixing
- Wort chilling
- Cooling source

The brainstorming resulted in multiple options for each of the sub-system components. These options will be vetted against the user requirements as well as cost. The choices made in designing these sub-system processes and components ultimately interact with the following overall considerations:

- Overall system layout
- The design of the processing containers
- Temperature regulation system

Discussion of progress

Temperature regulation was recognized as the first fundamental decision to be made. Extensive research into commercially available mash and wort temperature regulation systems showed that all such systems were too expensive for our budget. Knowing we cannot afford a commercial temperature regulation system, we explored less expensive alternatives such as modifying home heating thermostats along with designing and manufacturing our own heat exchangers.

The purchase of three commercially available mash and wort containers (15 gallon restaurant-grade stainless steel pots or specially designed brew pots) at more than $200 each would also exceed our budget. We are continuing to investigate buying used pots on eBay or worn out kegs.

Budget concerns

In the next two weeks, we will systematically evaluate the sub-system options generated over the past two weeks. Each option will be examined for both cost and compliance with user requirements. Inferior design concepts will be eliminated using a modified decision matrix to yield the best combination of system components as we move toward a final design. The existing Gantt chart will be updated. We anticipate that our final design will result in a quality mash and wort system which exceeds the quality of competing designs at a fraction of the cost.

Plan for subsequent reporting period

Attachments:
Concept Generation Listings
Generated Concept Figures

Attachments not included here due to space considerations

SUMMARY

In both industry and academia, innovative concepts lead to increased understanding, new products, and even new fields of study. Proposals are a standard mechanism for seeking support for implementing a project. A good proposal should anticipate the questions of the reviewers, provide a compelling justification for the need for the project, explain clearly the methods used in the project, demonstrate the expertise of the personnel, and provide mechanisms for assessing the outcomes and disseminating the project output.

Collaborative writing has become commonplace. For the collaboration to be successful, proper preparation and good communication are essential. Common vocabulary, notation, and formats need to be established and adhered to by all collaborators. A schedule should be established as well, and all collaborators need to commit to meeting their deadlines. Security is an issue in some forms of collaborative work. Make sure your collaborative writing plan satisfies the institutional or corporate security guidelines.

Progress reports need to be accurate, but they should also assure the funding organization that progress is being made. The funder should be convinced that the project is meeting its milestones as planned. If there has been a slight change, approval for deviation must be requested and granted.

Practicing engineers often struggle to document their accomplishments, but without documentation and dissemination of innovative and progressive ideas, these ideas will not affect the future in the way the engineer envisioned. Communication is an essential ingredient in any recipe for progress.

EXERCISES

1. In consultation with a faculty member, who may or may not be the writing instructor, select a topic of interest and prepare a preliminary research proposal to work with the professor on a project for a semester, a summer, or an entire academic year. This proposal may be a preliminary draft for a senior or honors thesis under the guidance of the faculty mentor.

2. Pick one of the topics presented in the Chapter 2 Exercises, and complete a survey of the current state of the art. Based on your analysis of the field, prepare a research proposal to fill a defined need in the field. Consider what expertise you possess and what additional expertise you might need to acquire or seek in a collaborator.

3. Prepare a progress report on any ongoing semester or multiweek project. The project may be course-related, or it may be a community service project or any other type of project. Establish a plan for completing one or more progress reports by a specific date. Use the progress report process to keep yourself on track for a timely completion of the project.

4. Develop a tentative master plan for your career, or perhaps even multiple master plans. For the master plan of your choosing, write a progress report and a plan for the timing of future progress reports. Use your progress reports as a means of performing self-assessment and of documenting milestones as you move toward your career objective.

5. With one or more classmates, select a topic from the list of topics in the Chapter 2 Exercises and prepare a collaborative report, focusing initially on the planning process. Write a short synopsis of the terminology, notation, and format you agreed upon and any mechanism you can think of for updating these plans as the need arises. Also describe in your synopsis the plan for sharing your drafts within your group. Select a group leader, split up the writing in an equitable way, and devise a timeline for completion. Prepare a group report that consists of the final edited manuscript, a brief list of the collaborators' contributions prior to editing, and a copy of the written synopsis of your initial planning agreement.

6. Search the Internet for a research proposal in an area of interest. Examine the proposal and read it to determine if it contains the components listed in Tables 8.2 to 8.4. Note the format of the proposal and list in outline form the main headings and subheadings in the document. Examine the proposal again to see if the author convinced you that the research is needed, the proposed methods are suitable, and the investigator is well qualified to carry out the work. Would you approve the proposal?

7. Reflect on your curriculum to this point in your academic career. Prepare a short one- to two-page proposal for curricular changes. Address your persuasive proposal to a hypothetical university administrator. Explain why the changes would be beneficial for your course of study. Research other institutions' curricula and compare them with your own.

8. Consider a leisure-time activity that you would like to pursue but cannot because there is no local group engaged in the activity. Write a proposal explaining exactly what the activity is, how you might recruit other participants, what the benefits would be for participants, and what financial support you would need to put your plans into motion.

REFERENCES

Allen, E.M. 1960. Why are research grant applications disapproved? Science, Vol. 132, 1532–1534. Also available online as DOI: 10.2307/1707045.

Foundation Center. 2012. Proposal writing short course. Accessed 8 May 2013 at <http://foundationcenter.org/getstarted/tutorials/shortcourse/index.html>.

Geever, J.C. 2007. The Foundation Center's guide to proposal writing, 5th ed. New York: The Foundation Center.

Glisson, A. 2004. Writing an effective proposal. Technologies of Writing, Vol. 2, No. 1. Accessed 24 February 2013 at <http://class.georgiasouthern.edu/writling/professional/TechWrite/2-1/glisson/index.html>.

Green, M.G. 2004. An empirical study of product functional families: Analyzing key performance metric trends to derive actionable design insights. Dissertation Proposal, Mechanical Engineering Department, University of Texas at Austin. Accessed 24 February 2013 at <https://webspace.utexas.edu/cherwitz/www/ie/sample_diss.html>.

Hilliard, M.D. 2005. A predictive model for aqueous potassium carbonate/piperazine/ethanolamine for carbon dioxide removal from flue gas. Dissertation Proposal, Chemical Engineering Department, University of Texas at Austin. Accessed 24 February 2013 at <https://webspace.utexas.edu/cherwitz/www/ie/sample_diss.html>.

Levine, J. 2011. Guide for writing a funding proposal. Accessed 24 February 2013 at <http://learnerassociates.net/proposal/>.

McGranaghan, M. 2012. Guidelines on writing a research proposal. Accessed 8 May 2013 at <http://www2.hawaii.edu/~matt%20/proposal.html>.

Montecino, V. and Williams, A. 2009. Collaborative writing. Accessed 24 February 2013 at <http://classweb.gmu.edu/nccwg/collab.htm>.

National Science Foundation (NSF) 2004. A guide for proposal writing. Accessed 24 February 2013 at <http://www.nsf.gov /pubs/2004/nsf04016/nsf04016.pdf>.

National Science Foundation (NSF). 2012a. Research performance progress report (RPPR), updated 1/27/12. Accessed 24 February 2013 at <http://www.nsf.gov/bfa/dias/policy/rppr/>.

National Science Foundation (NSF). 2012b. Draft format research performance progress report. Accessed 24 February 2013 at <http://www.nsf.gov/bfa/dias/policy/rppr/draftformat.pdf>.

National Sciences and Engineering Research Council of Canada (NSERC). 2012. Guidelines for the preparation and review of applications in engineering and the applied sciences. Accessed 24 February 2013 at <http://www.nserc-crsng.gc.ca/NSERC-CRSNG/Policies-Politiques/prepEngAS-prepGenSA_eng.asp>.

Research Proposal Guide. 2012. Research proposal guide. Accessed 24 February 2013 at <http://researchproposalguide.com/>.

Thackery, D. 2012. Why proposals are rejected, University of Michigan Proposal Writer's Guide. Accessed 24 February 2013 at <http://orsp.umich.edu/proposals/pwg/pwgrejected.html>.

University of New Brunswick. 2004. Grants and Contracts, Office of Research Services. Accessed 8 May 2013 at <http://www.unb.ca/research/ors/grants_contracts/grants.php>.

University of Pittsburgh Writing Websites. 2011. Selected proposal writing websites. Accessed 24 February 2013 at <http://www.pitt.edu/~offres/proposal/propwriting/websites.html>.

White, T. 2006. Principles of good research & research proposal guide. Policy, Performance and Quality Assurance Unit of London Borough Richmond upon Thames. Accessed 24 February 2013 at <http://www.richmond.gov.uk/research_proposal_guide.pdf>.

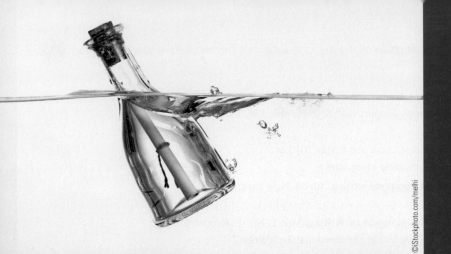

©iStockphoto.com/melhi

Chapter
9

SPECIFICATIONS, CONTRACTS, AND INTELLECTUAL PROPERTY

In many situations, engineers may not be the lead writers in documents with legal bearing, but they may be asked to contribute to such documents as specifications, contracts, and intellectual property registration (patents, trademarks, service marks, and patents). Especially in small and start-up companies, engineers might be asked to prepare drafts of these kinds of documents, which would then be reviewed and formulated in legal language by an attorney. In large organizations, junior engineers are often assigned the task of writing specifications.

The objective of this chapter is to provide engineers and other technical personnel with an overview of the purpose, format, and language of specifications, contracts, and intellectual property. At the conclusion of the chapter, the reader will have an understanding of the type of writing required, but will need further study and practice to become a skilled writer of these legalistic documents.

ENGINEERING SPECIFICATIONS

Specifications are legally binding agreements that define the requirements that a product, program, device, component, system, structure, or process (hereafter called "the object") must satisfy (Cockerill, 2005). Engineering input is usually required in the drafting of specifications to achieve the following objectives:

- Demonstrate how the object meets a client's needs
- Permit (initial) cost estimates
- Provide guidance for a contractor/developer who might undertake the project

TYPES OF SPECIFICATIONS

The word *specifications* has many different connotations depending, in part, on the industry. In general, however, all specifications address the following technical points:

- They describe an object (the product, structure, component, device, or program).
- They cite codes, standards, and regulations to justify why certain materials and methods are to be used.
- They reference and incorporate proprietary specifications.
- They specify the required and measurable performance of the object.

In this section, we will describe the purpose and content of specifications in selected disciplines and industries. We will begin with the familiar technical specifications that accompany consumer goods and then move on to new product development specifications used in manufacturing, electronic product requirements specifications, software development specifications, building construction specifications, process specifications, and then procurement specifications. Examples of a portion of a building specification and a process specification will serve to illustrate commonalities with regard to format, language, and grammatical structure. Finally, we will address the *writing* of specifications, in particular, word use and the characteristics of well-written specifications.

   This chapter may be read selectively or sequentially. Readers who work in a particular discipline may choose to focus on their discipline-specific specifications. Readers who are interested in a broader overview may prefer to read the chapter sequentially, comparing and contrasting specifications in different disciplines.

Product Technical Specifications

Products that come to market frequently come with a listing of performance characteristics. Most consumers, for instance, are familiar with the *technical specifications* that are found in user manuals of everything from microwave ovens to cameras. Quite often consumers use these same technical specifications to assess product quality and to select the appropriate product for their needs. Figure 9.1, for example, shows the technical specifications of a Canon MP270 all-in-one printer. Consumers considering the purchase of such a printer might compare features such as printing and scanning resolution, color or black-and-white scanning capability, copying and scanning size, and the requirements for the host computer.

General Specifications

Printing resolution (dpi)	4800* (horizontal) × 1200 (vertical) * Ink droplets can be placed with a pitch of 1/4800 inch at minimum.

Interface	USB Port:
	  Hi-Speed USB*1
	Direct Print Port: *2
	  PictBridge
	*1 A computer that complies with Hi-Speed USB standard is required. Since the Hi-Speed USB interface is fully upwardly compatible with USB 1.1, it can be used at USB 1.1.
	*2 MP270 series only

Print width	8 inches/203.2 mm (for Borderless Printing: 8.5 inches/216 mm)
Operating environment	Temperature: 5 to 35°C (41 to 95°F) Humidity: 10 to 90% RH (no condensation)
Storage environment	Temperature: 0 to 40°C (32 to 104°F) Humidity: 5 to 95% RH (no condensation)
Power supply	AC 100–240 V, 50/60 Hz

Power consumption	MP270 series:
	  Printing (Copy): Approx. 10W
	  Standby (minimum): Approx. 1.2W*
	  OFF: Approx. 0.5W*
	MP250 series:
	  Printing (Copy): Approx. 11W
	  Standby (minimum): Approx. 1.2W*
	  OFF: Approx. 0.5W*
	* USB connection to PC

External dimensions	MP270 series:
	  Approx. 17.8 (W) × 13.2 (D) × 6.1 (H) inches
	  Approx. 450 (W) × 335 (D) × 155 (H) mm
	MP250 series:
	  Approx. 17.5 (W) × 13.1 (D) × 6.1 (H) inches
	  Approx. 444 (W) × 331 (D) × 154 (H) mm
	* With the Paper Support and Paper Output Tray retracted.

Weight	MP270 series:
	  Approx. 5.6 kg (Approx. 12.3 lb)
	MP250 series:
	  Approx. 5.8 kg (Approx. 12.7 lb)
	* With the FINE Cartridges installed.

Canon FINE Cartridge	Black: 320 nozzles Color: 384 × 3 nozzles

Copy Specifications

Multiple copy	1–9, 20 pages
Reduction/Enlargement	Fit-to-page

Scan Specifications

Scanner driver	TWAIN/WIA (Windows Vista and Windows XP only)
Maximum scanning size	A4/Letter, 8.5" × 11.7"/216 × 297 mm

Scanning resolution	MP270 series:
	  Optical resolution (horizontal × vertical) max: 1200 dpi × 2400 dpi
	  Interpolated resolution max: 19200 dpi × 19200 dpi
	MP250 series:
	  Optical resolution (horizontal × vertical) max: 600 dpi × 1200 dpi
	  Interpolated resolution max: 19200 dpi × 19200 dpi

Gradation (Input/Output)	Gray: 16 bit/8 bit Color: 48 bit/24 bit (RGB each 16 bit/8 bit)

Minimum System Requirements

Conform to the operating system's requirements when higher than those given here.

	Windows	Macintosh
Operating System Processor RAM	Windows Vista, Vista SP1 1 GHz processor 512 MB Windows XP SP2, SP3 300 MHz processor 128 MB Windows 2000 Professional SP4 300 MHz processor 128 MB	Mac OS X v.10.5 Intel processor, PowerPC G5, PowerPC G4 (867 MHz or faster) 512 MB Mac OS X v.10.4 Intel processor, PowerPC G5, PowerPC G4, PowerPC G3 256 MB Mac OS X v.10.3.9 PowerPC G5, PowerPC G4, PowerPC G3 128 MB
Browser	Internet Explorer 6 or later	Safari
Hard Disk Space	750 MB Note: For bundled software installation.	800 MB Note: For bundled software installation.
CD-ROM Drive	Required	
Display	XGA 1024 × 768	

- Windows: Operation can only be guaranteed on a PC with Windows Vista, XP or 2000 pre-installed.
- To upgrade from Windows XP to Windows Vista, first uninstall software bundled with the Canon inkjet printer.
- Some functions may not be available with Windows Media Center.
- Macintosh: Hard Disk must be formatted as Mac OS Extended (Journaled) or Mac OS Extended.
- Internet Explorer 7 or later is required to install Easy-WebPrint EX.

Additional System Requirements for the on-screen manuals

Windows	Macintosh
Browser: Easy Guide Viewer Note: Internet Explorer 6 or later must be installed. The on-screen manual may not be displayed properly depending on your operating system or Internet Explorer version. We recommend that you keep your system up to date with Windows Update.	Browser: Easy Guide Viewer Note: The on-screen manual may not be displayed properly depending on your operating system or Safari version. We recommend that you keep your system up to date.

Specifications are subject to change without notice.

FIGURE 9.1 | Sample product technical specifications for an all-in-one printer-scanner-copier

Technical specifications are also prominent in marketing and advertising (for example, think of 4G versus 3G communication devices). There is an expectation that the device will perform as advertised. Failure to meet the product specifications may lead to a recall of the product, with the associated expenses. For this reason, manufacturers take random samples and test their products to ensure that they meet the product specifications.

Different specifications are used in the design and development of a *new product*. The role of these new product development specifications, described in the next section, is to establish the desired capabilities of a yet-to-be-created device and provide metrics against which design variants can be tested.

New Product Development

The first step in the development of a new product involves an individual or organization noting an explicit need for a device to do something that is not currently possible. This idea is then developed into *product design specifications* that describe what the product needs to be able to do and provides target values for the desired performance against which the product can be tested. Once the design specifications are laid out, the designers propose methods for achieving the requirements. Through an assessment of the economic and performance factors of perhaps multiple design suggestions, a final design evolves, which then forms the basis of the *manufacturing specifications*. **Manufacturing specifications** are the design drawings, the parts list, and the assembly procedures needed to manufacture the new product.

Product Design Specifications **Product design specifications** define the functions and performance characteristics that a particular component, device, or system (the object) must exhibit. Writing design specifications requires the engineer to consider the function of the object independent of any physical embodiment. How fast must it rotate? How much force must it transmit? Are there restrictions on the object's mass, volume, or conductivity? Does the object need to deliver a fixed voltage within tight tolerance, or can the output vary substantially? The engineer must define what the object is expected to do within defined tolerances. Careful consideration of the tolerances is needed because the tighter the tolerances, the higher the price of the object.

Each individual product design specification must meet the following criteria:

- Must be defined by a single metric and a value or range of values. The metric abstractly defines *what* the object will do in terms of a measurable quantity with a target value
- Does not consider *how* the object meets the metric
- Must be stated in complete, concise sentences that precisely define the performance without room for interpretation (i.e., it must be unambiguous)
- Must be traceable to customer needs. Numbering of each specification helps identify each need for purposes of traceability

The compilation of individual specifications constitutes the *product design specifications*.

Table 9.1 provides product design specifications for a folding sailboat mast from a senior design project. The first column lists the individual specifications based on the user's needs or requirements. These requirements are subdivided into basic (absolutely required), performance (better features likely to increase customer satisfaction), and excitement (even better features designed for ultimate satisfaction). Note that each specification is accompanied by a metric and a target value, and the rationale for choosing each target value is given. Adding justification for the product design specification makes the process more transparent because it allows design decisions to be traced back to the user's requirements. The analytical validation describes how performance will be calculated theoretically, and the physical validation explains how performance will be measured after the mast has been built.

The specifications are descriptive and performance-based. No particular design is implied, but rather the specifications serve as a guide for the design and provide metrics the final design must meet. The product design specification must highlight

TABLE 9.1 | Product design specifications for a collapsible sailboat mast

User Requirement	Origin of Requirement	Metric	Target Value or Range	Justification	Analytical Validation (pre-build)	Physical Validation (post-build)
Basic:						
1. The mast must be easily raised and collapsed by one individual.	Customer	Maximum sustained force required to raise mast	20 lb— 50 lb	If force to raise the mast is too high, a lone sailor could not use it	Working Model™ analysis and human strength capabilities	Have one person of average strength raise and lower the prototype
2. The mast must fit under bridges when collapsed.	Customer	Height of highest point of the mast above the waterline when collapsed	13 ft max	Passing beneath bridges is one of the reasons a collapsible mast is being created	Calculate collapsed height using the geometry of the mast	Measure the height of the collapsed prototype
3. The mast must be able to be transported on the boat trailer.	Customer	The distance the mast extends beyond the front of the boat when collapsed	2 ft max	There is not much space between the trailer and the back of the towing vehicle	Compare the length of the longest mast section to the length of the boat	Measure the length of the mast compared to boat length
4. The mast must be safe for use in usual boating conditions.	Customer	Maximum sustainable wind speed without failure	70 knots	The mast must be designed to withstand high wind speeds	Model mast in ANSYS™ to estimate stress relative to allowable stress	Test mast in structural testing laboratory
5. The mast must not collapse unintentionally under load.	Customer	Maximum compression load that moveable mast elements must withstand	2000 lb	Unintentional mast collapse would be dangerous to crew and poses a liability	Model mast in ANSYS™ to estimate stress relative to allowable stress	Test mast in structural testing laboratory

TABLE 9.1 | Continued

User Requirement	Origin of Requirement	Metric	Target Value or Range	Justification	Analytical Validation (pre-build)	Physical Validation (post-build)
Performance:						
6. The time required to raise or lower the mast must be reasonable.	Customer	Time required to raise mast (includes any setting up and stowing of auxiliary equipment)	30 min	User does not want to expend time erecting and collapsing mast	A higher required force to raise the mast will take longer for the user to erect the mast	One person raises and lowers the prototype within the allotted time
7. The mast must be at a competitive price.	Competition, Customer	Cost to retrofit on a boat with an existing noncollapsible mast	$1,000	Price is an overriding factor for acceptance of product	Create bill of materials to determine total part costs	Add up costs of parts and labor needed to manufacture the prototype
Excitement:						
8. The mast requires a power system for raising and lowering.	Customer	Power capability of motor for winch	2.5 hp max	A powered system makes it easier to raise the mast for users with strength limitations	Calculate power required from force and time required	Apply motor to winch and see if it can raise the mast

Based on Friday et al., 2008

performance and include quantitative values to provide criteria for selecting the final design from the various candidate designs that will be developed. There is obviously no single correct answer to the design problem, which may be disconcerting to engineering students who are used to solving problems specifically created to have a single correct answer. However, developing the ability to distinguish good from bad designs is part of becoming proficient in design. Additional information concerning the design process and methods to develop product design specifications can be found in numerous textbooks on mechanical, electrical, and computer design, such as Sclater and Traister (2003), Ford and Coulston (2008), and Ulrich and Eppinger (2011), along with their ancillary websites.

Manufacturing Specifications Once all stakeholders have reviewed, reformulated (if necessary), and agreed to the design specifications, the specifications are submitted to the designers. The designers' responsibility is to create a physical system (which includes any software required to operate the system) that delivers the required performance. (In other disciplines, the design may result in a software package, a chemical manufacturing process, or a maintenance or regulatory process.) After the designers have reached consensus on the preferred embodiment of the object and gained approval from the customer, they must produce a complete set of drawings, including dimensioned drawings of each component, with tolerances along with assembly drawings of the system. Written instructions accompany the drawings and describe the order and process of assembly. The drawings and written instructions together constitute the *manufacturing specifications* used by a machine shop to produce and assemble the parts. Any electrical and electronic components will need a complete set of schematics, layout of and production process for the mounting boards, as well as National Electrical Manufacturers Association (NEMA, 2013) specification of enclosures, including any required cooling system.

The standards for drawing, dimensioning, and tolerancing parts can be found in various professional society and regulatory agency publications and on their websites. It is common practice to cite these standards in the manufacturing specifications. Figure 9.2 shows an assembly drawing for the arm subassembly for a puck-shooting machine. Figure 9.3 shows the manufacturing drawing for the arm. There are usually a large number of drawings needed for even a relatively simple machine since a machinist must manufacture each part and then assemble the parts to make the device.

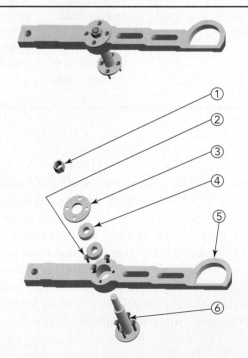

1	5810 1/2" − 13 Lock Nut
2	5820 1/4" − 20 × 3/8" Bolt
3	4832 Bearing Restraining Ring
4	4830 Ball Bearing
5	4810 Shooting Arm 1.75
6	4850 Bearing Support

Based on Kramer et al., 2010

FIGURE 9.2 | Assembly drawing for the projectile propulsion arm in a puck-shooting machine as a part of the manufacturing specification for the puck-shooting machine.

J. Spring Hangers with Vertical-Limit Stop <Insert drawing designation>: Combination coil-spring and elastomeric-insert hanger with spring and insert in compression and with a vertical-limit stop.
1. Frame: Steel, fabricated for connection to threaded hanger rods and to allow for a maximum of 30 degrees of angular hanger-rod misalignment without binding or reducing isolation efficiency.
2. Outside Spring Diameter: Not less than 80 percent of the compressed height of the spring at rated load.
3. Minimum Additional Travel: 50 percent of the required deflection at rated load.
4. Lateral Stiffness: More than 80 percent of rated vertical stiffness.
5. Overload Capacity: Support 200 percent of rated load, fully compressed, without deformation or failure.
6. Elastomeric Element: Molded, oil-resistant rubber or neoprene.
7. Adjustable Vertical Stop: Steel washer with neoprene washer "up-stop" on lower threaded rod.
8. Self-centering hanger rod cap to ensure concentricity between hanger rod and support spring coil.
K. Pipe Riser Resilient Support: All-directional, acoustical pipe anchor consisting of 2 steel tubes separated by a minimum of 1/2-inch- thick neoprene. Include steel and neoprene vertical-limit stops arranged to prevent vertical travel in both directions. Design support for a maximum load on the isolation material of 500 psig and for equal resistance in all directions.
L. Resilient Pipe Guides: Telescopic arrangement of 2 steel tubes or post and sleeve arrangement separated by a minimum of 1/2-inch- thick neoprene. Where clearances are not readily visible, a factory-set guide height with a shear pin to allow vertical motion due to pipe expansion and contraction shall be fitted. Shear pin shall be removable and reinsertable to allow for selection of pipe movement. Guides shall be capable of motion to meet location requirements.

The imperative is used to state an obligation on the part of the contractor (red highlight added)

PART 3 – EXECUTION

3.1 APPLICATIONS
A. Multiple Pipe Supports: Secure pipes to trapeze member with clamps approved for application by an agency acceptable to authorities having jurisdiction.
B. Hanger Rod Stiffeners: Install hanger rod stiffeners where indicated or scheduled on Drawings to receive them and where required to prevent buckling of hanger rods due to forces.
C. Strength of Support and Restraint Assemblies: Where not indicated, select sizes of components so strength will be adequate to carry present and future static loads within specified loading limits.

Contractor is obligated to undertake design work if sizes are not specified (red highlight added)

FIGURE 9.10 | Continued

3.2   VIBRATION-CONTROL DEVICE INSTALLATION
   A.   Comply with requirements in Division 7 Section "Roof Accessories" for installation of roof curbs, equipments, supports, and roof penetrations.
   B.   Equipment Restraints:
      1.   Install resilient bolt isolation washers on equipment anchor bolts where clearance between anchor and adjacent surface exceeds 0.125 inches.
   C.   Piping Restraints:
      1.   Comply with requirements in MSS SP-127.
      2.   Space lateral supports a maximum of 40 feet o.c., and longitudinal supports a maximum of 80 feet o.c.
      3.   Brace a change of direction longer than 12 feet.
   D.   Install cables so they do not bend across edges of adjacent equipment or building structure.
   E.   Install bushing assemblies for anchor bolts for floor-mounted equipment, arranged to provide resilient media between anchor bolt and mounting hole in concrete base.
   F.   Attachment to Structure: If specific attachment is not indicated, anchor bracing to structure at flanges of beams, at upper truss chords of bar joists, or at concrete members.
   G.   Drilled-in Anchors:
      1.   Identify position of reinforcing steel and other embedded items prior to drilling holes for anchors. Do not damage existing reinforcing or embedded items during coring or drilling. Notify the structural engineer if reinforcing steel or other embedded items are encountered during drilling. Locate and avoid prestressed tendons, electrical and telecommunications conduit, and gas lines.
      2.   Do not drill holes in concrete or masonry until concrete, mortar, or grout has achieved full design strength.
      3.   Wedge Anchors: Protect threads from damage during anchor installation. Heavy-duty sleeve anchors shall be installed with sleeve fully engaged in the structural element to which anchor is to be fastened.
      4.   Set anchors to manufacturer's recommended torque, using a torque wrench.
      5.   Install zinc-coated steel anchors for interior and stainless steel anchors for exterior applications.

3.3   FIELD QUALITY CONTROL
   A.   Perform tests and inspections.
   B.   Tests and Inspections:
      1.   Provide evidence of recent calibration of test equipment by a testing agency acceptable to authorities having jurisdiction.

FIGURE 9.10 | Continued

2. Test at least four of each type and size of installed anchors and fasteners selected by Architect.

3. Test to 90 percent of rated proof load of device.

4. Measure isolator restraint clearance.

5. Measure isolator deflection.

6. If a device fails test, modify all installations of same type and retest until satisfactory results are achieved.

C. Remove and replace malfunctioning units and retest as specified above.

D. Prepare test and inspection reports.

3.4 ADJUSTING

A. Adjust isolators after piping system is at operating weight.

B. Adjust limit stops on restrained spring isolators to mount equipment at normal operating height. After equipment installation is complete, adjust limit stops so they are out of contact during normal operation.

C. Adjust active height of spring isolators.

D. Adjust restraints to permit free movement of equipment within normal mode of operation.

END OF SECTION 15073

Specification provided courtesy of Bucknell University.

FIGURE 9.10 | Continued

While the specification format for building construction has been unified in the MasterFormat system, there are many other proprietary formats still in use. In highway construction, the Federal Highway Administration has guidelines, as does the Department of Transportation in most states. See FHWA (2010) and ODOT (2009) for examples of these guidelines. Entry-level engineers should familiarize themselves with the format and language of specification since the compilation of specifications is often assigned to junior members of an engineering firm.

As discussed previously, building and construction specifications prescribe how the construction process proceeds. There is far less reliance on end-user input, which is a hallmark of product technical specifications, new product development specifications, electronic product specifications, and software development specifications. Particularly in software and new product development, the output usually is new and unique, so the choices made to satisfy the end user are going to determine the success or failure of the software or product. In the building construction process, the end-user input is in the architectural design of the structure that precedes the building process.

In the next section, we will discuss process specifications. Similar to building specifications, process specifications are defined upstream of the process itself, and they delineate precisely how the process is to be carried out.

Process Specifications

Standardized specifications have many advantages. Most industries have adopted formatting and numbering systems analogous to those of MasterFormat® for easy identification and reference. As an example, Table 9.2 shows a process specification for the preventative maintenance of valves provided by one of the author's former students, now working in a corporate environment. The student had less than two years' experience when assigned the task of creating this specification. The annotations in the left column highlight some noteworthy characteristics of this specification. Minor typographical errors and inconsistencies in the original have been eliminated.

TABLE 9.2 | Sample engineering process specification supplied by an alumna. This specification is an example only and should not be interpreted as a validation for use as an actual specification for safety and relief valve maintenance or otherwise, and no warranty is being made in connection therewith.

Cover Page

ENGINEERING SPECIFICATION
XX-XX-XXXX
PREVENTIVE MAINTENANCE OF SAFETY AND RELIEF VALVES
REVISION 00
MAY 20XX

Contents Page

SECTION I – GENERAL AND TECHNICAL REQUIREMENTS
  1.0 SCOPE
  2.0 APPLICABLE STANDARDS AND REFERENCES
  3.0 QUALITY ASSURANCE
  4.0 GENERAL SAFETY AND ENVIRONMENTAL CONSIDERATIONS
  5.0 TEST, EQUIPMENT, AND CALIBRATION CRITERIA
  6.0 FREQUENCY
  7.0 PRECAUTIONS
  8.0 ACCEPTABLE METHODS
  9.0 DOCUMENTATION
  10.0 ACCEPTANCE CRITERIA
  11.0 RESTORATION
  12.0 ANALYSIS

SECTION II – ACCEPTANCE REQUIREMENTS AND SUPPLEMENTAL SPECIFICATIONS
Attachment 1 – Maintenance and Testing Intervals for Safety and Relief Valves
Attachment 2 – Tolerances for Pressure Lift Tests
Attachment 3 – NBIC In-service Inspection Requirements for Pressure Relief Valves
Attachment 4 – Certified Safety/Pressure Relief Valve Repair/Test Report (1 of 3)
Attachment 5 – NBIC Nameplate Requirements for 'VR' Repairs

TABLE 9.2 | Continued

SECTION I – GENERAL AND TECHNICAL REQUIREMENTS

1.0 **SCOPE**

  1.1 **Purpose** – This specification describes the recommended minimum schedule for the Preventative Maintenance (PM) of all safety and relief valves at the Company's Steam Generating Units. It is compiled as a guide for operating and maintenance personnel who are responsible for the optimum operation of this equipment.

    1.1.1 PM is defined as visually inspecting, pressure lift testing, repairing, or replacing the valve on a routine schedule.

  1.2 **Applicable Equipment** – This specification applies to any equipment in the computer system that falls under the Safety/Relief Valve asset category in the Generating Station.

  1.3 **General**

    1.3.1 The PM guidelines set forth in this specification have been formulated from legal requirements, industry standards, manufacturer's recommendations, and current Industry and Company practice. The guidelines are attached to this specification as Attachment 1 and provide the type of Preventative Maintenance to be performed based on the type of safety and relief valve.

    1.3.2 Contractors performing these services shall comply with these general requirements, Engineering Specification XX-XX-XXXX, and the purchase order – Company Standard Terms and Conditions of Service Contracts.

    1.3.3 The job specific work scope and list of valves to be examined, along with the technical requirements will be shown in the job specific specification.

2.0 **APPLICABLE STANDARDS AND REFERENCES**

> **NOTE**
> The latest edition of references applies, unless otherwise specified.

  2.1 American Petroleum Institute (API) 510: Pressure Vessel Inspection Code: In-Service Inspection, Rating, Repair, and Alteration.

  2.2 API 576: Inspection of Pressure-Relieving Devices.

  2.3 API 581: Risk Based Inspection Technology.

  2.4 American Society of Mechanical Engineers (ASME) Boiler and Pressure Vessel Code (BPVC) Sections I, VIII and IX.

  2.5 National Board Inspection Code (NBIC).

  2.6 National Fire Protection Association (NFPA) 25: Standard for the Inspection, Testing, and Maintenance of Water-Based Fire Protection.

  2.7 In addition to the above codes and standards, the Inspection Contractor shall comply with the federal, state and local laws, codes and all applicable specifications and standards including, but not limited to, those specified

First Page of Section I

Unlike an outline, the use of section number 1.1.1 without a subsequent 1.1.2 may be used

(Note: The original page breaks have been omitted.)

Industry codes and standards are cited

TABLE 9.2 | Continued

herein. Any conflicts that arise due to use of referenced specifications shall be identified and resolved with Engineering before starting or continuing work.

State laws are cited

2.7.1 Applicable State Law: [State] Industrial Code, Rule [Number]: Construction, Installation, Inspection, and Maintenance of High Pressure Boilers; Construction of Unfired Pressure Vessels.

3.0 QUALITY ASSURANCE

3.1 All safety and relief valve maintenance work shall be done in accordance with NBIC and ASME requirements.

3.2 All safety and relief valve maintenance work performed by contractors shall be done in accordance with Engineering Specification XX-XX-XXXX.

3.3 All welding procedures (WPSs and PQRs) to be used shall be certified in accordance with ASME Boiler and Pressure Vessel Code, Section IX.

3.4 All welders working on boilers and pressure vessels shall be qualified in accordance with ASME Boiler and Pressure Vessel Code, Section IX.

4.0 GENERAL SAFETY AND ENVIRONMENTAL CONSIDERATIONS

4.1 All Company personnel shall follow all applicable Company Environmental, Health and Safety policies, procedures and instructions. This includes but is not limited to the following:

Proprietary company specifications are cited

4.1.1 Company Specification XX.XXX – Hazard Assessment and Personal Protective Equipment

4.1.2 Company Specification XX.XXX – Personal Protective Equipment – Protective Clothing

4.1.3 Asbestos Management Manual

4.1.4 Company Specification XX.XXX – Oil Program Overview

4.1.5 Company Specification XX.XXX – Chemical Program Overview

4.1.6 Material Safety Data Sheets (MSDS) for any chemicals

Regulatory requirements are cited

4.2 All Contractor personnel shall follow all applicable OSHA rules and regulations, and all other federal, state and local laws, rules and regulations. This includes but is not limited to the following:

4.2.1 OSHA Subpart A – General

4.2.2 OSHA Subpart I – Personal Protective Equipment

5.0 TEST, EQUIPMENT, AND CALIBRATION CRITERIA

5.1 All pressure lift tests performed on safety and relief valves will be performed by a contractor.

5.2 Equipment Criteria

5.2.1 All vendor equipment will be certified under NBIC requirements.

5.2.2 When a lift test is performed, pressures must fall under the tolerances defined in ASME BPVC. See Attachment 2.

National standards are cited

5.3 Calibration Criteria – Equipment used shall be calibrated to a known standard traceable to the National Institute of Standards and Technology (NIST).

TABLE 9.2 | Continued

6.0 **FREQUENCY**

6.1 Maintenance and Testing Intervals for Safety and Relief Valves in Attachment 1 describes the recommended minimum PM schedule for safety and relief valves.

7.0 **PRECAUTIONS**

7.1 The instruction manuals for all safety and relief valves and equipment they are protecting shall be read and reviewed by the individuals performing all tests. All safety instructions within these manuals, as well as those identified in Section 4.0, General Safety and Environmental Considerations, shall be followed.

8.0 **ACCEPTABLE METHODS**

8.1 Visual inspections

8.1.1 Visual inspection will be performed by company personnel. Personnel will check that the valve meets the requirements in NBIC Part 2 – Inspection, Section 2.5.3. See Attachment 3.

8.2 Pressure lift tests

8.2.1 Pressure lift tests will be performed to check the set point of the safety and relief valves. For proper valve setting pressures and unit documentation, refer to job-specific specifications.

8.2.2 Pressure lift tests will be performed in accordance with Section 5.0, Test, Equipment and Calibration Criteria.

8.3 Repair

8.3.1 All repairs performed on safety and relief valves will be performed by a contractor.

8.3.2 Valves welded into the system shall be inspected, repaired and tested in place. All inspection, repair, and testing work shall be performed on-site unless otherwise authorized by the Station.

8.3.3 Flanged valves shall be removed and brought to the Contractor's Shop for inspection, repair and testing. All inspection, repair, and testing work shall be performed offsite unless otherwise authorized by the Station. Where it is not practical to remove the flanged valves, the Contractor may request permission from the Company, in writing, to perform the inspection, repair, and testing on site.

8.3.4 After reassembly of a repaired valve, valve lifting pressure will be set to the value specified on the nameplate and verified with a pressure lift test.

8.3.4.a If a valve was removed and repaired in the Contractors' shop, the valve set point shall be preset on a live boiler at the shop. If the contractors' boiler is not capable of code-specified pressure, then the safety valve shall be set at the Company station

8.3.5 Repair requirements shall be done in accordance with Engineering Specification XX-XX-XXXX.

TABLE 9.2 | Continued

8.3.6   Repair activities are subject to monitoring by the Company Field Representative.

8.3.7   A work permit must be issued by the station prior to starting any on-site work.

8.4   Replacement

8.4.1   Any new safety and relief valve must be designed to meet the capacity and maximum allowable working pressure (MAWP) of the system.

8.4.2   Refer to purchase order or job specification for specific design requirements.

8.4.3   A work permit must be issued by the station prior to starting any replacement work.

9.0   **DOCUMENTATION**

9.1   Visual inspections

9.1.1   Company personnel completing visual inspections must document that the safety and relief valve meets the requirements per NBIC Part 2 – Inspection, Section 2.5.3. See Attachment 3.

9.1.2   This documentation must be submitted to (name omitted) and retained until the next PM.

9.2   Pressure lift tests

9.2.1   The Contractor shall provide a "Certified Safety/Pressure Relief Valve Repair/Test Report" for each valve serviced. See Attachment 4. The Report shall document conditions found and work performed including certifications of testing. The Contractor shall document the following dimensional information in the report: "the critical minimum dimensions for the valve type," "As found" and "As refurbished". This information shall be available for Company review.

9.3   Repair

9.3.1   All repairs must be performed in accordance with NBIC.

9.3.2   The contractor shall apply an NBIC 'VR' stamp and NBIC nameplate. See Attachment 5.

9.4   Replacement

9.4.1   Not Applicable.

10.0   **ACCEPTANCE CRITERIA**

10.1   Contractor Work

10.1.1   Company shall confirm that all required work, tests, and inspections have been performed prior to Contractor's demobilization or acceptance of valves returned from the Contractor's shop.

11.0   **RESTORATION**

11.1   All safety and relief valves must meet code requirements before returning to service.

TABLE 9.2 | Continued

> 11.1.1 If a safety and relief valve does not meet the requirements of a visual inspection, the valve will be taken out of service immediately.
>
> 11.2 The work permit must be completed and signed before returning the valve to service.
>
> 12.0 **ANALYSIS**
>
> 12.1 Not Applicable.
>
> **SECTION II – ACCEPTANCE REQUIREMENTS AND SUPPLEMENTAL SPECIFICATIONS**
>
> Omitted here to save space.

There are several points worth noting in Table 9.2:

- Reference is made to industry standards. Specifications for preventative maintenance cite professional society (API, ASME, NBIC, NFPA) and regulatory agency (NIST, OSHA) standards, state laws, and other previously developed proprietary company standards.
- Inspection, testing, and repair procedures are stated in *general* terms. The *specific* procedures depend on the class of valves, the date of installation, the operating conditions, and other details defined in the attachments in Section II. Thus, the required performance of individual valves is specified by referencing the applicable codes and standards (Section 5) and all instruction manuals (Section 7).
- Terminology and abbreviations are defined and used consistently. Sentence structure is consistent in any given section. Under Section 2, notice how the abbreviations used later are all written out: American Petroleum Institute (API), American Society of Mechanical Engineers (ASME), and so on. Under Section 3, Quality Assurance, for example, notice the parallel sentence structure "subject *shall be* verb" for each specification.
- Requirements are stated without giving explicit explanations. Explanations are generally not given in specifications to keep them as concise as possible. Usually, the contractor does not want to know *why* the work is being done, just *what* needs to be done.

In the next section, we will discuss procurement specifications, which focus on the performance capabilities of devices and systems that are sought for purchase. Like process specifications, procurement specifications must provide characteristics that are essential for *performance*, leaving out characteristics that do not affect performance. Language that is too specific may result in no response from manufacturers; language that is too broad will result in a large number of unsuitable bids.

Procurement Specifications

Procurement specifications consist of a list of performance characteristics for items that a company wishes to acquire through a request for bids (also called a request for quotation, RFQ). In this example, assume an organization wants to replace 250 bench-top drill presses. Even when the company has a central purchasing office, the purchasing agents need to know precisely

BENCH TYPE DRILL PRESS	BENCH TYPE DRILL PRESS
General Detail	~~General Detail~~
1. Bench type drill press.	~~1. Bench type drill press.~~
2. Commonly used product from a recognized company	~~2. Commonly used product from a recognized company~~
Technical Detail	~~Technical Detail~~
3. Belt driven with 4 spindle speeds.	3. Belt driven with 4 spindle speeds.
4. Support table can be raised and lowered, rotated through an angle of 0–45°, and can swing sideways.	4. Support table can be raised and lowered, rotated through an angle of 0–45°, and can swing sideways.
5. Bench dimensions Length: 476 mm, Width: 178 mm	5. Bench dimensions ~~Length: 476 mm, Width: 178 mm~~ Length minimum: 470 mm, Width minimum: 150 mm
6. Distance from spindle to base is 632 mm	6. Distance from spindle to base ~~is 632 mm~~ minimum: 600 mm
7. Motor power 0.320 HP	7. Motor power ~~0.320 HP~~ minimum: 0.30 HP
8. Electrical supply is 110 V, 60 Hz, 1 phase.	8. ~~Electrical supply is 110 V, 60 Hz, 1 phase.~~
9. With chuck and chuck key.	9. With chuck and chuck key.
10. Complete with drill bits.	10. Complete with drill bits **(how many and sizes?)**
Option	~~Option~~
11. With 1 vise set.	11. With 1 vise set.
12. Complete with standard accessories and set up tools.	~~12. Complete with standard accessories and set up tools.~~
13. Minimum one year guarantee	~~13. Minimum one year guarantee~~ Belongs in contract

Based on Stout, 2004

FIGURE 9.11 | A flawed procurement specification on the left with a marked-up copy for revision. Revised specification is shown in Figure 9.12

what characteristics of the machines are essential. To enable the purchasing agent to submit the request for quotation, the engineer needs to develop the specification for the machines. In such cases, the engineer examines the various models that are available from manufacturers and vendors in order to ensure that the specified machine is physically realizable. The procurement specification should not enumerate exactly the characteristics of the product of any single manufacturer. Figure 9.11 shows a flawed specification with an annotated copy of the same specification, pointing out the flaws. A clean final version of the specification is shown in Figure 9.12.

According to Stout (2004), procurement specifications in a request for quotations (RFQ) should

- Promote fair competition among multiple vendors so that the company can get the best deal.
- Be clear and easily understood. Lack of clarity or unwarranted complexity will limit the number of bids a company will receive.
- Be complete with sufficient detail so that bidders have no doubt about the characteristics of the desired product.
- Be accurate with all numerical data and include all units on all data.
- Use a numbered list for easy reference. This facilitates communication between a company's purchasing agent and prospective vendors.

BENCH TYPE DRILL PRESS

1. Belt driven with 4 spindle speeds

2. Support table can be raised and lowered, rotated from 0 to 45°, and can swing sideways

3. Bench dimensions: Length minimum: 470 mm

   Width minimum: 150 mm

4. Distance from spindle to base minimum: 600 mm

5. Motor power minimum: 0.30 HP

6. With chuck and chuck key

7. Complete with 1 set of drill bits 7 pcs (1/8" increments)

8. With 1 vise set

© Cengage Learning 2015

FIGURE 9.12 | Edited procurement specifications from Figure 9.11

- Use a bulleted format or numbered list for ease of reading and checking for compliance. List one characteristic in each bullet point or number, with the most essential characteristics listed first.
- Avoid unnecessary details; list only those characteristics that affect the usability of the solicited product. For example, do not specify the weight of the drill press, since it does not affect performance.
- Use available product literature to make sure the specified machine exists.
- List model and manufacturer of existing equipment if compatibility is an issue. For example, specify Mac or PC compatibility for computer peripherals.
- List in detail any accessories that are to be included.

WORD USE IN ENGINEERING SPECIFICATIONS

When writing specifications, engineers must be especially careful with word use. To avoid legal disputes later on, the right words must be used to express unambiguously the obligations of the contractor, the manufacturer, or the software developer. When it comes to stating obligations and permissions in specifications, and by extension in contracts, there are three "language use schools." The *traditionalists* subscribe to the traditional legal use of the colloquially antiquated *shall* to express obligation and *may* to express permission. The *revisionists* prefer *must* and *will*. The *minimalists* suggest using the imperative (command) form of the verb in any specification, with the implication that the reader can supply the subject ("the contractor") if none is supplied. Geren (2012) and Fodden (2011) both offer arguments for making the language of specifications consistent with modern use. Geren (2012) favors the "minimalist" approach of including the subject of the action, replacing the verb *shall* with a colon, and then using the imperative form for the action that the contractor or manufacturer must carry out with the specified object, as shown below. This was the format used in the sample specification in Figure 9.10.

TRADITIONALIST: The Contractor *shall* tighten all bolts to 60 ft-lbf torque.

REVISIONIST: The Contractor must tighten all bolts to 60 ft-lbf torque.

MINIMALIST: All bolts: tighten to 60 ft-lbf torque.

The comments at the end of the Fodden (2011) article provide counterperspectives. For example, Swann argues that *must* is not a synonym for *shall* and extols the use of "the neat and precise pair, 'shall' and 'may', as exactly expressing either an obligation or a permission." Similarly, Fodden (2011) cites Adams' (2008) argument that "using 'shall' in contracts to express obligations imposed on the subject of the sentence is both the most rigorous approach and the most expedient approach."

Wolfe (1996), citing an earlier version of the CSI Manual (2011), strongly argues for continued use of *shall* to denote an obligation and *will* to denote a permissible action. He argues that there is no need for the word *must*.

Word Choice May Be Critical

When writing specifications, we always state requirements in the future tense using the emphatic form "shall." Hence, the finished product SHALL be, SHALL produce, SHALL consume . . . The weaker auxiliary verbs "will," "should" and "may" do not express a requirement. In the case of "will," the sentence places responsibility on the purchaser. "May" grants permission, and "should" states a preference. "Must" is ambiguous, since it may express a presumption instead of a requirement.

Oriel (2007)

Table 9.3 is an attempt to bring some order to the confusing use of words conveying obligations, prohibitions, and permissions. There may be additional uses of the entries in the table, which extend beyond the application to specifications and contracts.

Currently, in specifications and contracts, there is no single right way to express obligations, prohibitions, granting of permission, or denial of permission. The best advice is to follow the guidelines of your organization, your immediate supervisor, or your legal counsel.

TABLE 9.3 | The use of *shall*, *must*, *may*, and *will* in specifications and contracts

Condition	Traditionalist	Revisionist	Minimalist
Obligation	shall	must	: imperative command
Prohibition	shall not	must not	: negative imperative
Permission	may	may or will	may or will
Denial of permission (rarely used)	may not	may not or will not	may not or will not
Expectation or preference	[not used]	should	should
Expression of intent	will	will	will
Presumption or precondition (i.e., *"must obtain permits"*)	must	must	must

SIX CHARACTERISTICS OF GOOD SPECIFICATIONS

From both a technical and a writing standpoint, good specifications are

- Clearly written (unambiguous and understandable)
- Coherent and well organized
- Complete with no omissions
- Concise with no unnecessary words
- Consistent, both internally in terms of terminology, accuracy, and with any accompanying drawings, as well as externally with regard to constraints and customer needs
- Correct with no technical errors or grammatical errors

Clearly Written

When specifications are written clearly, the intended audience can easily understand what the object is supposed to do, how it should be built or programmed, and what codes, standards, and regulations are applicable to this object. To produce a clearly written specification, the writer must eliminate ambiguity, define potentially unfamiliar expressions, and rewrite complicated grammatical constructions in a form that is easier for the reader to process.

Ambiguity Due to a Technical Error
Consider the following specification:

TECHNICAL ERROR: Use bolts made of Grade 2 stainless steel.

This specification is ambiguous because of the juxtaposition of "Grade 2" and "stainless steel." Grade 2 bolts are defined as low-carbon steel, not stainless steel. The author's lack of familiarity with bolt standards results in conflicting instructions to the contractor.

REVISION: Use bolts made of 18-8 stainless steel.

Ambiguity Due to a Punctuation Error
Often ambiguity arises when the grammatical structure is sufficiently complex that a single comma might change the interpretation of the text.

MISSING COMMA: Use screws, nuts and bolts of 18-8 stainless steel.

Which components need to be made of 18-8 stainless steel: "screws, nuts, and bolts" or just "nuts and bolts"? In a product with many screws, the cost savings could be substantial if the contractor/manufacturer substituted carbon steel screws for the intended stainless components. To specify unambiguously that all connecting components be made of stainless steel, add a comma after the word *nuts*:

REVISION 1: Use screws, nuts, and bolts made of 18-8 stainless steel.

To make the meaning even more unambiguous, insert a comma and the word *all*.

REVISION 2: Use screws, nuts, and bolts, all made of 18-8 stainless steel.

Ambiguity Due to Word Use
Review the "Frequently confused word pairs" section in Chapter 5 to make sure you are using these words correctly. In general, whenever you are uncertain about the meaning of a word, look it up in a dictionary or ask your supervisor to explain the proper context and use.

Ambiguous Pronoun Reference

Is it possible to identify in the following specification who is responsible for the delays?

UNCLEAR PRONOUN REFERENCE:	The Contractor may hire a subcontractor for all masonry work; he will be liable for all delays unless he demonstrates due diligence, and the delay results from an act of God.

The ambiguity is easily removed by replacing the pronoun *he* with the noun *the Contractor* and adding the word *however* for emphasis:

REVISION:	The Contractor may hire a subcontractor for all masonry work; *however, the Contractor* will be liable for all delays unless he demonstrates due diligence, and the delay results from an act of God.

See Chapter 5 for other grammatical issues related to ambiguity.

Ambiguous Expressions to Avoid

Good specifications state requirements and provide well-defined performance measures with no further clarification needed. For this reason, adjectives and adverbs are seldom warranted. In fact, adjectives and adverbs such as *adequate* or *adequately*, *desirable* or *desirably*, *normal* or *normally*, *excessive* or *excessively*, *pleasing* or *pleasingly*, *sufficient* or *sufficiently*, among many others, are rather subjective and a difference of opinion between the customer and the contractor may spell trouble in the final product. FHWA (2010) and Oriel (2007) provide many examples of adjectives and adverbs that are particularly inappropriate in specifications.

Other expressions that can lead to ambiguous interpretations are listed in Table 9.4. Experienced technical writers avoid these expressions in specifications and in contracts.

TABLE 9.4 | Ambiguous expressions to avoid

Expression	Explanation
and/or	This combined conjunction permits the reader to choose the more advantageous of the two options. If you mean *and*, use *and*. If you mean *or*, use *or*.
any	This word has many possible meanings, ranging from one item to all items.
as agreed as allowed as appropriate as approved as authorized as directed as indicated as identified as required as necessary as permitted as specified as stipulated	All such expressions are incomplete and require someone or something to indicate *what* is allowed by *whom*, *what* is appropriate, and so on. Avoid these expressions and state every requirement clearly.

TABLE 9.4 | Continued

Expression	Explanation
as a minimum not limited to at least not less than	All such "inclusively limiting phrases" make a specification ambiguous. State with precision what is required.
as well as	This phrase is commonly used to avoid repeating the word *and*. In specifications, "Do A as well as B" can be construed to require that actions A and B must be done equally well (or poorly), but does not mean explicitly that either must be done.
because	This conjunction is used to provide an explanation. Because specifications should not provide explanations, just state the requirements and avoid using the word *because*.
capable	This word has many meanings, none of which are particularly applicable to specifications and contracts. A piece of equipment capable of some performance may require substantial additional assembly, auxiliary components, or control systems before the capability can be realized. Specifying capability is equivalent to saying "deliver and leave in the original packaging." There is no requirement to make the system functional.
comprise	This verb means *to include* or *contain*. The whole comprises the parts. Comprise is frequently used incorrectly as in . . . *is comprised of* . . . .Avoid this expression if you cannot use it correctly. See Grammar Girl (2009).
critical	This adjective has over 30 synonyms. If you cannot make the meaning unambiguous, reformulate the requirement without using the word *critical*.
etc. or et cetera	This Latin phrase means *and the rest, and so on, and so forth*. In a specification, there is no obligation for a contractor to infer what was intended, so it is best to avoid this phrase.
in a workmanlike manner	The contractor and the customer may have different opinions of what constitutes *workmanlike*.
include(s)	This word has too many possibly ambiguous meanings.
shall function as intended	This expression places the customer at the mercy of the contractor. The contractor cannot be expected to infer the customer's intent.
to the satisfaction of the customer's representative	Customer satisfaction should be strived for, but it is not a measurable requirement. Instead, describe precisely the function of the device, product, or software that will result in acceptable performance.
up to	*Up to* could mean the inclusive range of all values between the specified minimum and the specified maximum. Or it could mean only one value within the range. The requirements must be stated precisely.
slash (virgule), "/"	Using the slash in a specification permits the manufacturer, contractor, or developer to choose one of the options separated by the slash. If both options are intended, use the word *and* instead of the slash.

Based on Clendining, 2009; FHWA, 2010

Other words to consider carefully before using are those whose meanings in legal contexts are substantially different from those in common use. These include *consideration*, *cause*, *substantive*, *cure*, *material* (as an adjective), and *harmless*, among others (Oriel, 2007).

Jargon

Jargon is the specialized language used by experts in a particular discipline. The appropriate use of jargon makes communication among knowledgeable people more efficient. When using jargon in specifications, keep in mind that engineers are not the only stakeholders who will read the specifications. To make sure all interested parties will be able to understand the jargon, define terms in the text where they are first used.

Acronyms

Analogous to jargon, some acronyms are accepted as a part of the language. For example, the acronym SCUBA (meaning **S**elf-**C**ontained **U**nderwater **B**reathing **A**pparatus) is so well recognized that it has a meaning of its own and is seldom written out. Other acronyms, such as USB (universal serial bus) and HTML (hyper-text markup language), are familiar to most computer users. As with jargon, it is advantageous to use commonly understood acronyms in specifications because they shorten the text. To enhance readability, use acronyms judiciously and define them when they are first mentioned.

Nominalizations

Some writers believe that using long instead of short expressions makes their writing seem more sophisticated. One technique they use is to add a suffix such as *–(a)tion*, *–ance*, *–ence*, *–ity*, *–ment*, *–ness*, and *–ing* to a verb, changing the verb into a noun (which is called a **nominalization**). Compare the readability of a sentence with nominalizations to that containing a mixture of nouns and verbs:

NOMINALIZATION: Repetitive use of nominalizations makes the processing of writing more difficult.

REVISION: Text is harder to read when verbs are nominalized repeatedly. *or*

Repeatedly using the noun form of a verb makes the text harder to read.

Notice also that the revised sentences are shorter than the sentence containing nominalizations. Conciseness is another attribute of good specifications.

Noun Stacks

In some instances, a stack of nouns can be used adjectively to describe another noun, as in the following example:

NOUN STACK: aircraft wing-body junction vortex generation

An experienced aerodynamicist would readily understand that the four-noun stack *aircraft wing-body junction* modifies the last two nouns *vortex generation*. For those less familiar with aerodynamics, however, the phrase could be made more comprehensible by breaking up the noun stack with prepositions as follows:

REVISION: vortex generation at the wing-body junction of an aircraft

For ease of reading, limit noun stacks to three or fewer nouns.

Long Sentences

As discussed in Chapter 5, sentence length should be guided by the complexity of the idea being described. Many long sentences may actually be run-on (fused) sentences that contain more than one independent clause. To enhance readability, consider reformulating long sentences as two or more shorter sentences. Then use connecting words such as *furthermore* and *however* (see Table 5.4 for different categories of transitional words and phrases along with examples) to show how the shorter sentences are related to each other.

Coherent and Well Organized

Good specifications have an easily recognizable structure. For example, the construction industry has adopted MasterFormat® with its logical numerical framework for construction specifications. Many organizations use a format in which the specification document is divided into sections (Oriel, 2007), for example,

- Section 1: Scope and background information
- Section 2: Relevant documents
- Section 3: Requirements list
- Section 4: Testing and verification
- Section 5: Packaging and shipping
- Section 6: Related information, references, and notes (if applicable)

The process specification in Table 9.2, for example, is divided into sections. The sections are organized in such a way that the information flows from general to specific, allowing contractors to identify easily the steps needed to inspect and certify safety valves. Patterns of paragraph organization, such as general to specific, chronological, and cause and effect, can also be applied to specifications (see Chapter 5).

The absence of text, in other words, the white space on a page, also helps readers identify the document structure. Use headings and subheadings, short paragraphs, bulleted or numbered lists, and visual elements, such as tables and figures, to break up large blocks of text. The white space created using these techniques helps the reader to process subconsciously where one topic ends and the next one begins.

Complete

To produce a readily usable and easily implemented specification, the writer must think carefully through the steps needed to produce, construct, manufacture, or develop the object of the specification. Failure to describe all aspects pertaining to object development will lead to customer dissatisfaction and to potential litigation. Keep in mind that the contractor, manufacturer, or developer is not responsible for pointing out and correcting any omissions, errors, inconsistencies, or ambiguities. It is the writer's job to make sure the specifications and the drawings are consistent, complete, and accurate.

In addition, provide an appendix for referenced sources and other details that help make the specifications clearer.

Concise

Specifications can be made concise by trimming "deadwood," which includes

- Redundant words and phrases (see, for example, Table 5.5)
- Empty phrases that can be replaced with concise alternatives (see, for example, Table 5.6)
- Vague and ill-defined adjectives and adverbs
- Unnecessarily complex sentences

Other ways to remove extraneous information from specifications include:

- Searching the document for the words *because*, *as a result*, *thus*, and *therefore*, words that are frequently related to explanations. Specifications typically do not explain or justify requirements. Instead, the requirements are stated in a way that allows performance of the finished product to be traced back to the customer's original requirements.

• Making sure a requirement is not repeated in different sections of the specifications.

Consistent

When preparing specifications, and especially when preparing specifications collaboratively, make sure that

• The terminology used is consistent throughout the document.
• The dimensions and annotations on any drawing are consistent with in-text citations.
• The terminology and symbols in the specifications are consistent with their use in cited codes and standards.

See Chapter 8 for suggestions on proper preparation for collaborative writing.

Correct

First, the *technical details* must be correct.

• The cited codes and standards must be relevant to the object of the specification.
• The specified function and performance measures must be appropriate, of correct magnitude, and attainable. If a metric is unattainable, a contractor, manufacturer, or developer may be excused from meeting the requirement.

Second, the *writing* itself must be grammatically correct to avoid misunderstandings that could lead to unpleasant surprises in the finished product. Oriel (2007) defines three classes of grammatical errors:

• Trivial errors
• Unintelligibility errors
• Ambiguity errors

Trivial errors are those in which a word is used incorrectly, but the meaning of the sentence is still clear. For example:

> **WRONG**
> **COMPARATIVE FORM:**   The system shall have three or *less* moving parts.
>
> **REVISION:**   The system shall have three or *fewer* moving parts.

Even though the wrong comparative is used, the only possible meaning is that the number of moving parts cannot exceed three.

   Unintelligibility errors are grammatical errors that render the sentence incomprehensible. Usually, these kinds of errors result from a lack of proofreading and editing. An unintelligibility error may be as simple as an omitted word.

> **OMITTED WORD:**   The control system shall be supplied within three of awarding the contract.

Clearly, the time duration has been omitted. Should it be days, weeks, months, or years?

> **REVISION:**   The control system shall be supplied within three *months* of awarding the contract.

In most cases, unintelligibility errors will be detected before execution of the contract. If they are not, the contractor may ignore any requirement that is unintelligible. Correcting an unintelligibility error in a signed contract will result in the need for an engineering change proposal (see the subsequent section on contracts) and increased cost.

Ambiguity errors result when a phrase can be interpreted in more than one way. Different types of ambiguity errors are discussed in the "Clearly Written" section in this chapter and in Chapter 5. When proofreading specifications for ambiguity, play the devil's advocate. In other words, look for sentence constructions that would allow you to get out of having to comply with a particular requirement. Then rewrite the sentence to remove all doubt about what is intended.

SPECIFICATION CHECKLIST

As you write the specifications for a project, it is prudent to maintain a checklist of what is needed in a specification. During the editing stage, consult the checklist to ensure that you included all essential requirements and avoided all common errors. A sample checklist is given in Table 9.5. Other checklists are available on the FHWA (2010) website.

Most specifications documents will be written by a team. As discussed in the section on collaborative writing in Chapter 8, the keys to effective collaborative writing are

- Proper planning
- Disciplined attention to assignments and deadlines
- Good recordkeeping of multiple revisions of the document
- Use of consistent terminology, notation, and writing formats

Several useful sources about specification writing are available on the Internet: Clendining (2009), FHWA (2010), Oriel (2007), and ODOT (2009), among many others.

TABLE 9.5 | Checklist for specifications draft

Characteristic	Detail	Yes	No
Clear and unambiguous	Are all requirements unambiguous with only one possible interpretation?	❑	❑
	Is the meaning of each requirement straightforward?	❑	❑
	Have grammatical constructions been simplified?	❑	❑
	Have all acronyms, abbreviations, and jargon been defined, and are they being used appropriately?	❑	❑
	Are noun stacks limited to three nouns?	❑	❑
	Do most sentences consist of no more than 12–18 words?	❑	❑
Coherent and Organized	Is there a coherent logic to the document structure?	❑	❑
	Is good use made of white space, headings, visual elements, and lists?	❑	❑

TABLE 9.5 | Continued

Characteristic	Detail	Yes	No
Complete: Product Design Specifications or Software Design Specifications	Are all user requirements addressed without specifying any design embodiment?	❏	❏
	Are all requirements testable?	❏	❏
	Are all requirements feasible?	❏	❏
	Are all requirements traceable to user needs?	❏	❏
	Are all external references readily available?	❏	❏
Complete: Construction Specifications or Manufacturing Specifications	Have all dimensions and tolerances been specified correctly and consistently?	❏	❏
	Are all requirements testable?	❏	❏
	Are all requirements feasible?	❏	❏
	Are all requirements traceable to user needs?	❏	❏
	Are all external references readily available?	❏	❏
Concise	Has all deadwood been trimmed?	❏	❏
	Have unnecessary adjectives and adverbs been eliminated?	❏	❏
	Is each requirement specified only once?	❏	❏
Consistent	Is all terminology consistent internally within the specifications and externally with cited codes and standards?	❏	❏
	Does any requirement conflict with any other aspect of the specification?	❏	❏
Correct	Are all requirements technically correct?	❏	❏
	Are all sentences grammatically correct and unambiguous?	❏	❏

CONTRACTS

A **contract** is a legally binding agreement between two people or organizations (usually called the *parties* to the contract), which involves an exchange of entities that have some value (legally called *considerations*) to each of the parties. In a familiar context, employees sign contracts that state they agree to perform services that their employers value in exchange for wages or salaries that the employees value. When the considerations are complex, specifications are required to provide a detailed accounting of these considerations. Because the specifications that engineers write are invariably incorporated into contracts, engineers need to be cognizant of some basic aspects of contracts.

The contract is considered valid and unchangeable when it is executed, which means both parties sign the contract attesting to their agreement to the terms. If there is an omission or an error, or one party wishes to change any aspect of the contract, an additional agreement will need to be negotiated. Such a supplemental agreement, called an **Engineering Change Proposal** (ECP) or something similar, requests equitable adjustment (claims) and modifications to recently purchased items (mods), usually obligating the party that is receiving the services to supply additional (frequently substantial) compensation to the party supplying the services. ECPs and mods are initiated by the party *receiving* the services, while claims are initiated by the party *supplying* the services. In both cases, changes need to be made to the contract.

The need for such ECPs, claims, and mods should, in theory, be a rare occurrence since there is usually a very explicit and detailed set of specifications for the project, which both parties are expected to read and understand before signing the contract. In reality, projects often may be so complex that neither party truly understands the full nature of the project until it is already underway. Consider the specifications in Figure 9.10, and realize that the entire specification is at least two orders of magnitude longer. If the interpretation of the service provider differs from that of the receiving party, the legal system usually sides with the service provider, as long as the provider's interpretation is reasonable. This need to resolve differences after the contract has been signed is usually expensive for the party receiving the services. To avoid costly ECPs, claims, and mods, it is essential that all specifications be correct, unambiguous, and consistent with codes, standards, and any pertinent regulations. Unscrupulous service providers may intentionally complete a contract according to their interpretation, even though they know it is wrong. The party receiving the service must then pay for the first incorrect execution of the specification, the cost of undoing the incorrect work, and finally the cost to do it correctly. Further details on engineering contracts can be found in a variety of sources, including Oriel (2007), Fisk (2000), Clough and Sears (1994), Williams (1992), and Bockrath (1986).

INTELLECTUAL PROPERTY—PATENTS, TRADEMARKS, AND COPYRIGHTS

Intellectual property is any work of your intellect (your mind) which, when incorporated into some physical embodiment, takes on commercial value. If the physical embodiment of your thought process is

- A new machine or device
- A new material
- A new manufactured product
- A new process for manufacturing or
- An improvement to any of these items

then a *patent* gives you a government-regulated right to a monopoly on using your invention for 20 years from the date from which you file your patent application. Patents can also be obtained for new varieties of plants (predominantly genetically engineered varieties) and for the particular design of a product (i.e., its "look and feel"). Diegel (2012) provides an easy-to-read introduction to intellectual property and detailed considerations on when and why to consider patenting an invention.

Warning

In 2011, the United States changed its patenting system from one that required proof that the applicant was the first to *invent* to one that awarded the patent to the first inventor to *file*. This change to the "first-inventor-to-file" system brought the United States in line with most other countries. Keep this change in mind when you read pre-2011 sources concerning U.S. patents.

If the physical embodiment is

- A book
- A poem
- A musical composition
- A work of art
- A photograph or other image
- Any combination of these items

then a *copyright* provides the author protection from unauthorized use of the author's intellectual output by other people.

If the intellectual embodiment is a device, a process, or a chemical formula that can be kept secret, and if there is little chance of it being **reverse-engineered** (another person taking apart the device and then reconstructing it to learn how it works), then many companies prefer to keep it as a *trade secret*.

A *trademark* and a *service mark* are similar to each other. Both can be a word, phrase, logo, picture, or visual image that is used to distinguish a manufactured product (trademark) or a specified process (a trademark or service mark) from those of others. Other people cannot use a listed trademark to label their product. A trademark, however, does not stop others from making a similar product; they just cannot call it by the same name or use the same logo or marketing phrase. Patents, trademarks, service marks, and copyrights will be discussed in greater detail in the following sections.

PATENTS

What exactly is a patent, and why would an engineer want to apply for one? The United States Patent and Trademark Office provides a wealth of resources on this topic (USPTO, 2012a). Ma (2011) gives a good introduction to patents for scientists and engineers. Pressman (2011) has written what many consider to be the single best, most authoritative guide to the processes of patent preparation and patent filing. Nolo's Guide to Provisional Patent Applications (Nolo PPA, 2009), which is available as a free download, is also very useful. Diamond (2012), Goldman (2011), Dahl (2010), Ehrlickman (2010), Poltorak (2009), and Calvert (2010) are less comprehensive, but these sources are readily available on the Internet.

According to the USPTO (2012a) website, a **patent** is defined as "a property right granted by the Government of the United States of America to an inventor 'to exclude others from making, using, offering for sale, or selling the invention throughout the United States or importing the invention into the United States' for a limited time in exchange for public disclosure of the invention when the patent is granted." What this means to an inventor is that the inventor's rights to exclude others from manufacturing, using, or selling an invention will be protected for a specified time, as long as the inventor declares the invention in writing to the U.S. government.

In 2011, the U.S. patent law changed when the Leahy-Smith America Invents Act was signed (Goldman, 2011; Diamond, 2012). The America Invents Act (AIA) changed the U.S. patent application process from a "first-to-invent" system to a "first-inventor-to-file" system. Inventors no longer have to prove that they were the first to invent a given product, but they still have to prove that they did actually invent the product. The requirement that the inventor is the only person permitted to file a patent application remains. Under the old law, if an individual invented a product, but did not file a patent application, this first inventor could have successfully challenged the granting of the patent to an individual who invented the same product later, but filed the application first. To do so, the first inventor would only need to show proof that the date of his invention predated the second inventor's invention. Today, the patent right belongs to the first inventor who files a patent application, regardless of whether someone else may have invented the same device at an earlier date. This change brings U.S. patent processes into alignment with those in almost all other countries.

What Can and Cannot Be Patented?

A *utility patent* can be issued for a(n)

- Process
- Machine
- Article of manufacture
- Composition of matter
- Improvement of any of the above

provided it can be shown that the device is "new, non-obvious, and useful" (USPTO, 2012a). The USPTO will also consider issuing a patent on the design of an article of manufacture (a *design patent*) or on asexually reproduced plant varieties (a *plant patent*). Software can currently be patented, but there are substantial efforts afoot to limit the use of "functional claims," in which the inventor claims the rights to all similar problem solutions instead of the specific algorithm used to solve the problem. Software patent laws are evolving, so it is difficult to provide more definitive information.

The USPTO will not consider any application claiming rights to

- Laws of nature
- Physical phenomena
- Abstract ideas
- Nonuseful and infeasible devices (such as perpetual motion machines)
- Works of art (literary, musical, or artistic—all of which can be copyrighted)
- Any invention that is offensive to public morality

The requirements (USPTO, 2012a) for awarding a patent are that the invention be

- Useful
- Novel

- Nonobvious
- Adequately described
  » to enable someone skilled in the field to manufacture the invention
  » to demonstrate that the applicant possesses full knowledge of the invention
  » to provide for one or more embodiment(s) of the invention
- Claimed by the inventor in clear and definite terms

Prior to filing an application for a utility, design, or plant patent, an inventor may choose to file a *provisional patent application* (PPA). Calvert (2010), Diegel (2012), and Nolo's Guide (Nolo PPA, 2009) all explain that the provisional patent application allows an inventor a one-year window in which to decide on filing the actual, nonprovisional patent application. In this year, the inventor can refine the invention, undertake a market analysis, secure financial backing, or reconsider the advisability of filing the formal patent application. A PPA is a record of invention and sets a filing date that can be used retroactively if the inventor does file the actual patent application on the same invention within the one-year window. All details in the invention must be included in the provisional patent to use the PPA filing date. Because the Patent Office does not examine the provisional patent application, the cost of filing a provisional patent is lower, and the fees for legal advice from a patent attorney are also substantially lower, than those associated with a formal patent application. If the inventor decides not to file a formal application, there is no public disclosure of the invention. An added advantage of filing a provisional patent application is that the inventor can use the description "patent pending" for the object of the patent and can publicly disclose the invention without jeopardizing the patentability of the invention. Currently, the main disadvantage of the PPA is that the date for filing a formal patent cannot be extended beyond a year; if the inventor misses the deadline, the PPA filing date no longer applies for purposes of "first-inventor-to-file." More details on PPAs can be found in Calvert (2010), Nolo PPA (2009), and Diegel (2012).

How to Prepare a Patent Application

Before an inventor submits a PPA or a formal patent application, some important preliminary criteria must be satisfied:

- The inventor must ensure that the invention fits the USPTO definition of a patentable invention.
- The inventor must prove an invention is "new" or "novel" by reviewing the state of the art (**prior art** in patent terminology) in the particular field of the invention. This review must take into account previous patents, any printed text (now most likely including electronic formats as well) that was publicly available before the filing date of the new invention, any existing invention that existed at least one year prior to the filing date of the new invention, and any public disclosure or commercial sale of the new invention prior to the filing date. The United States permits disclosure within one year of filing; other countries do not.
- The inventor no longer needs to prove that the invention predates a nondisclosed invention of a competitor, only that the filing date predates that of any competitor's invention ("first-inventor-to-file" process in the 2011 AIA).
- The inventor should consider the merits of filing a provisional patent application to nail down the filing date while carrying out further refinement of the invention and preparation of the formal patent application.

The review of the prior art must include a patent search. In the past (approximately before the early to mid-1990s), a patent search would entail going page-by-page through thousands of pages of patent records over a one- or two-week-long sojourn at the Patent Office in Washington, D.C. An experienced patent attorney could conduct this search much more thoroughly and expediently than the inventor. For this reason, many inventors still assume a patent attorney must be engaged to undertake a patent search. Today, however, essentially the entire patent database is available as a digital record, and a patent search can be undertaken anywhere with Internet access. To save money, an inventor can undertake the patent search without engaging a patent attorney. A full search is estimated to consume 20 to 30 hours. The USPTO provides keyword searches of U.S. patents back to 1976 and by patent number and classification searches back to 1790. To use this service, go to <www.uspto.gov/patft/index.html>. A tutorial on searching the USPTO patent database is provided on the University of Texas at Austin library website, <http://www.lib.utexas.edu/engin/patent-tutorial/index.htm>. Pressman (2011) identifies other sources of patent databases. For international patents, the World Intellectual Property Organization (WIPO, <http://www.wipo.int/portal/index.html.en>) has search features at their website.

A patent application includes a cover sheet with information about the inventor, the assignee (the entity to whom the patent rights belong), and the filing date. A payment voucher and an administrative action form follow. The remainder of the application consists of a narrative section and a claims section. The narrative part, also called the **specification** (yet another use of the word), explains the prior art, the problem solved by the invention, any additional background on the invention, the way it differs from prior art, and—very importantly—gives sufficient information to enable a person knowledgeable in the field to manufacture the invention. The narrative should also explain the best physical embodiment of the invention or several alternatives.

The **claims** comprise the core of the patent application. They define the legal coverage conferred by the patent. Writing patent claims requires both an understanding of the technology of the invention and the arcane wording of these claims required by the USPTO. Although the inventor may choose to undertake the patent search, a patent attorney should be retained to prepare, or at least review, the claims. The claims in the patent application specify the exact features that are being patented. Each independent claim must stand alone and yet be tied to the concept of the patent. Claims that are too broad will be rejected; claims that are too narrow will not provide sufficient protection for your ideas, and your competition may find ways to use your ideas without infringing on your patent rights. The patent examiner may require claims that describe more than one feature to be split into multiple claims. The number of individual claims determines the final issuing fee for the patent.

Office actions (suggested or required deletions recommended by the patent examiner) may remove claims, split claims, or otherwise result in modifications of the original application. *No new material may be added to a patent application.*

Figure 9.13 shows the first page of an issued patent. It includes information about the inventors, the assignee, the prior art (if any) that was considered by the patent examiner, the classification of the patent (the area assigned by USPTO), and the abstract. It also shows a figure to explain the problem solved by the invention. Figure 9.14 shows the first page of the narrative, while Figures 9.15 and 9.16 show the claims for this patent.

‖‖‖‖‖‖‖‖‖‖‖‖‖‖‖‖‖‖‖‖‖‖‖‖
US 20080156242A1

(19) **United States**

(12) **Patent Application Publication**   (10) Pub. No.: **US 2008/0156242 A1**
Knisely et al.   (43) **Pub. Date: Jul. 3, 2008**

(54)   **FOLDABLE MAST ASSEMBLY FOR A SAILING VESSEL**

(75)   Inventors: **Charles W. Knisely**, Lewisburg, PA (US); **Henry M. Baylor**, Lewisburg, PA (US); **Matthew A. O'Rourke**, Bridgewater, NJ (US); **Thomas C. Walker**, Lewisburg, PA (US); **Jacob C.D. Clark**, Downingtown, PA (US)

Correspondence Address:
ALLEMAN HALL MCCOY RUSSELL & TUTTLE LLP
806 SW BROADWAY, SUITE 600
PORTLAND, OR 97205-3335

(73)   Assignee: **Susquehanna Yacht Manufacturing, Inc.**, Lewisburg, PA (US)

(21)   Appl. No.:   **11/969,137**

(22)   Filed: **Jan. 3, 2008**

Related U.S. Application Data

(60)   Provisional application No. 60/883,321, filed on Jan. 3, 2007.

Publication Classification

(51)   Int. Cl.
       B63B 15/00   (2006.01)
(52)   U.S. Cl. **114/90**; 114/97; 114/112

(57)   **ABSTRACT**

As one non-limiting example, a foldable mast assembly for a sailing vessel is provided. The foldable mast assembly includes a lower mast section; an intermediate mast section having a lower end foldably coupled to an upper end of the lower mast section; an upper mast section having a lower end foldably coupled to an upper end of the intermediate mast section; and a boom coupled to the lower mast section. A locking device internal the mast assembly is also provided to inhibit folding of the mast assembly.

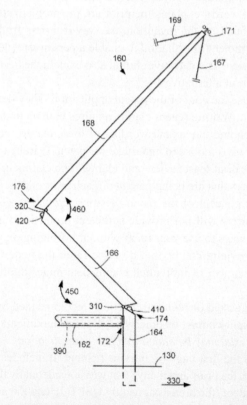

FIGURE 9.13 | The first page of patent US 2008/0156242 A1

US 2008/0156242 A1   Jul. 3, 2008

1

FOLDABLE MAST ASSEMBLY FOR A SAILING VESSEL

[0001] The present application claims priority to U.S. Provisional Patent Application No. 60/883,321, filed Jan. 3, 2007, and entitled "Reconfigurable Watercraft", the entire contents of which are incorporated herein by reference.

BACKGROUND AND SUMMARY

[0002] Sailing vessels can utilize a mast for supporting one or more sails. In some examples, the size and/or configuration of the mast can impose limitations on the use or transportability of the watercraft. As one example, a sailing vessel may be transported by trailer between two bodies of water or between a body of water and a storage location. During transportation, the mast may be removed or unstepped from the deck or hull where it may be secured along the length of the hull to reduce the height of the sailing vessel. However, where the mast is of greater length than the hull, the mast may extend considerably beyond the hull profile, thereby making transportation of the sailing vessel more difficult. Furthermore, during operation of the sailing vessel, the height of the mast may also limit the ability of the sailing vessel to pass under low lying structures, such as bridges or wires that are located at a relatively low height above the water.

[0003] The process of stepping and unstepping the mast can also be difficult, work intensive, and time consuming. For example, some masts may require the assistance of multiple people to complete the stepping process due to the size and/or weight of the mast. Further still, mast stays or guy wires may be adjusted, reattached, or disassembled as a consequence of the stepping or unstepping process, thereby further complicating the stepping process.

[0004] U.S. Pat. No. 4,112,861 (Lewis) provides one approach for addressing some of the above issues. Lewis describes a mast stepping and unstepping structure that enables the mast to be stepped or unstepped without requiring that the stays and shrouds be tuned.

[0005] However, the inventors have recognized some issues with the above approach. As one example, the inventors have recognized that the approach of Lewis does not address how a boom structure may be treated during the mast stepping and unstepping process. Furthermore, the inventors have recognized that the approach of Lewis hinges two mast sections on the sail, which can reduce the continuity of the track across a joint or interface between two mast sections or obstruct the track. Further still, the approach by Lewis still requires stepping and unstepping of the mast for purposes of transportation and storage.

[0006] To address at least some of the above issues, the inventors have provided, as one example, a foldable mast assembly for a sailing vessel, comprising: a lower mast section; an intermediate mast section having a lower end foldably coupled to an upper end of the lower mast section; an upper mast section having a lower end foldably coupled to an upper end of the intermediate mast section; and a boom coupled to the lower mast section. By coupling the boom to the lower section of the mast assembly, the inventors have recognized that the intermediate and upper mast sections may be more easily folded, without necessarily requiring reconfiguration of the boom.

[0007] As another example, the inventors have provided a foldable mast assembly for a sailing vessel, comprising: an upper mast section including a first track segment at a stern side of the upper mast section, the first track segment being adapted to guide a luff edge of a sail between a raised configuration and a lowered configuration of the sail; an intermediate mast section including a second track segment at a stern side of the intermediate mast section, the second track segment being adapted to guide the luff edge of the sail between the raised configuration and lowered configuration; a first hinge assembly foldably coupling a lower end of the upper mast section at a bow side of the upper mast section to an upper end of the intermediate mast section at a bow side of the intermediate mast section. Thus, the inventors have recognized that in some examples, placing the hinge on an opposite of the mast assembly from the track can reduce discontinuities between two track sections that are located at different foldable mast sections.

[0008] As yet another example, the inventors have provided a foldable mast assembly for a sailing vessel providing at least an erected position and a folded position, comprising: a lower mast section; a stepping support fixedly coupling the lower mast section to a sailing vessel; an intermediate mast section having a lower end rotationally coupled to an upper end the lower mast section by a first hinge assembly that permits an upper end of the intermediate mast section to rotate from the erected position toward the stern of the sailing vessel and into the folded position without unstepping the lower mast section from the stepping support; and an upper mast section having a lower end rotationally coupled to an upper end of the intermediate mast section by a second hinge assembly that permits an upper end of the upper mast section to rotate from the erected position toward the bow of the sailing vessel and into the folded position without unstepping the lower mast section from the stepping support. In this way, the mast assembly can be folded or erected without requiring the mast assembly to be unstepped from the stepping support.

BRIEF DESCRIPTION OF THE DRAWINGS

[0009] FIG. 1 illustrates an example watercraft having a sailing boat configuration.

[0010] FIG. 2 illustrates the watercraft of FIG. 1 in a power boat or transportable configuration.

[0011] FIGS. 3A, 3B, 3C, and 4 illustrate an example mast assembly that is reconfigurable between folded and erected configurations.

[0012] FIG. 5A illustrates a detailed view example interface between intermediate and upper mast sections of the mast assembly.

[0013] FIG. 5B illustrates a detailed view of an example interface between lower and intermediate mast sections of the mast assembly.

[0014] FIGS. 6A, 6B, and 6C illustrate examples of alternative hinge assemblies that may be used at an interface between two mast sections of the mast assembly.

[0015] FIGS. 7A-7C illustrate an example approach for locking two mast sections of the mast assembly.

[0016] FIGS. 8A-8D illustrate several example approaches depicting how an insert within the mast assembly can be translated between locked and unlocked positions.

[0017] FIGS. 9A, 9B, and 9C illustrate alternative examples for an interface between two mast sections.

[0018] FIGS. 10A, 10B, and 10C illustrate example cross-sections of a mast assembly.

[0019] FIG. 11 shows an example of a mast stepping support for the mast assembly.

United States Patent Application Publication

FIGURE 9.14 | The first page of the narrative section of patent US 2008/0156242 A1

[0120] Alternatively or in addition to the air injection system of FIGS. 20 and 21, one or more doors or shields may be utilized to at least partially cover the depressed region for receiving the keel bulb when fully retracted. For example, FIGS. 22 and 23 illustrate how a door indicated at 2210 can be moved from a stored position above water line 120 to cover the depressed region. For example, each side of the watercraft can include one or more doors that seal around keel arm 184 to provide a more hydrodynamic hull. Further, as illustrated in FIG. 23, one or more doors 2210 can be configured to provide different hull configurations, for example, as indicated at 2310 and 2320. For example, the bottom surface of bulb 182 can provide a first hull configuration when fully retracted and one or more doors 2210 can provide a second hull configuration. As one non-limiting example, the bulb of the keel can provide a planing hull configuration when fully retracted while the doors can provide a displacement hull configuration when utilized to cover a portion of the hull opening. In this manner, the hull configuration may be adjusted to provide different hull characteristics.

[0121] Doors 2210 can be deployed to cover a depressed region of the hull using any suitable approach. As one example, doors 2210 can be moveably coupled to the hull as indicated at 2220 above the water line so that the doors do not add additional drag to the hull when not in use. As the keel is deployed, the doors can be translated downward to at least partially cover the hull. For example, the doors may be coupled to the hull via a track that enables the doors to translate between the deployed configuration below water line and the stowed configuration above waterline. As yet another example, one or more doors may be stored within the depressed region, where they may be deployed when the keel is deployed. However, it should be appreciated that these are just examples of the various approaches that may be used to reduce drag on the hull by the application of one or more doors.

[0122] Note that the example control routines included herein, for example, with reference to FIGS. 13 and 16, can be used with various watercraft configurations. The specific rou-

We claim:
1. A foldable mast assembly for a sailing vessel, comprising:
   a lower mast section;
   an intermediate mast section having a lower end foldably coupled to an upper end of the lower mast section;
   an upper mast section having a lower end foldably coupled to an upper end of the intermediate mast section; and
   a boom coupled to the lower mast section.
2. The mast assembly of claim 1, wherein the boom includes a boom furler for furling and unfurling a sail.
3. The mast assembly of claim 1, further comprising a mast stepping support adapted to receive and rigidly couple a lower end of the lower mast section to a sailing vessel.
4. The mast assembly of claim 1, wherein the upper end of the lower mast section is foldably coupled to the lower end of the intermediate mast section by a first hinge assembly, wherein the first hinge assembly is arranged on a stern side of the mast assembly.
5. The mast assembly of claim 4, wherein the upper end of the intermediate mast section is foldably coupled to the lower end of the upper mast section by a second hinge assembly, wherein the second hinge assembly is arranged on a bow side of the mast assembly.
6. The mast assembly of claim 1, wherein the upper mast section is longer than the intermediate mast section; and wherein the upper mast section is longer than the lower mast section.
7. The mast assembly of claim 1, wherein an upper end of the upper mast section folds from an erected position toward the bow side of the mast assembly and wherein the upper end of the intermediate mast section folds from the erected position toward the stern side of the mast assembly.
8. The mast assembly of claim 1, further comprising:
   a first locking device configured to:
      inhibit folding of the upper mast section relative to the intermediate mast section in a locked position; and
      permit folding of the upper mast section relative to the intermediate mast section in an unlocked position;

United States Patent Application Publication

FIGURE 9.15 | A page from patent US 2008/0156242 A1 where the listing of claims begins

It is possible for an inventor to prepare and submit a patent application without an attorney (Pressman, 2011), but the hours spent learning how to prepare claims may take time away from other tasks. To decide whether or not to retain an attorney, study related patents and pay attention to the details of the language, the figure references, the reference to other descriptions of your invention, and particularly the language and structure of the claims. Further guidance on the patent process can be found on the USPTO (2012a) website under "Patent Information." Diegel (2012) also presents a well-thought-out process for deciding on the advisability of filing a patent application.

For students and others interested in a career as a patent agent, a patent attorney, or a patent examiner, Oppenheimer (2008) provides a fairly concise overview of how to enter the field.

US 2008/0156242 A1   Jul. 3, 2008

13

wherein the first locking device includes a first moveable insert internal the upper and intermediate mast sections, whereby the first moveable insert is moveable between the locked and unlocked positions by adjusting an amount of overlap between the moveable insert and at least one of the intermediate and upper mast sections.

9. The foldable mast assembly of claim 8, further comprising a second locking device configured to:

inhibit folding of the intermediate mast section relative to the lower mast section in a locked position; and

permit folding of the intermediate mast section relative to the lower mast section in an unlocked position;

wherein the second locking device includes a second moveable insert internal the intermediate and lower mast sections, whereby the second moveable insert is moveable between the locked and unlocked positions by adjusting an amount of overlap between the second moveable insert and at least one of the intermediate and lower mast sections.

10. The foldable mast assembly of claim 9, wherein the first and second moveable inserts each include an elongate element that is tapered along its length.

11. A foldable mast assembly for a sailing vessel, comprising:

an upper mast section including a first track segment at a stern side of the upper mast section, the first track segment being adapted to guide a luff edge of a sail between a raised configuration and a lowered configuration of the sail;

an intermediate mast section including a second track segment at a stern side of the intermediate mast section, the second track segment being adapted to guide the luff edge of the sail between the raised configuration and lowered configuration;

a first hinge assembly foldably coupling a lower end of the upper mast section at a bow side of the upper mast section to an upper end of the intermediate mast section at a bow side of the intermediate mast section.

12. The foldable mast assembly of claim 11, wherein the first and second track segments are aligned to guide the luff edge of the sail between the first track segment and the second track segment.

13. The foldable mast assembly of claim 11, further comprising:

a lower mast section having a lower end adapted to be received by a stepping support of the sailing vessel; and

a second hinge assembly foldably coupling a lower end of the intermediate mast section at the stern side of the intermediate mast section to an upper end of the lower mast section at a stern side of the lower mast section.

14. The foldable mast assembly of claim 13, wherein the second track segment has an upper end that extends to the upper end of the intermediate mast section and a lower end that terminates a distance from the lower end of the intermediate mast section before reaching the second hinge assembly.

15. The foldable mast assembly of claim 14, wherein the first track segment has a lower end that extends to the lower end of the upper mast section and an upper end that terminates near the upper end of the upper mast section.

16. The foldable mast assembly of claim 13, further comprising a boom coupled to the lower mast section between.

17. The foldable mast assembly of claim 13, wherein the second hinge assembly includes:

a first hinge, including a first hinge half coupled to the intermediate mast section and a second hinge half coupled to the lower mast section; and

a second hinge, including a third hinge half coupled to the intermediate mast section and a fourth hinge half coupled to the lower mast section;

wherein the second track segment is located along the length of the stern side of the intermediate mast section and passes between the first and third hinge halves at the lower end of the intermediate mast section;

wherein the lower mast section further includes a third track segment at the stern side of the lower mast section, the third track segment being adapted to guide the luff edge of the sail between the raised configuration and the lowered configuration;

wherein the third track segment and the second track segment are aligned to guide the luff edge of the sail between the second track segment and the third track segment.

18. A foldable mast assembly for a sailing vessel providing at least an erected position and a folded position, comprising:

a lower mast section;

a stepping support fixedly coupling the lower mast section to a sailing vessel;

an intermediate mast section having a lower end rotationally coupled to an upper end the lower mast section by a first hinge assembly that permits an upper end of the intermediate mast section to rotate from the erected position toward the stern of the sailing vessel and into the folded position without unstepping the lower mast section from the stepping support; and

an upper mast section having a lower end rotationally coupled to an upper end of the intermediate mast section by a second hinge assembly that permits an upper end of the upper mast section to rotate from the erected position toward the bow of the sailing vessel and into the folded position without unstepping the lower mast section from the stepping support.

19. The foldable mast assembly of claim 18, further comprising a boom coupled to the lower mast section.

20. The foldable mast assembly of claim 19, further comprising a substantially continuous track system for guiding a luff edge of a main sail along a length of the mast assembly, the track system including at least:

a first track segment at a stern side of the intermediate mast section that has a first end that extends to an upper end of the intermediate mast section; and

a second track segment at a stern side of the upper mast section that has a first end that extends to a lower end of the upper mast section;

wherein the first end of the first track segment and the first end of the second track segment are substantially aligned.

* * * * *

United States Patent Application Publication

FIGURE 9.16 | The final page of patent US 2008/0156242 A1 showing the remainder of the claims

COPYRIGHTS

In the United States, any creative work, including work in written, musical, dramatic, or art form, may be protected, subject to the constraints of Sections 107 to 118 of the copyright law, by registering the work with the United States Copyright Office (2012). A copyright protects the physical manifestation (electronic, written, and audio or visual format), but not the idea of the creative work. Copyright registration carries with it the right to control the following:

- Who may copy or otherwise reproduce and disseminate the creative work
- Who may prepare works derived from the registered work
- Who may perform a creative work such as a play or musical composition

One of the most significant exceptions to reproducing copyrighted works without authorization falls under Section 107 of the copyright law and applies to "fair use." According to Section 107, the following factors must be considered (quoted verbatim; see United States Copyright Office, Fair Use, 2013):

- The purpose and character of the use, including whether such use is of commercial nature or is for nonprofit educational purposes
- The nature of the copyrighted work
- The amount and substantiality of the portion used in relation to the copyrighted work as a whole
- The effect of the use upon the potential market for, or value of, the copyrighted work

For specific details on recent interpretations of these factors and other questions on fair use, please consult with a librarian or another information technology professional.

Any author of a creative work has an inherent copyright upon production of the work, regardless of whether the work has been registered for copyright. This copyright applies to works produced by students, such as homework assignments, reports, exam solutions, and notes.

There is a small fee for registering a copyright. Currently, the government copyright website indicates a three-month period for approval of electronic submissions and a ten-month backlog in approving paper submissions.

TRADEMARKS

According to the USPTO (2012b), "a **trademark or service mark** includes any word, name, symbol, device, or any combination, used or intended to be used to identify and distinguish the goods/services of one seller or provider from those of others, and to indicate the source of the goods/services." A trademark is used by businesses that supply goods; a service mark is used by businesses that supply services. Federally registered trademarks and service marks are followed by the ® symbol. The symbols ™ and ℠ indicate that someone is claiming ownership of the mark, but these marks are not federally registered.

Benefits of Federal Registration of a Trademark

Trademarks and service marks are marketing tools designed to help the public remember and identify a particular company and its goods or services. Marks registered with the USPTO give notice to the public that the registrant claims ownership of the mark and that the owner alone is entitled to use the mark throughout the United States to promote the specific goods or services described in the registration. Registration also provides legal rights in federal courts. Registering your trademark in the United States is a prerequisite for registering your trademark in other countries. Registration protects domestic goods from imported items that bear the same trademark.

Registration Process

To register a trademark, follow the instructions for filing an application with the USPTO (2012b). While it is not mandatory to hire an attorney, doing so may make the process more efficient. Usually the USPTO takes about three months to examine a trademark application, but it may take longer depending on the volume of applications (see USPTO, 2012b for the current expected time for completion).

Table 9.6 summarizes the different types of intellectual property discussed, along with the protection conferred, some familiar examples, and the duration of the protection.

TABLE 9.6 | An overview of intellectual property

Protection	What Is Protected	Examples	Duration
Utility patent	Process, machine, article of manufacture, composition of matter, improvement of any of the above	Amazon online retail process; sailboat mast; fax machine	20 years from priority date (filing date)
Design patent	A unique design; "look and feel"	US patent D48,160 for the original Coca-Cola bottle; US patent D11,023 for the Statue of Liberty	14 years
Plant patent	A new asexually propagated species	Monsanto corn varieties	20 years from priority date
Copyright, ©	Text, poems, screenplays, music, photos, and works of art and sculpture	This textbook, the songs of Bob Dylan, the poems of T.S. Eliot	70 years after the author's death
Registered trademark or service mark, ®	A logo, symbol, word, or phrase identifying a product or service	Coca-Cola®, Dell®, Asus®, ANSYS®, Roto-Rooter®	As long as it remains in use or until it becomes generic
Common law trademark, ™	An unregistered trademark used frequently enough to be recognized	Android™, Google Play™, ANSYS® Mechanical™	As long as it remains in use
Common law service mark, ℠	An unregistered service mark used frequently enough to be recognized	Local service suppliers, plumbers, delivery services, lawn services; Cross Cut℠ Lawn Service	As long as it remains in use
Trade secret	A process or formula that a company carefully guards	The recipe for Coca-Cola; the process for making Bayer aspirin	As long as the company can keep the secret

Based on Nolo PPA, 2009

SUMMARY

The material in this chapter is intended to introduce engineers and other technical personnel to legal documents such as specifications and contracts as well as different types of intellectual property (patents, trademarks, service marks, and copyrights). Specifications describe an object (a product, program, device, component, system, structure, or process) and the object's performance, referencing the codes and standards used to design, develop, or manufacture the object. We described different kinds of specifications: technical specifications that accompany consumer goods; product design and manufacturing specifications used for new product development; electronic product requirements specifications; software development specifications; and construction specifications. The standardized MasterFormat® adopted by the construction industry was used as a springboard for the discussion of specification format. We then addressed the actual *writing* of specifications, in particular, how to choose words and grammatical constructions to produce a clear, concise, consistent, accurate, well-organized, and complete specification. The last part of this chapter provided a brief overview of intellectual property, in particular, patents, trademarks and service marks, and copyrights. The final preparation of these kinds of legalistic documents may be done by an attorney, but engineers are likely to provide technical input and should therefore be familiar with their content and structure.

ACKNOWLEDGMENTS

For this particular chapter, the authors are indebted to several colleagues who helped us better understand their field's use of specifications: Dan Hyde (computer science professor), Stu Thompson (electrical engineering professor), Charles Kim (mechanical engineering professor), Mike Toole (civil engineering professor), and Dave Zartman (construction company owner).

EXERCISES

1. Consider a homework assignment in one of your courses. Restructure the problem statement as a specification. Use the structure of a specification to itemize the required and discretionary work to be delivered. Make careful use of the words *shall, will, may, should,* and *must* to indicate the required work and discretionary work.

2. Consider the product design specifications given in Table 9.1. Assume you have been asked to proofread the specification. What kind of information has been omitted? Was any information included that should not be there? Were any phrases ambiguous? Can you find any internal inconsistencies? Correct any grammatical and formatting errors.

3. Use search techniques from Chapter 2 to find online samples of specifications used in your field of study (construction specifications, new product design specifications, manufacturing specifications, process specifications, or software requirement specifications). Use the specifications checklist in Table 9.5 to evaluate the specifications you find. Some questions to consider are:
   - Is the meaning of each requirement clear and unambiguous?
   - Are acronyms and abbreviations defined and used appropriately?

- Is the terminology used consistently?
- Are the specifications complete?
- Can you suggest ways to make the document more concise?

4. Think of a common device such as a household appliance, a hand tool, or a power tool; a device used in computing, recording music, or making videos; a software package or computer app; or another, more specialized device. Brainstorm how you might improve the function, manufacture for cost, manufacture for sustainability, or appearance of the device. Translate your brainstorming ideas into product design specifications.

5. After developing your product design specifications in the preceding exercise, use resources at the United States Patent and Trademark Office website to determine if your "invention" could meet the USPTO requirement for "novelty."

6. Think of a common device such as a household appliance, a hand tool, or a power tool; a device used in computing, recording music, or making videos; a software package or computer app; or another, more specialized device. Use the USPTO (or another) patent database to trace the patent history of the device.

7. Find all patents related to the development of the electric guitar. What is the expiration date of the various patents? Which patents are still in force, and which are considered public domain? Write a synopsis of how the designs vary from one patent to the next.

8. What plant (the botanical variety) patents have produced the greatest return for the inventor/assignee? Explore the criteria used to define whether a plant can be patented. Provide an overview of the plant patenting process for a fellow student or an entry-level engineer.

9. Investigate the development of the global positioning system (GPS). Are there any GPS patents that are still in force? Are there any expired GPS patents? If there are no patents, can you find the letters GPS as a registered trademark? Explain who, if anybody, owns the right to use the letters GPS. Compare your results with the use of the letters UPS?

10. Several registered trademarks have become common words and have lost their trademark status. Some examples include nylon, band-aid, frisbee, elevator, and escalator. Write an expository essay on how a company can prevent their trademark from becoming a common, generic word. There are Internet resources available on this topic.

REFERENCES

Adams, K.A. 2008. Manual of style for contract drafting, 2nd ed. Chicago: American Bar Association.

Ambler, S.W. 2009. Introduction to user stories, <http://www.agilemodeling.com/artifacts/userStory.htm>, accessed 4 May 2013.

Beck, K., Beedle, M., van Bennekum, A., Cockburn, A., Cunningham, W., Fowler, M., Grenning, J., Highsmith, J., Hunt, A., Jeffries, R., Kern, J., Marick, B., Martin, R.C., Mellor, S., Schwaber, K., Sutherland, J., and Thomas, D. 2001. Manifesto for agile software development. <http://agilemanifesto.org/>, accessed 2 March 2013.

Beck, K., and Andres, C. 2004. Extreme programming explained: Embrace change, 2nd ed. Boston: Addison-Wesley.

The Bible. 1996. (New Living Translation) Genesis 6:14–16. Wheaton, IL: Tyndale House Publishers. Accessed 2 March 2013 at <http://www.biblestudytools.com/nlt/genesis/6.html>.

Bockrath, J.1986. Dunham and Young's contracts, specifications, and law for engineers, 4th sub-edition. New York: McGraw-Hill.

Boehm, B.W. 1988. A spiral model of software development and enhancement. Computer, 21(5), 61–72.

Calvert, J. 2010. Provisional patent applications: What you need to know. USPTO Inventors Assistance Program, InventorsEye. Accessed 4 May 2013 at <http://www.uspto.gov/inventors /independent/eye/201004/provisional.html>.

Clendining, W. 2008. Writing Engineering Specifications-Vocabulary in Specifications. Accessed 28 August 2013 at <http://www.technical-expressions.com/engineering/writing-specs/vocab/cover .html>.

Clendining, W. 2009. An overview seminar of MasterFormat 2004. Accessed 2 March 2013 at <http://www.technical-expressions.com/mf/intro/MF04-Overview.pdf>.

Clough, R.H., and Sears, G.A. 1994. Construction contracting, 6th ed. New York: John Wiley & Sons.

Cockerill, A.W. 2005. Writing specifications. Accessed 8 May 2012 at <http://www.achart.ca/articles /publications/words9.htm>.

Copeland, L. 2001. Extreme programming. Accessed 2 March 2013 at <http://www.computerworld .com/s/article /66192/Extreme_Programming?taxonomyId=063>.

CSI 2010 MasterFormat. 2010 MasterFormat® 2010 Update. Washington, DC: Construction Specifications Institute.

CSI Guide. 2011. The CSI specifications practice guide. Construction Specifications Institute. Hoboken, NJ: John Wiley & Sons.

CSI. 2012. MasterFormat® Numbers & Titles, April 2012. Accessed 28 August 2013 at <http://csinet .org/numbersandtitles>.

Dahl, D. 2010. Patents: 5 Things you need to know, AOL Small Business. Accessed 2 March 2013 at <http://smallbusiness.aol.com/2010/05/10/how-to-file-a-patent/>.

Diamond, P. 2012. Patent reform update: sweeping changes to US patent law enacted, Accessed 2 March 2013 at <http://ezinearticles.com/?Patent-Reform-Update:-Sweeping-Changes-to-US -Patent-Law-Enacted&id=6930617>.

Diegel, O. 2012. Intellectual property. Everything you wanted to know but were afraid of being told…, Creative Industries Research Institute, Auckland University of Technology. Accessed 2 March 2013 at <http://www.ciri.org.nz/downloads/Intellectual%20Property.pdf>.

Donaldson, T.L., and Glen, W.C. 1860. Handbook of specifications: Or, practical guide to the architect, engineer, surveyor, and builder, in drawing up specifications and contracts for works and constructions. London: Lockwood and Co.

Ehrlickman R. 2010. Tips and tricks—Patents: 10 things engineers should know, Control Engineering. Accessed 2 March 2013 at <http://www.controleng.com/index.php?id=483&cHash=081010&tx _ttnews[tt_news]=40486>.

FHWA. 2010. The Federal Highway Administration Technical Advisory: Development and Review of Specifications, HIAM-20. Accessed 4 May 2013 at <http://www.fhwa.dot.gov/construction /specreview.cfm>.

Fisk, E.R. 2000. Construction project administration, 6th ed. Upper Saddle River, NJ: Prentice Hall.

Fodden, S. 2011. Shall we keep using "shall" or must we stop? Accessed 2 March 2013 at <http://www.slaw.ca/2011 /05/26/shall-we-keep-using-shall-or-must-we-stop/>.

Ford, R.M., and Coulston, C.S. 2008. Design for electrical and computer engineers. New York: McGraw-Hill.

Friday, B., Lindgren, S., Shannon, S., and Stern, D. 2008. Folding sailboat mast—Final report. Senior design report, Mechanical Engineering Department, Bucknell University.

Geren, R.L. 2012. Must We Use "Shall" In Our Specifications?! Accessed 4 May 2013 at <http://goo.gl/RJOIh> or <http://www.specsandcodes.com/Articles/Must%20We%20Use%20Shall%20In%20Our%20Specifications.pdf>.

Goldman, D. 2011. Patent reform: What inventors need to know. Accessed 2 March 2013 at <http://www.legalzoom.com/intellectual-property-rights/patents/patent-reform-what-inventors-need>.

Grammar Girl. 2009. Quick and dirty tricks for better writing, "Comprise" versus "Compose", episode 188. Accessed 2 March 2013 at <http://grammar.quickanddirtytips.com/comprise-versus-compose.aspx>.

IEEE Std 830-1998. 1998 IEEE Recommended practice for software requirements specifications. New York: Institute of Electrical and Electronics Engineers.

Kramer, B., Saunders, C., Schramke, B, and Weaver, D. 2010. Third generation puck-shooter final report. Senior design report, Mechanical Engineering Department, Bucknell University.

Kruchten, P. 2001. From waterfall to iterative development—A challenging transition for project managers, The Rational Edge, Accessed 2 March 2013 at <http://www.ibm.com/developerworks/rational/library/content/RationalEdge/dec00/FromWaterfalltoIterativeDevelopmentDec00.pdf>.

Kruchten, P. 2004. Going over the waterfall with the RUP, developerWorks. Accessed 4 May 2013 at <http://www.ibm.com/developerworks/rational/library/4626.html>.

LeVie, D. 2010. Writing software requirements specifications (SRS). Accessed 4 May 2013 at <http://techwhirl.com/writing-software-requirements-specifications/>.

Ma, M.Y. 2011. Fundamentals of patenting and licensing for scientists and engineers. Hackensack, NJ: World Scientific Publishing Company.

MasterFormat®. 2012. Accessed 18 June 2012 at <www.masterformat.com>.

NEMA. 2013. National Electrical Manufacturers Association Standards. Accessed 17 February 2013 at <http://www.nema.org/Standards/Pages/All-Standards.aspx>.

Nolo PPA. 2009. Nolo's guide to provisional patent application. Accessed 4 May 2013 at <http://nolonow.nolo.com/noe/popup/provisional_patent_application_guide.pdf>.

ODOT. 2009. Oregon Department of Transportation, Specification writing and style manual, Accessed 2 March 2013 at <http://www.oregon.gov/ODOT/HWY/SPECS/docs/forms_manuals/Specification_Writing_Style_Manual.pdf?ga=t>.

Oppenheimer, S.C. 2008. Patent careers for technical writers and scientific, engineering, and medical specialists. Accessed 2 March 2013 at <http://www.opatent.com/Patent%20Careers%20for%20Tech%20Writers,%20Engineers,%20and%20Scientists.pdf>.

Oriel, J. 2007. Guide to specification writing for U.S. Government Engineers, Naval Air Warfare Center Training Systems Division (NAWCTSD). Accessed 4 May 2013 at <http://nawctsd.navair.navy.mil/Resources/Library/Acqguide/spec.htm>.

Poltorak, A.I. 2009. What you need to know about patents and their value. Accessed 2 March 2013 at <http://ezinearticles.com/?What-You-Need-to-Know-About-Patents-and-Their-Value&id=1982047>.

Pressman, D. 2011. Patent it yourself, 15th ed. Berkeley, CA: Nolo Press.

Robbins, J. 2012. ICS 121: Specification vs. Design. Accessed 4 May 2013 at <http://www.jrobbins .org/ics121f03/lesson-spec-design.html>.

Royce, W. 1970. Managing the development of large software systems. Proceedings of IEEE WESCON, pp. 328–338. Accessed 28 August 2013 at <http://leadinganswers.typepad.com /leading_answers/files/original_waterfall_paper_winston_royce.pdf>.

Sclater, N., and Traister, J.E. 2003. Handbook of electrical design details, 2nd ed. New York: McGraw-Hill.

Stout, P. 2004. Equipment specification writing guide. Accessed 4 May 2013 at <http://www.peterstout.com/pdfs/tech_specs_detailed.pdf>.

Ulrich, K.T., and Eppinger, S.D. 2011. Product design and development, 5th ed. New York: McGraw-Hill/Irwin.

United States Copyright Office. 2012. Accessed 2 March 2013 at <http://www.copyright.gov/>.

United States Copyright Office, Fair Use. 2013. Copyright fair use. Accessed 4 May 2013 at <http://www.copyright.gov/fls/fl102.html>.

USPTO.2012a. Patents, United States Patent and Trademark Office. Accessed 4 May 2013 at <http://www.uspto.gov/patents/index.jsp>.

USPTO. 2012b. Trademarks, United States Patent and Trademark Office. Accessed 2 March 2013 at <http://www.uspto.gov/trademarks/index.jsp>.

Waine, P. 2007. Definition of functional specification. Accessed 2 March 2013 at <http://searchsoftwarequality.techtarget.com/definition/functional-specification>.

Williams, K. 1992. Civil engineering contracts. Chicester, England: Simon & Schuster/Ellis Horwood.

Wolfe, S. 1996. When shall I use will? Accessed 4 May 2013 at <http://www.northstarcsi.com /writing/spex-shallwill.htm>.

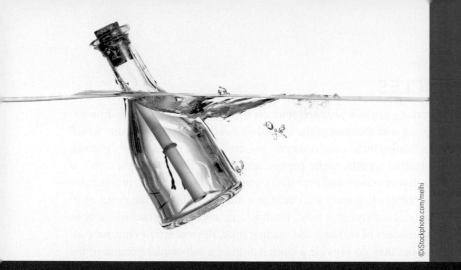

Chapter 10

OVERVIEW REPORTS

Some forms of technical writing require the writer to summarize and synthesize information from a variety of sources. In this chapter, we will describe the following types of "overview" writing:

- Review articles (or literature reviews in preparation for academic theses)
- Site visit reports that summarize findings of visits to locations of interest
- White papers that identify customer problems and provide solutions, frequently suggesting the use of the sponsoring company's products
- Technical trade journal articles that provide concisely focused presentations of technical updates for readers interested in practical applications

Understanding the format of these reports and articles will help the reader more quickly assimilate the information presented. Knowing what belongs in these types of communications will prepare students and entry-level engineers to undertake this type of writing if, and when, they are asked to do so. Studying the phrasing, word choices, tone, and structure of these communications will also further guide the development of each engineer's professional voice.

REVIEW ARTICLES

Engineers who begin working in a new area and students who undertake a thesis or dissertation project must be familiar with the state of the art in their field. Often the literature, which includes journal articles, books, theses and dissertations, conference proceedings, patents, government reports, industrial reports, white papers, and many other Internet sources, is scattered across many different venues and over many years. On the one hand, producing a comprehensive review of the existing literature could be a full-time effort for several weeks, depending on the maturity of the research field. Finding a relatively recent review article, on the other hand, can save hours of research and reading time. Review articles are not considered primary sources, but they do provide a compilation of a substantial portion of the relevant primary literature. Authors of review articles summarize and synthesize the current state of the art, point out any inconsistencies in the literature, identify any unresolved questions, and suggest possible paths and related topics for future investigation. Reading a good review article is a highly effective way to familiarize yourself with a new area of study.

A review article is often written as a precursor to a new research project. The article may provide the background and justification for the project, which may be required in the funding proposal (see Chapter 8). Similarly, graduate students prepare literature reviews before beginning their theses for the following reasons:

- To determine what is already known in the field
- To understand the research methods and instrumentation used in the field
- To assimilate and synthesize the existing literature
- To integrate additional research techniques from allied fields of study
- To identify the inconsistencies and gaps in knowledge that justify their research
- To identify unresolved issues in the field
- To identify ultimately the niche into which their research fits

A literature review must be more than an annotated bibliography. References cited should be directly relevant to the thesis topic and should be discussed thematically in the context of the arguments being made in the thesis. A review is not a chronological listing of *all* the available literature, but rather a critical analysis of the work that is *relevant* to the project at hand. The review should integrate past knowledge, synthesize the relevant information, and provide the reader with a perspective not found in any of the individual papers reviewed.

Audience

As with any form of communication, the review article or literature review must be tailored to the audience. Why are you writing the review, and who will be interested in what you write? You may write a review for one of the following reasons:

- It is intended for a general audience interested in some aspect of technology.
- It is required by a professor as a stand-alone paper for a course.
- It is requested by managers and other stakeholders who have an agenda for a particular project and who may have decision-making power over your thesis or proposal.
- It is targeted at specialists in the field as a comprehensive overview of the state of the art.
- It is geared to reviewers of a proposal for project funding (in-house, extramural funding agency, venture capitalist, or another source).

Your answers to the questions why and for whom you are writing will dictate the level of detail and the precision with which you search the literature in the field. Figure 10.1 illustrates the scope of some projects in which a literature review might be undertaken.

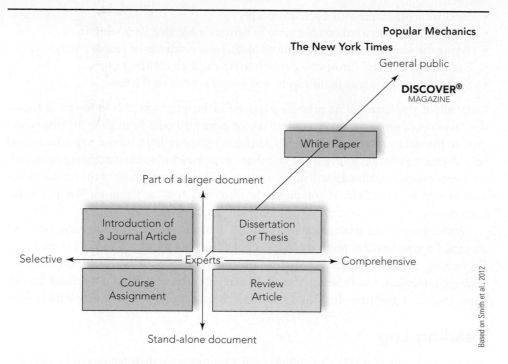

FIGURE 10.1 | A literature review can range from selective to comprehensive. It can stand alone or be part of a larger document. Reviews can be written for audiences with widely different interests

As shown by the *x*-axis, the review may be relatively selective, as in the introduction to a journal article, which focuses on one aspect of a specific topic, or in an assignment for a particular course. At the other end of the spectrum, the review may be comprehensive, as required for a dissertation or thesis or a review article for a journal.

The *y*-axis reflects the contribution of the literature review to a given project. The introduction section of a journal article or a dissertation, for example, is just one part of that particular paper, but course assignments and review articles typically stand alone.

The *z*-axis describes the level of the audience's technical sophistication, ranging from experts in the field to the general public. Experts would be interested in reading journal articles, dissertations, and review articles, whereas interested lay people would be likely to read articles in the popular press. Somewhere in the middle of this axis would be a mixed-bag audience of technical personnel, managers, and business people who read white papers (to be discussed in detail later in this chapter) because they are looking for solutions to their problems or wish to stay on top of developments in their field.

References for a Literature Review

The most efficient way to find references that are relevant to your topic is to search article databases using the strategies described in Chapter 2. In short, these strategies involve the following steps:

• Gaining a basic understanding of the topic
• Narrowing down your research goals

Planned changes for the facility include expansion of the firing range on the south side of Building C. This expansion would be below grade so that the outdoor grade level space will still be available for parking, which is scarce. This project is in the early planning stage, and no budget has been set aside for it.

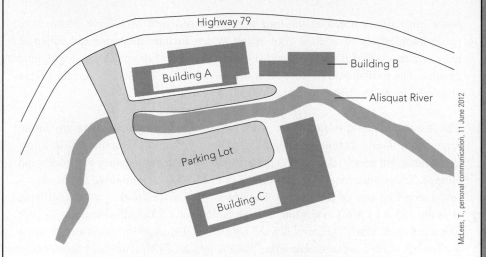

Sketch of Fremont Firearms property

Another major change involves the storage of methanol, a flammable liquid, on the south side of Building A. Presently storage is in four 1,000 gallon independent single-use tanks. The tanks are fenced and sit in a concrete containment basin adjacent to the south wall of the building. This wall is of noncombustible construction and has several glass casement windows approximately eight feet above ground that are typically kept open. Mr. Apple has planned to replace the four tanks with one new bulk storage tank. His plans include grounding, bonding, and providing a non-freeze automatic sprinkler system for protection for the new tank. Considering fire protection features in the planning stages of such a project, especially when flammable liquids are involved, is an excellent loss prevention policy of Mr. Apple's. Plans for any changes or additions to the fire protection systems should be submitted to Timothy McLees, Jr. for review and comment prior to the installation or fabrication of any materials. In the meantime, the present setup for the methanol tanks is still an exposure, and it is recommended that automatic sprinkler protection be installed, given the tank replacement project does not have a project timeframe established.

Recommendation 1

Another recommendation developed as a result of the tour included adding one valve to the weekly fire protection valve checklist. One quarter-turn 1-inch shutoff valve for the Stock Office in Building C should be added to the list and confirmed to be open.

Recommendation 2

FIGURE 10.6 | Continued

Recommendation 3

The last recommendation from this visit involves the deluge sprinkler system protecting the "Truck Fill Area" attached to Building B. This is an empty shed; the deluge system has not been trip tested in over five years. The system should either be trip tested annually or simply capped off and removed from service. If the system is capped off, the area that was protected should remain free of all combustibles.

Suggested procedural revision

Some other issues noted on the tour were discussed with Mr. Apple. Plans were made immediately to address roof issues which include moderate quantities of debris on the roof of Building A and several cases of roof drains clogged with pine needles. The roof should be kept clear of all equipment and debris, and the drains should be inspected and cleaned periodically.

Information for underwriters

The flammable liquid storage room was also identified as an area where any future renovations should include additional fire protection features. The room as it is now is tolerable, but poorly designed and not capable of accommodating any additional storage. There were four 55 gallon drums of Flash3000, a flammable liquid, stored in the receiving area of Building B. The drums were moved to the east side of Bldg B in the 205 ft by 80 ft extension, and the pallet load of 55 gallon drums was kept along the north wall. This area is used for steel rod storage on racks and presents a tolerable storage arrangement since there is a lack of continuity of combustibles in this area and it is away from the forklift traffic pattern in this area. The use of this space could become a more serious issue if the receiving area becomes the alternative option as drums of flammable liquids arrive and need to be stockpiled.

Information for underwriters

The final issue that Mr. Apple immediately addressed regarded the outdoor sprinkler heads protecting the transformer in front of Building A. Four sidewall sprinkler heads are located against the noncombustible building wall, the heads are installed backwards, and there are no heat collectors over the heads so they probably would not operate. For these reasons, these heads would not control a transformer fire, but the protection is not necessary in the first place. The noncombustible building walls and sufficient space reduces the exposure of a transformer fire having any effect on the building. Mr. Apple will leave the system as it is, and if there is a chance, have the maintenance staff turn the heads around. This is acceptable, and it is not necessary to make any modifications to the system.

A closing conference was held with Mr. Apple, Mr. Carrot, and Ms. Avocado to discuss the results of the visit.

Occupancy

Information for underwriters

Fremont Firearms is one of the nation's leading manufacturers of high-quality firearms for the commercial sporting market. This location is the primary manufacturing, assembly, testing, and distribution point for over seventy varieties

FIGURE 10.6 | Continued

of pistols, rifles, and shotguns. Production is presently at 4,000 guns per day, with an expected increase to 5,000 within the coming year. Fremont Firearms has had strong business over the past few years, even with cyclical hunting season demand changes. XYZ Castings, a division of Fremont Firearms, occupies Building A at this location and produces high-precision steel investment castings both for Fremont Firearms products and for other customers.

The three buildings at this location are designated Buildings A, B, and C. There are over 1,000 employees working three shifts, six days per week. There is a facility shutdown for two weeks in July. First shift has approximately 625 employees and runs from 5:00 AM to 3:00 PM. Second shift has approximately 275 employees and runs from 3:00 PM to 1:00 AM. Third shift has minimal staff and completes the work day overlapping with the main production shifts. Buildings A and C are occupied 24 hours per day. Building B is not occupied during the third shift.

Building A houses XYZ Castings and includes a variety of casting operations as well as a machine tool shop, some storage areas, and small office space. Wax injection molding uses up to 22 injection molding machines. Mold patterns are stored onsite for all current and previous parts. Ceramic shell molding uses a series of both robotically and manually operated dipping lines. Humidity is carefully controlled. Heat treatment of molds is accomplished using four electrical internal quench atmosphere furnaces, four vacuum annealing furnaces, four tempering furnaces, and one electric salt-bath martempering and austempering heat treating system. The steel foundry has three 350 kW induction melting units, heating steel for pouring into the molds. The castings are freed from the molds using a variety of steps including grinding, acid pickling, welding, grit blasting, drilling and milling. There is an inspection station and metallurgical laboratory that accompany the casting operations. In the rear of Building A, located directly outside, there are four caged 1,000 gallon methanol storage tanks, one bulk carbon dioxide tank, one bulk liquid nitrogen tank, and one bulk argon tank. There are two 12,000 gallon underground fuel oil tanks and 50 feet from the building, two caged 25,000 gallon propane tanks.

Building B has the facility's woodworking operations as well as receiving and storage of raw materials, including wood, oil, steel, and packaging materials. Woodworking includes the production of gun stocks, from machining and sanding to finishing and painting. There are spray booths and a flammable liquids storage room in Building B. The flammable liquids storage room is tolerable in its current setup but is poorly designed and limited in its ability to accommodate additional flammable liquid storage. If that area or the surrounding dispensing or spray areas are ever upgraded, the flammable liquid storage room should be modified to include all appropriate fire protection features.

FIGURE 10.6 | Continued

Building C is located on the south side of the Alisquat River and is accessible by a single bridge connecting its parking lot to those of Buildings A and B. This building is largely dedicated to the final manufacturing and assembly steps for the firearms. There are hundreds of individual tooling stations using both computer and manually operated machines. There are two designated welding areas supporting the maintenance of all the machines. The building also includes a test range for completed products, offices, a computer room, and shipping area for finished goods. Completed firearms are shipped to the finished goods warehouse located at 638 St. Margarite Street, Fremont, Pennsylvania. There is also a caged area for ammunition storage; ammunition is used explicitly for firing in the test range, and none is produced onsite.

Fremont Firearms has one other assembly facility in Los Angeles, California. That location only assembles semi-automatic handguns, one of the product lines. All castings for both locations are done at the Fremont, Pennsylvania location.

McLees, T., personal communication, 11 June 2012

FIGURE 10.6 | Continued

WHITE PAPERS

Another genre of technical writing is the white paper. Kemp (2005) defines **white papers** as "authoritative business communications that achieve marketing goals by explaining technical ideas clearly with a compelling presentation of business value." The white paper is much more focused on presenting a solution to a common business or industry problem than it is on developing advanced technology. White papers should not be confused with technical conference papers. Conference papers are written and presented by the engineers who carried out the work; their audience consists of fellow technical personnel who are attending the conference for professional development purposes (corporate engineers) or because the research is relevant to their own work (engineers in academia). White papers, on the other hand, are usually prepared by corporate engineers working with professional writers for a target audience that may include corporate managers, engineers, and technicians. White papers are often used as marketing tools to educate readers on the technical advantages and business value of selecting products from the sponsoring companies to solve a specific technical or business problem.

What Is a White Paper?

In the British government, long government reports have traditionally been bound into books with blue covers, known as "Blue Books." Reports that focused on providing background and defining or arguing for a particular position or policy, and that were too short for binding, were stapled together with a white cover page and became known as "White Papers." White papers (some argue for continued capitalization of the words) are still used in government and politics. Politicians and government agencies frequently commission the preparation of political white papers by experts in specialized fields.

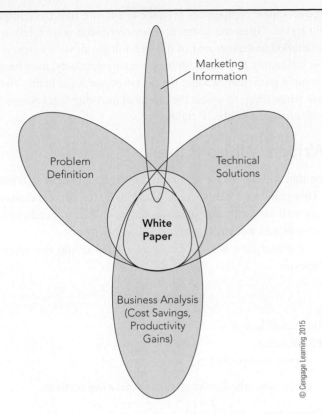

FIGURE 10.7 | A white paper lies at the "sweet spot," defined by the intersection of the problem definition, technical solutions, and the business analysis, with just a sliver of marketing.

A technical white paper is primarily an educational and informational document with only a very subtle underlying marketing aspect. It is a focused review of a particular, commonly occurring, sector-specific industrial or managerial problem. The white paper is written in relatively plain English to allow engineers, engineering managers, and executives who often hold decision-making powers to understand the content readily. A white paper, as shown schematically in Figure 10.7, defines a specific problem, suggests alternative technical solutions, and provides insight into the business advantages of each solution. The white paper, when done well with an objective presentation of alternatives, enhances the credibility of the sponsoring organization and often increases the likelihood that potential customers will investigate the product offerings of the sponsoring organization.

In the past several years, the generation of technical white papers has become an increasingly common business practice. The sponsor is a manufacturer or vendor who is marketing a particular product effective in solving the problem addressed in the white paper. White papers appear with regular, and seemingly increasing, frequency via links connected with electronic technical trade journals and electronic trade news magazines. Without making any concerted effort to do so, other than responding to email solicitations, we have been able to obtain access to over 100 white papers in less than a year.

White papers are also used for internal company purposes. Companies with substantial intellectual property (IP) to protect must be vigilant to protect their rights. They must monitor the products their competitors produce to be sure that the competition is not infringing on their rights. Often the distinction between what is a violation and what is not requires a very detailed understanding of the technology. In such cases, a senior company engineer, who is intimately familiar with the company products, may be asked to produce a white paper about a particular product for the corporate legal team. The legal team then uses the in-house white paper to assess the merits of pursuing legal action against competitors suspected of infringing on the IP rights.

How to Write a White Paper

The instructions that follow are based largely on Kemp's excellent *White Paper Writing Guide* (2005). This guide provides specific instructions for anyone contemplating writing a white paper, as well as myths and mistakes to avoid. The appendices in Kemp's guide describe the strengths and weaknesses of sample white papers.

The process of preparing a white paper can be divided into five stages (adapted from Kemp's nine stages):

- Planning
- Researching
- Designing the overall layout
- Writing and illustrating
- Revising, reviewing, and publishing

Each of these stages is described in detail in the following sections.

Stage 1—Planning
Planning begins by defining the goals of the white paper and identifying the target audience as well as the needs of its constituent groups. According to Kemp (2005), overt goals (those specifically mentioned in the white paper) may include

- Stating a problem and how the technology solves this problem
- Explaining how the technology works
- Establishing a company's expertise
- Convincing investors to provide financial support for the project

Some covert goals (those not specifically mentioned in the white paper) may be to persuade technical personnel to buy the product or service, to convince businesspeople of the advantages of the technology, and to demonstrate the superiority of the product or service over that of the competition.

The second critical part of planning is identifying the audience. As shown in Table 10.3, the audience may consist of a diverse group of people, each with different professional

---

TABLE 10.3 | Prospective readers of white papers

Venture capitalists	R & D directors	Business analysts	Production supervisors
Industry analysts	Engineering managers	Service technicians	Shop supervisors
Senior executives	Engineers	Purchasing agents	End users

interests and backgrounds. Some are primarily business-oriented, others are technically in-clined, and still others work in administration. Consider the barriers that might prevent them from being interested in or understanding the white paper and reasons why they would be hesitant to accept the paper's conclusions. For example, senior executives are motivated by the cost savings or return on the investment of a new technology, but may be less familiar with how the technology works. Engineers might be excited about the technology itself, but probably would be wary of promotional literature. End users want to know how the technology will make their jobs easier, but are concerned with how much training would be involved.

After the goals and the audience for the white paper have been established, a strategy must be developed for describing the subject matter in a way that supports the goals and is understandable to the audience. Always consider the readers' perspective when writing about the following topics in relation to the product or service highlighted in a white paper:

- What is it?
- How does it work?
- How does it improve performance?
- How does it improve productivity?
- How much training is involved in learning how to use it?
- What are the financial benefits?
- What other options are available, and what are their advantages and disadvantages?

When the audience is diverse, it may be advisable to generate multiple white papers to tar-get the interests and backgrounds of each constituent group. Regardless, keep the interests of the audience members as the first priority, focus on their needs, and anticipate the issues that might make them reluctant to acquire the product or service.

Another aspect of planning pertains to logistics such as

- How the white paper will be used to sell and market the technology
- Media (print or electronic) used to disseminate the white paper
- Production team (experts from the sales, marketing, and technical divisions; writer(s); illustrator(s); and reviewers)
- Desktop publishing and drawing programs
- Estimate of production costs (labor, software, printing, website development, and so on)
- Schedule for producing the final product

Throughout the planning stage, it is prudent to ask for feedback from those who commis-sioned the white paper. It makes more sense to make sure the project is on the right track before investing heavily in the researching, designing the layout, writing and illustrating, and revising stages that are yet to come.

Stage 2—Researching
Researching is acquiring information that pertains to the goals of the white paper, the featured technology, and the interests and needs of the target au-dience. Take advantage of human as well as printed or electronic resources. Human resources include members of the target audience and so-called subject matter experts such as sales-people, marketing personnel, and technical staff. When possible, ask questions in person rather than via email. Face-to-face meetings make communication more efficient because they provide input on the other person's true feelings through the display of body language and permit the exploration of tangentially related topics that evolve during the conversation. Other ways to make good use of human resources is to brainstorm with other members of the production team and write down the group's collective ideas for further consideration.

Printed or electronic resources include

- In-house technical documents, marketing brochures, and sales presentations
- Trade journal articles
- Specifications and brochures for competitors' products
- Social media (forums, blogs, and so on) and other information available on the Internet

After collecting as much information as possible, analyze and organize it with the original goals and the defined audience in mind. Make a mind map and use it to construct a rough outline with no more than two or three levels. Try to make the main topics in the outline approximately the same length. Think about the topic that is of greatest interest to the target audience, and put this "attention grabbing" topic first. Arrange subsequent topics in a logical order.

Stage 3—Designing the Overall Layout
The goals of the white paper and the audience drive the content, and the content drives the layout design. Consider the following variables:

- Portrait or landscape orientation
- One column or two
- Color or black and white
- Typography
- White space
- Headings and figure captions
- Desirability of interactive features and multimedia presentation

In portrait format, the long edge of the page is vertical. This orientation is preferred when the white paper is a printed document. Kemp (2005) recommends using two text columns or one column about two-thirds the width of the page for text and another column for visual elements such as tables, figures, illustrations, sidebars, pull quotes, and so on. In landscape format, the long edge of the page is horizontal. This orientation should be used when readers read the document on their computer screen. Both one- and two-text column formats are easy to read in landscape orientation.

The color scheme of the white paper is determined in part by the dissemination media. A professionally printed document in color makes a good first impression and is more likely to attract reader attention than a drab black and white publication. On the other hand, if the audience is going to print an electronic white paper on a black and white printer, the benefit of color is lost. Whether the design is in color or black and white, make sure there is good contrast between the text and the background to ensure that the text is legible.

The term **typography** refers to typeface, font size, line length, and leading (the space between the lines). Choose a typeface that gives your white paper character and a font size that is large enough to be read easily. Some authorities on page layout suggest that shorter line lengths promote comprehension, while longer lines can be read faster. The leading and the white space (the space not occupied by text and visual elements) are important, because they help create a cleaner, less cluttered look and make the paper easier to read. White space makes the white paper more attractive and is also a great way to separate sections.

The layout of a white paper should make it easy to scan and find information quickly. To facilitate browsing, make headings, subheadings, and figure captions descriptive to bear out the most important points. Use visual elements to break up blocks of text and to enhance readers' understanding of the content.

If multimedia and interactive features are desirable, then the white paper must be designed from the start as an online document. Used in appropriate measure, interactive features and links to multimedia presentations can make knowledge acquisition more efficient for online viewers. Done poorly or in excess, these same features can drive away some of the more traditional senior executives with substantial decision-making authority. Always consider the needs and preferences of the target audience.

Stage 4—Writing and Illustrating

White papers, like other types of short-form writing, will ideally follow the 3-30-3 rule (Eight Rules, 2005). According to this rule, readers take 3 seconds to read the title to gain a first impression. If the title catches their interest, then they spend another 30 seconds reading the introduction. If readers are still interested, they devote 3 minutes to reading about how the featured product or process will solve their technology problem. The bottom line for writers of white papers is to get to the point sooner rather than later.

The title should contain the words "White Paper" and a short, catchy description of what readers can expect to get from reading the paper. Here are a few titles from the literature:

- Six Best Practices for Effective Wind Farm Operation Using Industrial Ethernet Switches to Assure Maximum Uptime. White Paper (a bit long but very compelling if you run a wind farm)
- The Ten Commandments of USB
- The Benefits of Quad-Core CPUs in Mobile Devices
- White Paper: Save Energy in Your Supermarket with a CO_2 Refrigeration System

---

Unfortunately, not all documents labeled *white papers* are truly white papers. Often a technical report will be passed off as a white paper. Occasionally, a marketing brochure will also be called a white paper. Be careful and be aware! White papers are typically 4 to 12 pages long, focus on a particular problem, and inform readers of a proposed solution to the problem. The primary objective is education, not self-promotion.

After reading the title, readers skim the introduction. To keep readers' attention, the first few paragraphs must address their needs and present a solution that makes good sense from technical and business perspectives. When writing the introduction to a white paper, follow the tips given in Table 10.4.

After the introduction, describe the topics in the order formulated in the outline. Explain how the technology works and why the reader would want to acquire the technology, and provide quantitative data that showcases the technology's effectiveness. Also demonstrate that the technology makes good sense from the business perspective, highlighting increased productivity, decreased energy cost, or other savings that enhance the customer's bottom line. Although the primary goal of white papers is to educate, the clarity and logic of a well-written white paper may subtly persuade readers that the featured technology is the obvious choice to solve their problem. The last part of the white paper is typically an understated acknowledgment of the company's sponsorship of the white paper and the products they offer for solving the problem addressed in the paper.

TABLE 10.4 | Attention-keeping measures for the introduction of a white paper

Do	Do not
Describe a problem that people in this business face	Give a lengthy account of the history of this problem
Give readers a reason to be interested in this problem	Explain the technology in detail before convincing the reader that the problem is serious enough to require a solution
Show how the current technology can solve the problem	Discuss the problem or the solution at a level that is either too elementary or too complex

Based on Kemp, 2005

Writing the content of a white paper can be challenging when the intent is to explain a technically complex subject on a level that can be understood by a diverse audience. To help the target audience understand technical concepts, Kemp (2005) has the following suggestions:

- Write at an appropriate level, using the language of the reader.
- Define concepts and acronyms.
- Use terminology consistently.
- Explain how each component works and show how it fits into the process; this helps the reader understand why the technology is important.
- Use visual elements to complement the narrative.
- Give examples and use analogies to help the reader make connections.

Well-chosen visual elements are critical to the success of the white paper. Graphs, drawings, flowcharts, photographs, images, tables, sidebars, and pull quotes break up blocks of text and add to the paper's visual appeal. These elements also speed comprehension, as we know from the adage "a picture is worth a thousand words." In order for visual elements to be effective, however, their resolution must be appropriate for the selected media. Images in professionally printed documents must have a high resolution, as grainy, blurry pictures ruin an otherwise well-written paper. High-resolution images on webpages, however, are a handicap because they take longer to display. If readers will download and print PDF files, the resolution is limited by the printer's output, typically given as 300 dots per inch (dpi).

Stage 5—Revising, Reviewing, and Publishing

Once the first draft is complete, focus on the content. Do the concepts fit together? Does one topic transition smoothly to the next? Is the amount of detail appropriate, not too much and not too little? Are more examples or hard data needed? Do the visual elements complement the text and enhance reader understanding? In a second pass, revise the paper for grammatical and typographical errors. After the second review, pass the draft to the assigned technical reviewer to check the facts. Finally, if possible, get feedback from a select group of potential customers, or at least from someone in the company's sales force, to make sure the white paper is addressing the needs of the audience. Revise accordingly.

Once the revisions are finished, the company that commissioned the white paper will authorize its release, printing, and dissemination. The initial distribution is already planned, but do not be afraid to expand the readership by handing out the white paper at conferences and trade shows, submitting it to professional trade journals, and posting it on the corporate website.

Final Thoughts on White Papers

Technical white papers are a common form of communication in the corporate world, and many examples are available on the Internet. In academia, however, this genre of technical writing is underrepresented. Most undergraduate engineering students have never heard of white papers, yet when they work in industry, they need to read and write them.

Many Internet sources provide guidance on writing white papers. We have cited Kemp's (2005) detailed writing guide frequently in our discussion; other useful resources include Kemp (2010), Canright (2011), Eight Rules (2005), and Stelzner (2005). In closing, we provide advice from Canright (2011), listed here as Table 10.5, concerning what a writer should and should not do when writing a white paper.

Sample White Paper

Many documents labeled as white papers are actually marketing brochures. Others are technical papers focused solely on the technical fundamentals behind a device or process, without defining the problem that the device or process solves. To illustrate the characteristics of good white papers, we include some sample extracts in Figures 10.8 and 10.9. Note how these white papers promote the greater good (not just the sponsoring company's products) by educating and informing us about the effects of various refrigerants on the environment and the need to consider seasonal performance.

TABLE 10.5 | Suggestions for what to do and what not to do when writing a white paper

Recommended	Not Recommended
• Know the material about which you write	• Assume that the reader knows anything about the topic
• Keep the reader's education as first priority	• Use a style that is too technical or too simplistic, too formal or too informal
• Show real-world applications and provide sources	• Give rambling, unnecessarily detailed explanations
• Remember that short is good; shorter without loss of concept is better	• Present incorrect or biased information
• Provide readers with a variety of solutions to their problems, not just your own product	• Promote only one product that can solve the problem
• Find a balance between formal writing and street talk; use language familiar to your audience	• Push your product on the reader
	• Take any available report or article and call it a white paper

Based on Canright, 2011

In 1997 the Kyoto Protocol, signed and ratified by many nations around the world, focused attention on the impact of human activity on climate change. As a result, there is now more attention on global warming. Although the Kyoto Protocol does not apply to the United States, our industry has worked to lower the impact of refrigerants on climate change with higher-efficiency refrigerants and system designs.

In 1997 the Air-Conditioning and Refrigeration Institute (ARI) finished a major international testing program entitled the Alternative Refrigerants Evaluation Program (AREP). The AREP report indentified several suitable HFC replacements for HCFC R-22. In the USA and Europe, these HFC replacements are already being widely used. Some of these replacement refrigerants have different operating characteristics than HCFC R-22, but they all eliminate chlorine and potential ozone depletion, leaving climate change as the focus for future regulations and control.

Environmental drivers

There are two factors important to the discussion of the environmental impact of refrigerants: ozone depletion and global warming.

Ozone depletion

The ozone layer surrounding the earth is a reactive form of oxygen 25 miles above the surface. It is essential for planetary life, as it filters out dangerous ultraviolet light rays from the sun. Depleted ozone allows more ultraviolet light to reach the surface, negatively affecting the quality of human, plant, animal and marine life.

Scientific data verifies that the earth's ozone layer has been depleted. The data also verify that a major contribution to ozone depletion is chlorine, much of which has come from the CFCs used in refrigerants and cleaning agents.

Research has shown that even the chlorine found in R-22 refrigerants can be harmful to the ozone layer. The need to protect the earth's ozone has resulted in new government regulations and HFC refrigerants. Since HFCs are chlorine free, they will not damage the ozone layer.

Climate Change

According to the National Academy of Scientists, the temperature of the earth's surface has risen by about one degree Fahrenheit (0.5 degree Kelvin) in the past century[2]. There is evidence that suggests that much of the warming during the last 50 years is because of greenhouse gases, many of which are the by-product of human activities. Greenhouse gases include water vapor, carbon dioxide, methane and nitrous oxide, and some refrigerants. When these gases build up in the atmosphere, they trap heat. The natural greenhouse effect is needed for life on earth, but scientists believe that too much greenhouse effect will lead to climate change. **Figure 1** shows the mechanism of this global warming process.

The greenhouse effect

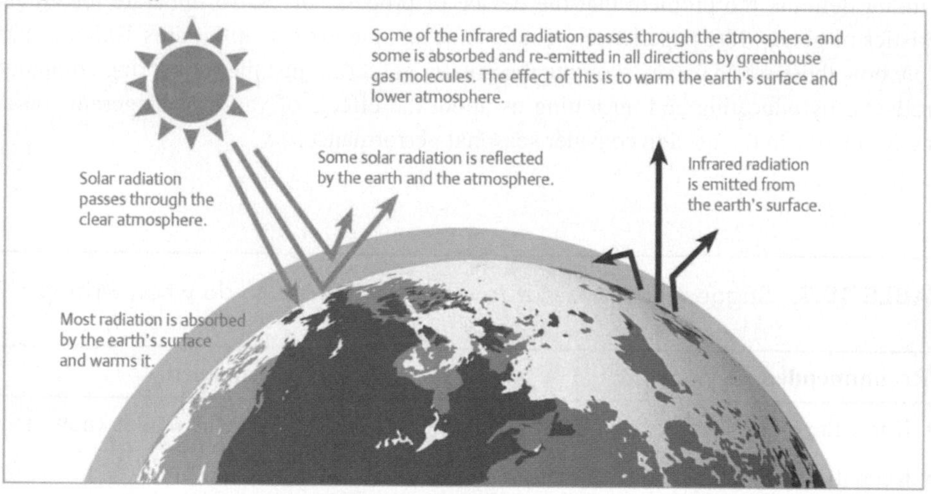

Some of the infrared radiation passes through the atmosphere, and some is absorbed and re-emitted in all directions by greenhouse gas molecules. The effect of this is to warm the earth's surface and lower atmosphere.

Some solar radiation is reflected by the earth and the atmosphere.

Infrared radiation is emitted from the earth's surface.

Solar radiation passes through the clear atmosphere.

Most radiation is absorbed by the earth's surface and warms it.

Figure 1

FIGURE 10.8 | Sample page from a white paper discussing the role of refrigerants in global warming

White paper	Seasonal performances and compression technology

By Jean de Bernardi[1] and Jean-François Le Coat[2]

Background of air conditioning measurement standards

The pioneers in this domain were the US during the 1980s. They established the 3 main founding measurement standards which have been guiding our profession for the last 30 years. The first of these was the ARI 210/240 introduced by the US Department of Energy (DOE) which included for the first time the concept of seasonal COP as applied to the most popular local domestic units (air conditioning units-air/air-12/19 kW, and packaged centralised systems). Three new parameters were in fact introduced at this time. These were the SEER for the cooling mode (Seasonal Energy Efficiency Rating), the HSPF for the heating mode (Heating Seasonal Performance Factor) and the APF which constitutes an annual summary of the two preceding parameters (Annual Performance Factor). At this stage, capacity variation has not been taken into account.

The ARI (Air-Conditioning and Refrigeration Institute) then introduced the ARI 340/360 in response to the need for a measurement standard for larger air/air air conditioning units of over 19 kW, by at the same time creating a first coefficient allowing for the adaptation of the need to the demand to be taken into account with the AIR IPLV (Integrated Part Load Value).
The only thing missing was a measurement standard for water systems. The ARI remedied this situation by introducing the ARI 550/590 for air-cooled or water-cooled water chillers. It then made sense to consolidate an AIR IPLV and introduce a WATER IPLV.

It was only in 2004 that other parts of the world decided to adapt these 3 large US standards to local applications, in particular by taking into account climate differences and the most popular systems in each geographical area.
This is what EUROPE has done with the EUROVENT standard which was mainly built on the foundations of the ARI 550/590 for air-cooled and water-cooled water chillers. A COP exclusively for the cooling mode was introduced: the ESEER based on the WATER IPLV.
Japan and Korea reacted at the same time and were inspired by the ARI 210/240 for domestic air/air reversible systems (<10 kW) and compact units (10 to 28 kW). They made two seasonal performance criteria available to manufacturers: the SEER and the HSPF.

China waited until 2006 to adapt the ARI 340/360 within the framework of air/air exchanges, for either compact units, multi-split units or VRV (Variable Refrigerant Volume) systems. A new annual COP based on the AIR IPLV then emerged.

Labelling, measuring standards and impact on sales

The energy labels which are applied to air conditioning systems all rely on measuring standards to define a performance scale. Therefore it follows that, depending on the standard chosen, which in turn depends on the geographical area, the labels mainly concern the cooling mode, while few of them take into account the heating and cooling modes.

Although energy labels were initially used on a voluntary basis, they are increasingly becoming mandatory all over the world. Some additional voluntary labels also exist to distinguish the best

products (for example the Energy Star label in the USA identifies heat pumps which are 8% more energy efficient than the market average, and 20% more efficient than the models installed). The figure below gives an idea of the main labels encountered in daily life.

FIGURE 10.9 | Sample page from a white paper concerning seasonal ratings of refrigeration systems and compression technologies

PROFESSIONAL TRADE JOURNAL ARTICLES

Another type of writing that is somewhat similar to the background section of a white paper is an informative article for a trade journal. A vast number of professional trade groups publish magazines for their members, a sampling of which is given in Table 10.6.

The audience for a professional trade journal consists of practitioners who are primarily interested in *implementing* an idea rather than expanding the overall body of knowledge about a subject. Like academics, readers of trade journals have sufficient background

TABLE 10.6 | Sampling of professional trade journals with news articles, new product information, and emerging trends in various industries

Name	Coverage	URL
Global Processing Magazine	Worldwide process industry news and product information	<http://www.globalprocessingenews.com/>
Processing Magazine	Domestic process industry news and product information	<http://www.processingmagazine.com/>
Water/Waste Processing Magazine	Water resource and waste process industry news and product information	<http://www.waterwaste.com/>
Power Engineering	Worldwide power industry news and product information	<http://www.power-eng.com/>
Nuclear Power International	Worldwide nuclear power industry news and product information	<http://www.nuclearpowerinternational.com/>
Sustainable Plant	Worldwide news and product information on sustainability issues in manufacturing	<http://www.sustainableplant.com/>
Flow Control Magazine	Technical articles and new product information for fluid handling engineers	<http://www.flowcontrolnetwork.com/>
Industrial Info Reports (IIR)	Industrial project reports across 12 vertical markets, including capital and maintenance project activities, key contacts, unit details, and planned units	<http://www.industrialinfo.com>
Electronic Design Magazine	Internationally focused information, news, and commentary on electronic design solutions	<http://electronicdesign.com/>

to understand the basic terminology without lengthy explanation. In informative articles, company names can be mentioned, but the focus should be on the technology and not on the company product. Other sections of the trade magazine are reserved for review of new products.

In the sample trade article given in Figure 10.10, the focus is on advances in and applications of signal compression technology. A novice will have difficulty understanding the technical details in the article, but with a bit of background research, most engineering students should be able to extract a fair understanding of why signal compression is needed.

Compression Reduces Memory Bottlenecks, From Supercomputing To Video

<u>Al Wegener</u>
June 05, 2012

In 1899, a writer for the popular magazine *Punch's Almanack* famously wrote, "Everything that can be invented has been invented." In 1943, IBM founder Thomas J. Watson allegedly said, "I think there is a world market for maybe five computers." And in 1981, Bill Gates notoriously predicted, "640 kbytes ought to be enough for anybody." Today we all laugh at these off-the-mark predictions, as they demonstrate the difficulty of predicting technology's future.

Which of today's technical prognostications might fall short? We engineers are confidently told about the end of Moore's Law, the impossibility of automating software to extract parallelism, and (my personal favorite) the futility of expecting commercially useful results from new compression research. I'd like to refute that last prediction with two examples that illustrate how a novel compression technology is yielding big payoffs in two radically different applications.

Super Supercomputing

Supercomputing is facing the proverbial memory wall as increasing multicore CPU and GPU core counts strain already overloaded DDRx memory, PCI Express, and InfiniBand bus bottlenecks *(see "<u>HPC and 'Big Data' Apps Tap Floating-Point Number Compression</u>")*. Let's first examine how the memory and bus bandwidth walls are challenging one of supercomputing's key semiconductor components, graphics processing units (GPUs).

In the mid-2000s, Nvidia repurposed GPUs that it originally developed for gaming and video applications toward high-performance computing (HPC). Using Nvidia's CUDA or OpenCL software frameworks, HPC algorithms coded in ANSI C can be parallelized to utilize hundreds of floating-point rendering engines per GPU. The industry calls this technology general-purpose GPU (GPGPU) computing.

GPGPU is used for such diverse HPC applications as finite element analysis, computational fluid dynamics, seismic processing, and weather forecasting. GPGPU

FIGURE 10.10 | A sample of a technical article from a professional trade magazine. Note that while there appears to be no references, the actual article has live links to the referenced paper. Please see Wegener (2012) to access the original website.

technology has also become a key enabler in HPC's push towards Exascale (1018 floating-point operations per second), supercomputing's target goal for 2018.

To achieve Exascale performance, GPGPU technology must overcome daunting I/O challenges both to GDDR5 memory and PCI Express bus bandwidth. A 2008 DARPA report identified the need for a 16x improvement in memory bandwidth and a 100x increase in bus bandwidth if Exascale performance is to be achieved. Such order-of-magnitude I/O improvements will not be achieved by 2018 simply with evolutionary memory and bus standards, such as the 2x increase in memory bandwidth from DDR3 to DDR4, or the 2x increase in PCI Express bus bandwidth from Gen2 to Gen3.

Disturbingly, GPU memory and bus bandwidth per core has actually decreased by a factor of six since 2003, because the number of floating-point GPU cores rose by 24x (from 64 to 1536) while memory and bus rates only increased by a factor of four. Similar technology forces kept CPU per-core I/O rates flat since the mid-2000s for multicore CPUs like Intel's Sandy Bridge and AMD's Opteron. A novel compression approach that compensates for this missing I/O acceleration factor of 2x to 6x could accelerate struggling CPU and GPU interfaces that carry HPC's floating-point operands.

Compression's improvements to memory and bus bandwidth can be quantified. In 1948, Bell Labs researcher Claude Shannon succinctly described a theory of information content illustrating the tradeoffs between compression ratio and distortion using rate-distortion (RD) curves. RD curves succinctly quantify where compression algorithms reach their limits. However, only certain RD curves are affordable in real-time applications where compression and decompression must operate at tens or hundreds of megahertz. Thus, RD curves have an implicit third axis: complexity.

Consumer Compression

In today's consumer electronics devices, video compression algorithms must simultaneously exceed a target compression ratio with acceptable distortion while implementation complexity meets a silicon or MIPS constraint. The H.264 compression algorithm is broadly used for video distribution across wireless networks, both for downloads of streaming video and uploads of user-captured video. H.264 compression exhibits excellent quality at 20:1 to 30:1 compression by using inter-frame correlations between groups of pixels, typically 16- by 16-pixel units called macroblocks (MBs).

During compression, a memory-intensive process called Motion Estimation (ME) compares the current MB to hundreds of MBs stored in previous frames. The most similar MB in previous frames (and also interpolated by a factor of four) is then encoded using a motion vector and the quantized transform coefficients of the MB

FIGURE 10.10 | Continued

error signal. The corresponding process during H.264 decompression is called Motion Compensation (MC).

For 1080p HD video displayed at 60 frames/s, H.264 DDR memory bandwidth requirements exceed 3 Gbytes/s. Rather than accelerating video signal processing operations, recent H.264 research has tried to optimize cache memory access to megabytes using special software or hardware techniques. As in supercomputing, I/O is now the critical bottleneck in H.264 video compression. Could the same I/O acceleration technology that provides a 2x to 6x speed-up for supercomputing be used to accelerate H.264 memory bottlenecks?

A new Samplify compression technique called APAX ("APplication AXceleration") efficiently reduces memory bottlenecks for both supercomputing and video. For HPC applications, APAX floating-point compression reduces memory and storage bandwidth by 2x to 6x while generating the same results for climate modeling, weather forecasting, seismic processing, and 3D rendering. For video and graphics applications such as H.264 and rendering, APAX's integer compression technology reduces frame buffer, texture, and mesh traffic across memory and bus interfaces by a user-selected factor, from lossless to 8:1, with visually imperceptible results.

In conclusion, an emerging compression technology that accelerates supercomputing's push toward Exascale computing also accelerates video decoding, one of compression's most well-researched and ubiquitous application areas.

Note: Al Wegener is the CTO and founder of Samplify Systems, a startup in Santa Clara, Calif. He holds 17 patents and is named on additional Samplify patent applications. He earned a BSEE from Bucknell University and an MSCS from Stanford University.

Used by permission of Electronic Design, a Penton Media publication

FIGURE 10.10 | Continued

SUMMARY

The types of writing explored in this chapter are common professional assignments for engineers working both in industry and in academia. Review articles, site visit reports, white papers, and articles in professional trade journals all require the writer to summarize and synthesize information from a variety of sources. The writing requires a clear understanding of the needs of the audience and the ability to write objectively and persuasively at the same time. Not every engineer will write all of these types of reports. However, gaining familiarity with these types of writing as a student is good preparation for the workplace. Studying the writing style used in these communications will also contribute to the development of an engineer's professional voice, especially if sufficient attention is given to the purpose and audience of these genres.

EXERCISES

1. Select a topic appropriate for your background in your field of study (ask for your instructor's guidance). Write out a question about this topic to investigate. Read relevant sections of your textbook or other textbooks in the library to gain background. Revise your topic as needed. Find 6 to 10 primary sources on this topic, with at least 2 print references and at least 2 online references. Evaluate each source using the reading log template in Figure 10.2. Based on the reading log assessment, develop an appropriate method of grouping the sources.

2. For the topic selected in question 1, or another topic of your choice, use the Web of Science to find a relevant reference. Then use this reference to find newer, related references in which the first reference was cited. Now look at the cited references in the initial reference and work backwards to assemble a list of older sources for further investigation.

3. Prepare a literature review of a topic agreed upon by you and your instructor.

4. Use a computer to access the site visit report found at <http://www.cpwr.com/pdfs /Perini/d000887.pdf>. Read the first page of the report. Is it clear to you why the site visit was undertaken? Would knowing that the CPWR is affiliated with the AFL-CIO (trade union) help you understand the purpose of the visit? Now read page 3 and answer the same question: What was the purpose of the visit? Read the remainder of the report and comment upon the tone and what you believe was the intent of the authors of this site visit report.

5. Go to the site <http://www.nvidia.com/object/white-papers.html> and select a white paper to read. How well does the selected white paper fit the format described in the section in this chapter about writing white papers? Does the paper start by identifying a problem and explaining why it is a real problem? Is there an objective discussion of alternative solutions, and is there any evidence of economic return for the company? Comment on any differences in format from the white paper format presented in this chapter and suggest possible reasons.

6. Download a technical trade journal article from one of the sites listed in Table 10.6 or from another professional trade publication site. Often these articles follow the same format as white papers. How well does the format of the selected article fit that described in the section about writing white papers? Does the article start by identifying a problem and explaining why it is a real problem? Is there an objective discussion of alternative solutions and is there any evidence of economic return for the company? Comment on any differences in format from white paper format and suggest possible reasons.

7. Access a technical trade journal article from one of the sites listed in Table 10.6 or from another professional trade publication site. Try to read the article. How much of it did you understand? Wait a day and then try to re-read the article. How much of it were you able to understand after a day of not looking at it? Now find the definition of words and acronyms that are prohibiting your understanding. Finally, read the article a third time and paraphrase the contents of the article in a three-to-four-paragraph report.

REFERENCES

AAM. 2013. American Association of Museums Peer review manual. Washington, DC: American Alliance of Museums. Accessed 6 May 2013 at <http://www.aam-us.org/docs/peer-review/prmanual.pdf?sfvrsn=0>.

AIT. 2005. Writing a literature review, Language Center, Asian Institute of Technology. Accessed 3 March 2013 at <http://web.pdx.edu/%7Ebertini/pdf/literature_review.pdf>.

Canright, C. 2011. White paper basics: The dos, don'ts, whys, whats, and hows of white papers (pdf), Canright Communications, Chicago. Accessed 3 March 2013 at <http://www.canrightcommunications.com/samples/Canright_WhitePaperBasics.pdf >.

de Bernardi, J., and Le Coat, J.F. 2010. White paper: Seasonal performances and compression technology, Danfoss A/S (RA Marketing). Accessed 3 March 2013 at <http://variablespeed.danfoss.com/NR/rdonlyres/8B522814-BD3E-4530-9CC8-8A1232FE7BFF/0/FRCCPE001A102Seasonalperformances.pdf>.

Eight Rules. 2005. Eight rules for creating great white papers, KnowledgeStorm, The Content Factor, Marietta, GA. Accessed 6 May 2013 at <http://www.idemployee.id.tue.nl/g.w.m.rauterberg/lecturenotes/Eight-Rules-for-Writing-Great-White-Papers.pdf>.

Emerson Climate. 2005. Refrigerants for commercial refrigeration applications, Emerson Climate Technologies, Inc. Accessed 6 May 2013 at <http://www.emersonclimate.com/en-us/WhitePapers/ECT2005ECT162.pdf>.

Finio, B. 2007. Performance characteristics of a vortexing jet pump, Undergraduate Honors Thesis, Department of Mechanical Engineering, Bucknell University.

Gittleman, J., Ellenberger, D., Stafford, P., Gillen, M., and Kiefer, M. 2008. Worksite assessment team site visit report for City Center and Cosmopolitan construction projects, Las Vegas, Nevada. Accessed 3 March 2013 at <http://www.cpwr.com/pdfs/Perini/d000887.pdf >.

Kemp, A. 2005. White paper writing guide (pdf), Impact Technical Publications. Accessed 3 March 2013 at <www.impactonthenet.com/wp-guide.pdf>.

Kemp, J. 2010. Creating effective white papers, Kemp Copywriting. Accessed 6 May 2013 at <http://kempcopywriting.com/creating-effective-white-papers.html>.

McAuliffe, S. 2002. Sample site visit worksheet, The David and Lucile Packard Foundation. Accessed 3 March 2013 at <http://www.cof.org/files/Documents/WebNotebook/July2003/Grants/Sample_Site_Visit_Worksheet.pdf>.

Monash University. 2013. Writing literature reviews, Monash University, Language and Learning Online. Accessed 6 May 2013 at <http://www.monash.edu.au/lls/llonline/writing/general/lit-reviews/index.xml>.

Mongan-Rallis, H. 2006. Guideline for writing a literature review. Accessed 6 May 2013 at <http://www.duluth.umn.edu/~hrallis/guides/researching/litreview.html>.

Smith, E., Duckett, K., Bankston, S., Classes, J. Orphanides, A., and Baker, S. 2012. Literature reviews: An overview for graduate students, North Carolina State University Libraries. Accessed 3 March 2013 at <http://www.lib.ncsu.edu/tutorials/lit-review/>.

Soledad, A.-B. 2011. Writing Chapter 2: Review of the literature, NOVA Southeastern University, Abraham S. Fischler School of Education. Accessed 6 May 2013 at <http://www.fischlerschool.nova.edu/Resources/uploads/app/35/files/ARC_Doc/writing_chpt2_Litreview.pdf>.

Stelzner, M.A. 2007. Writing white papers. Poway, CA: WhitePaperSource Publishing.

University of North Carolina. 2012. Literature review. University of North Carolina Writing Center. Accessed 3 March 2013 at <http://writingcenter.unc.edu/handouts/literature-reviews/>.

University of Melbourne. 2012. Site visit reports for engineers. The University of Melbourne, Academic Skills Unit. Accessed 3 March 2013 at <http://services.unimelb.edu.au/academicskills /writing/?a=454073>.

University of Ottawa. 2007. Writing a literature review. Academic Writing Help Centre, Graduate Writing, University of Ottawa. Accessed 3 March 2013 at <http://www.sass.uottawa.ca/writing/kit /grad-literature-review.pdf >.

University of Wisconsin. 2011. Learn how to write a review of literature, University of Wisconsin -Madison, The Writing Center. Accessed 3 March 2013 at <http://writing.wisc.edu/Handbook /ReviewofLiterature.html>.

Wegener, A. 2012. Compression reduces memory bottlenecks, from supercomputing to video, Electronic Design Magazine. Accessed 3 March 2013 at <http://electronicdesign.com/article/digital /compression-reduces-memory-bottlenecks-supercomputing-video-74003>.

Part 4

ORAL PRESENTATIONS AND POSTER PREPARATION

©iStockphoto.com/melhi

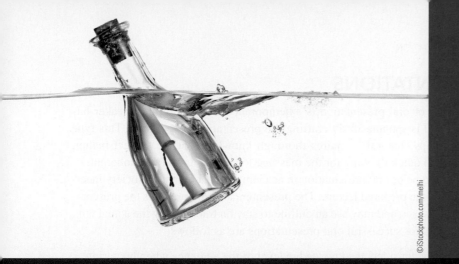

Chapter
11

ORAL PRESENTATIONS AND RUNNING MEETINGS

Several formats for the written communication of technical information have been discussed in earlier chapters, including the technical report (Chapter 4), correspondence (Chapter 7), proposals (Chapter 8), specifications (Chapter 9), and overview reports (Chapter 10). In these written formats, readers can scan, skip, and re-read sections of the report at their own pace, at any time. On the other hand, there is no interaction between the reader and the writer; the only connection between them is the text on the page.

Oral presentations differ from written communications; the audience relies heavily on the speaker's delivery to provide information. The audience cannot skip to the conclusions to assess the value of the presentation, nor can they return to topics covered previously if they missed some key information. Therefore, not only must the speaker be knowledgeable about the subject matter, he or she must also be skilled in public speaking for the presentation to be successful.

Parts of this chapter have been adapted from *A Student Handbook for Writing in Biology*, Chapter 8. Used with permission from the publishers, Sinauer Associates, Inc. and W. H. Freeman & Co.

PRESENTATIONS

The most natural form of oral presentation is extemporaneous, whereby the speaker has no prepared outline and is spontaneously crafting the presentation in real time. This type of oratory, while perhaps "natural," requires thorough knowledge of the subject matter, substantial practice, and quick thinking for the message to be meaningful and coherent.

The most common type of oral presentation in academia and at technical society meetings and conferences is the prepared lecture. The presenter carefully prepares the presentation beforehand, rehearses it, and may use an outline to stay on track during the actual talk. Some general principles for successful oral presentations are as follows:

- The speaker carefully considers the knowledge level of the audience when preparing the visuals and the delivery.
- The speaker establishes a good rapport with the audience.
- The presentation is well organized, coherent, and focused.
- The presentation leaves listeners satisfied that they learned something new.
- The visuals are simple and legible, and they help listeners focus on the important points.

Organization

Oral presentations are not unlike technical papers in their structure, but they are much more selective in their content. As in a technical paper, the **introduction** captures the audience's attention, explains what motivated the work, provides background information on the subject, and clarifies the objectives of the work (Table 11.1). The **body** is a condensed version of the Apparatus and Procedures and Results sections, which contains only the details necessary to emphasize and support the speaker's conclusions. If the focus of the talk is on the results, then the speaker spends less time on the methods and more time on visuals that highlight the findings.

The visuals may be simpler than those prepared for a journal article because the audience may only have a minute or two to digest the material in the visual display (in contrast to a journal article, where the reader can re-read the paper as often as desired). Finally, the **closing** is comparable to the Discussion and Conclusions sections because here the speaker summarizes the objectives and results, states conclusions, and emphasizes the take-home message for the audience.

TABLE 11.1 | Comparison of the structure of an oral presentation and a journal article

Oral Presentation	Journal Article
	Abstract
Introduction	Introduction
Body	Apparatus and Procedures, Results
Closing	Discussion and Conclusions
	References

There is no distinct Abstract or References section in an oral presentation, but published work should be cited on the slides. If the presentation is part of a meeting, an abstract may be provided in the program. If listeners are interested in the references, they should contact the speaker after the talk.

After the closing, speakers may include an acknowledgments slide in which they recognize sources of funding, their research adviser, and others who helped them with their work.

Preparation of a Presentation

In preparing a presentation, some speakers will write out the entire presentation word for word, while most experienced speakers will work from an outline of bulleted points. Preparing both the written and oral parts of a presentation is an iterative process that requires multiple rounds of drafts and revisions before arriving at the final polished form.

Plan Ahead   Before you start writing the content of your presentation, make sure you know the following:

- Your audience. What do they know? What do they want to know?
- Why you are giving the talk, not just why you did the experiment. Are you informing the audience, or are you trying to persuade them to adopt your point of view?
- The speaking environment. How large is the audience? What audiovisual equipment is available (chalkboard, overhead projector, computer with projection equipment)?
- How much time do you have? A one-hour presentation will certainly require greater preparation and attention to detail than a 10- to 15-minute talk.

Write the Text   An oral presentation will actually take you longer to prepare than a paper. D'Arcy (1998) breaks down the steps as follows:

- Procrastination   25%
- Research   30%
- Writing and creating visuals   40%
- Rehearsal   5%

To overcome procrastination, follow the same strategy as for writing a technical report: Write the body first (Apparatus and Procedures and Results sections), then the introduction, and finally the closing.

- Write the main points of each section in outline form.
- Use the primary literature (journal articles) as well as secondary sources such as textbooks to provide background in the introduction and supporting references in the closing.
- Transfer your outline to note cards. Use only one side of the note cards, and limit yourself to one main topic per note card. Note where you plan to use visual aids.
- If you plan to use a presentation graphics program such as Microsoft PowerPoint in addition to note cards, make sure each slide is just an outline of what you plan to say. Use the slides to *support* what you say, not *repeat* what you say.
- Keep it short and simple. Tell your audience what you are *going* to tell them, *tell* them, and then tell them what you *told* them. Focus on your main take-away message.

Prepare the Visuals   Effective visual aids help the audience to visualize the data and to see relationships among variables. They clarify information that would otherwise involve lengthy descriptions.

Effective visuals are simple, legible, and organized logically. They capture the listener's interest, increase comprehension, and focus attention on the important findings. When deciding on which visuals to use, keep in mind that most audiences prefer figures to tables. For guidelines on what kind of figure to use, see Chapter 6. For instructions on how to make different kinds of graphs in Microsoft Excel, see Appendix II. Photos, diagrams, and drawings are effective ways to show physical rather than numerical results. Mechanisms are much easier to visualize when the verbal description is accompanied by a drawing.

Because the audience will have only a limited time to process the information on a visual, it is best to use the simplest possible graphics to convey the message. Furthermore, too much detail may make the visual illegible to people in the back of the room.

Take the time to make good visuals for your oral presentation. We receive more than 83% of our information through sight and only 11% through hearing. Without visuals, your listeners will forget most of your presentation within 8 hours. These statistics help explain why Microsoft PowerPoint and similar programs have become so popular. PowerPoint allows you to create a slide show containing text, graphics, and even animations. Power-Point has the same advantages as word processing programs: It is easy to make revisions and to send an electronic copy of the presentation to a colleague to review (or to yourself as a backup).

Because PowerPoint is so easy to use, you may be tempted to write every word of your presentation on the slides. Do not write out every word! Follow these tips for preparing visuals:

- **Do not put too much information on any one slide or transparency.** A good rule of thumb is to cover only one concept per slide or transparency. For more complicated concepts, consider using the animations feature in PowerPoint, which makes it possible to display individual components on a slide sequentially rather than all at once (see Appendix III). When using transparencies, overlays serve a similar function.
- **Keep the wording simple.** Instead of using full sentences, bullet keywords or key points (Figure 11.1). Limit the number of key points. Some authorities suggest a maximum of three bullet points per slide. You may use a few more, but certainly no more than six.
- **Make sure the lettering can be read from anywhere in the room.** Use no smaller than 18–24 point (abbreviated as pt) sans serif font and black type on a white background for best contrast. The light background makes the room much brighter, which has the dual advantage of keeping the audience awake and allowing you to see your listener's faces. Use color sparingly in text slides; a bright color or a larger font are good ways to emphasize a point.
- **Use the same font and format for each slide for a consistent look.** Avoid "cute" graphics that might detract from the technical message.
- **Keep graphs simple.** Make the numbers and the lettering on the axis labels large and legible. Choose symbols that are easy to distinguish. Instead of following the conventions used in journal articles, modify the figure format as follows (compare Figures 11.2 and 11.3):
  » Position the title *above* the figure and *do not* number it. The title should reflect your most important finding.
  » Label each plotted data set instead of identifying the data symbols in a legend. The labels may be made using text boxes in Excel, Word, or PowerPoint.

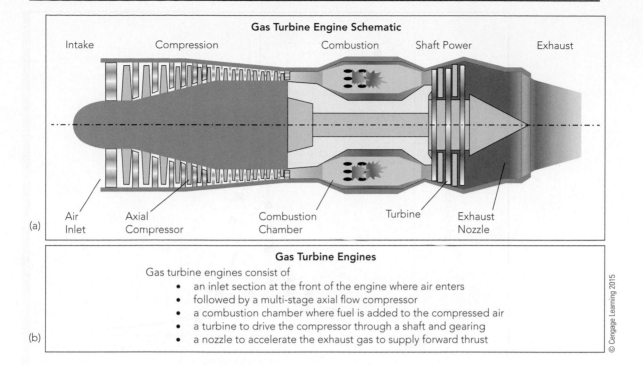

Gas Turbine Engine Schematic

Intake   Compression   Combustion   Shaft Power   Exhaust

(a) Air Inlet   Axial Compressor   Combustion Chamber   Turbine   Exhaust Nozzle

Gas Turbine Engines

Gas turbine engines consist of

- an inlet section at the front of the engine where air enters
- followed by a multi-stage axial flow compressor
- a combustion chamber where fuel is added to the compressed air
- a turbine to drive the compressor through a shaft and gearing
- a nozzle to accelerate the exhaust gas to supply forward thrust

(b)

© Cengage Learning 2015

FIGURE 11.1 | **(a)** An effective slide with a labeled diagram of gas turbine engine components and **(b)** a less effective list of the components of a gas turbine engine. The large amount of text on Slide b will have the audience reading instead of listening to the speaker; the absence of a diagram in Slide b slows comprehension.

Both of these modifications make it easier for the listener to digest the information within the short time the graph is displayed during the presentation.

- **Use symbols consistently in all of your graphs.** For example, if you use a triangle to represent "27.1% damping" in the first graph, use the same symbol for the same condition in all other graphs.
- **Appeal to your listeners' multiple senses.** Use images from the Internet, recordings of machine, traffic, or flow noises (as appropriate), and animations or movie clips if they help make your point. Make sure these "extras" do not detract from your take-home message and that they fit within the time limit of your presentation.
- **Allow at least 20 seconds for each slide,** more time for complicated slides. Many experienced presenters plan 1 minute or more per slide, because they will have substantial commentary for each slide. Complex slides may require up to 5 minutes of explanation.
- In some venues, the audience may not be able to see the bottom third of the screen, for example, because all the seats are on the same level and viewers' heads in front obstruct the view of viewers in the back. In such circumstances, be prepared to take extra time to explain the graphics for those who have only a partial view.

Rehearsal   After you have written the presentation and prepared the visuals, you must practice your delivery. Give yourself plenty of time so that you can run through your

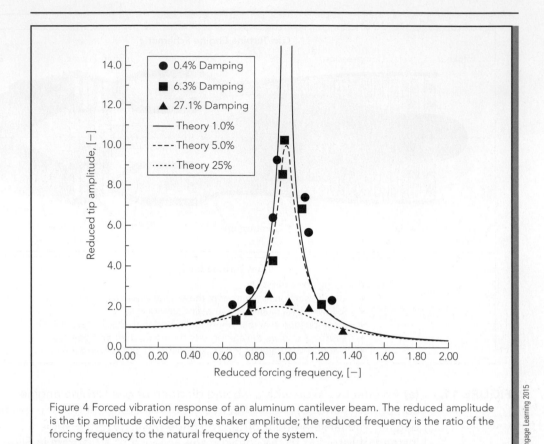

Figure 4 Forced vibration response of an aluminum cantilever beam. The reduced amplitude is the tip amplitude divided by the shaker amplitude; the reduced frequency is the ratio of the forcing frequency to the natural frequency of the system.

© Cengage Learning 2015

FIGURE 11.2 | Example of a figure formatted for a technical report or paper. A legend is needed for more than one data set. The caption with figure number is positioned below the graph. All lines and symbols are black for best contrast in a black and white publication.

presentation several times and, if possible, do a practice presentation in the same room where you will hold the actual presentation. Other tips to consider are as follows:

- Go to a place where you can be alone and undisturbed. Read your presentation aloud, paying attention to the organization. Does one topic flow into the next, or are there awkward transitions? Revise as needed.
- Time yourself. Make sure your presentation does not exceed the time limit.
- After you are satisfied with the organization, flow, and length, ask a friend to listen to your presentation. Ask him or her to evaluate your poise, posture, voice (clarity, volume, and rate), gestures and mannerisms, and interaction with the audience (eye contact, ability to recognize if the audience is following your talk). Use an Evaluation Form (see Tables 11.2a and 11.2b, for example) as a guide.

It is natural to be nervous when you begin speaking to an audience, even an audience of your classmates. Adequate knowledge of the subject, good preparation, and sufficient rehearsal can all help to reduce your nervousness and enhance your self-assurance.

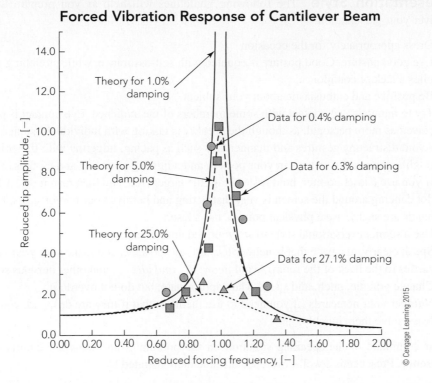

Forced Vibration Response of Cantilever Beam

FIGURE 11.3 | Re-formatted graph for an oral presentation. An abbreviated title is positioned above the graph, and the caption including the figure number is omitted. The legend has been eliminated, and each data set and theoretical curve is labeled for quick identification. The colored symbols make the visual more appealing.

Delivery The importance of the delivery cannot be overstated: Listeners pay more attention to body language (50%) and voice (30%) than to the content (20%) (Fegert et al., 2002). Remember that you must establish a good rapport with the audience in order for your oral presentation to be successful.

Different parts of your presentation may warrant changes in inflection, pace, and sentence length. Using these techniques advantageously keeps the audience engaged. Deliver your talk with a high-energy level and enthusiasm for your topic. Use hand gestures and facial expressions to accent your words, as both are natural accompaniments of speech.

Often, presenters experience anxiety or stage fright. Recognize that anxiety can be both beneficial and detrimental. A bit of anxiety keeps you at a high-energy level, but too much prevents you from performing effectively. **Stage fright** is a fear of inadequacy when facing an audience. The obvious antidote for stage fright is to prepare your talk thoroughly, fine-tune it for its purpose, and, when possible, get to know members of the audience beforehand. Immediately before your talk, use relaxation techniques such as listening to music, breathing slowly and deeply for a minute or two, or visualizing a pleasant scene.

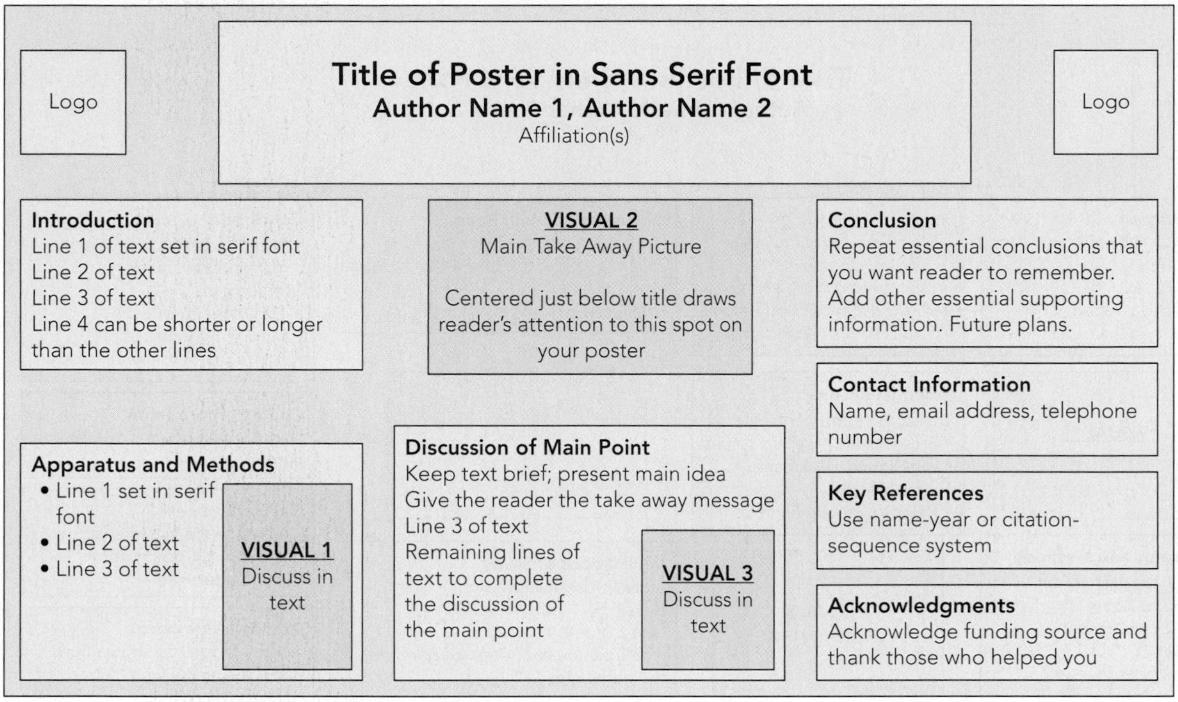

Title of Poster in Sans Serif Font
Author Name 1, Author Name 2
Affiliation(s)

Logo   Logo

Introduction
Line 1 of text set in serif font
Line 2 of text
Line 3 of text
Line 4 can be shorter or longer
than the other lines

VISUAL 2
Main Take Away Picture

Centered just below title draws
reader's attention to this spot on
your poster

Conclusion
Repeat essential conclusions that
you want reader to remember.
Add other essential supporting
information. Future plans.

Contact Information
Name, email address, telephone
number

Apparatus and Methods
• Line 1 set in serif
  font
• Line 2 of text
• Line 3 of text

VISUAL 1
Discuss in
text

Discussion of Main Point
Keep text brief; present main idea
Give the reader the take away message
Line 3 of text
Remaining lines of
text to complete
the discussion of
the main point

VISUAL 3
Discuss in
text

Key References
Use name-year or citation-
sequence system

Acknowledgments
Acknowledge funding source and
thank those who helped you

© Cengage Learning 2015

FIGURE 12.1 | One possible layout for a three-column landscape format poster with a pastel background. Individual elements may be lightly framed or unframed. Keep the use of framing consistent.

narrower portrait orientation because they can fit more posters per horizontal distance on an available mounting space. A poster might also be presented on a projection screen or a large-format TV in a classroom environment.

Appearance

The success of a poster presentation depends on its ability to attract people from across the room. Interesting graphics and a pleasant color scheme are good attention-getters, but avoid "cute" gimmicks. Posters should be presented in a serious and professional manner so that participants will take the work seriously.

Organize the sections so that information flows from top left to bottom right. Left align text rather than centering it or using full justification. The smooth left edge provides the reader with a strong visual guide through the material. Columns of text facilitate quick scanning of the text, similar to the format found in newspapers and news magazines.

Avoid crowding. Large blocks of text turn off viewers; instead, use bulleted lists to present your objectives and conclusions clearly and concisely. Use blank space to separate sections and to organize the poster for optimal flow from one section to the next.

Use appropriate graphics that communicate the data clearly. Use three-dimensional graphs *only when necessary* for three-dimensional data. Proofread figure legends and running text carefully and correct spelling and grammatical errors.

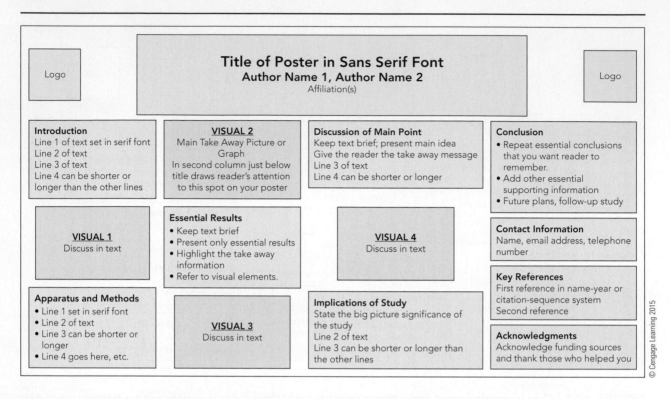

FIGURE 12.2 | One possible layout for a four-column landscape format poster on a white background. Individual elements may be lightly framed or unframed, but keep framing consistent.

Colored borders around graphics and text enhance contrast, but keep framing to a minimum. **Framing** is the technique whereby the printed material is mounted on a piece of colored paper, which is mounted on a piece of different-colored paper to produce colorful borders. Use borders judiciously so that they do not distract from the poster content. Heavy borders and thick framing lines disrupt the visual flow.

If the poster is prepared by gluing multiple pieces of paper with text and graphics onto a poster board, use adhesive spray or a glue stick. These tend to have fewer globs and bulges than liquid glues, and the edges adhere better.

Font (Type Size and Appearance)

Remember that most readers of the poster will be at least 3 feet and up to 10 feet away, so the print must be large and legible. Sans serif fonts like Arial or Calibri, are good for titles, but serif fonts like Times, Palatino, and Garamond are much easier to read in extended blocks of text. The **serifs** (small strokes that embellish the character at the top and bottom) create a strong horizontal emphasis, which helps the eye scan lines of text more easily. Select one sans serif font for titles and headings and one serif font for the text and captions. Not all experts agree on this suggestion, but it appears to be a majority view. The font sizes you select should be scaled in proportion to the poster width. Recommended font sizes are given in Table 12.1.

Make the title Mixed Type or ALL CAPS in 72 to 100 point **bold**, as shown below. Mixed type has the advantages of being easier to read and taking up less space than all caps.

TABLE 12.1 | Range of fonts and styles for poster components. Preferred values are bolded and underlined.

Poster Width (inch/cm) Font →	Title Font and Size (Pt) BOLD	Author Font and Size (Pt) BOLD	Affiliation Font and Size (Pt) BOLD	Header Font and Size (Pt) Regular	Body Font and Size (Pt) Regular	Caption Font and Size (Pt) *Italic*
96/244	Sans serif **Mixed** or ALL CAPS 100–120	Serif or sans serif Mixed 60–78	**Sans serif** or serif Mixed 54–70	Sans serif **ALL CAPS** or Mixed 44–60	Serif Sentence case 36–48	Serif Sentence case 32–44
56/142	Sans serif **Mixed** or ALL CAPS 86–102	Serif or sans serif Mixed 52–66	**Sans serif** or serif Mixed 48–62	Sans serif **ALL CAPS** or Mixed 40–52	Serif Sentence case 32–42	Serif Sentence case 30–40
44/112	Sans serif **Mixed** or ALL CAPS 78–94	Serif or sans serif Mixed 48–60	**Sans serif** or serif Mixed 44–54	Sans serif **ALL CAPS** or Mixed 36–46	Serif Sentence case 30–38	Serif Sentence case 28–36
32/81	Sans serif **Mixed** or ALL CAPS 72–86	Serif or sans serif Mixed 42–56	**Sans serif** or serif Mixed 40–50	Sans serif **ALL CAPS** or Mixed 34–44	Serif Sentence case 26–34	Serif Sentence case 24–32

Do not use all caps if there are case-sensitive words in the title, such as pH or Ph.D. Limit title lines to 65 or fewer characters. Taking into account the paper size and title length, select a font size for the title so that the title occupies between half and two-thirds of the poster width.

Calibri 72 Pt
CALIBRI
Arial 72 Pt
ARIAL

Authors' names should be 36 to 56 point bold, sans serif or serif font, mixed type:

Times Roman 48 Pt

Authors' affiliations can be slightly smaller than or the same size as the authors' names. The section headings can be 28 to 48 point **bold**, sans serif or serif font, mixed type:

Lucida Sans 28 Pt

Garamond 28 Pt

The text itself should be no smaller than 24 point, serif font, mixed type, and not bold face:

Times 24 Pt

When choosing a font size from the range given in Table 12.1, be consistent in keeping all elements near the maximum, near the minimum, or around a constant average value between the minimum and the maximum. The scaling in Table 12.1 is similar to the scaling algorithm used in PowerPoint. This scaling can be easily demonstrated: Create a poster in PowerPoint by choosing a poster size and creating the poster content. When resizing the poster, the font size changes in proportion to the poster size.

Mounting Options

Ask the conference organizer (or your instructor) about how posters will be displayed at the poster session. Some possibilities include a pinch clamp on a pole, an easel, a table for self-standing posters, large white boards for taping on posters, and cork bulletin boards to which posters are affixed with pushpins. Plan accordingly. Some experienced poster presenters always carry a roll of strong, clear tape as a backup if other display options do not materialize.

Transporting your poster also poses challenges. Plastic or cardboard mailing tubes are practical, but they may delay the security inspection at airports. Some conference attendees mail their posters to the conference venue. Other experienced presenters email their poster files to an office supply store near the conference venue and then pick up the printed poster upon their arrival at the venue. Be aware that most poster-size plotters use water-soluble ink, so laminating your poster provides an added level of insurance that the poster will look as intended upon arrival at the conference.

MAKING A POSTER IN MICROSOFT POWERPOINT 2010

The vast majority of posters for professional meetings and conferences are prepared using PowerPoint. Some reasons for PowerPoint's widespread use are

- Availability (it comes with the Microsoft Office Suite)
- User familiarity with the commands (from creating slide shows for oral presentations–see Chapter 11)
- Ease of customizing the design and layout
- Ease of revising the content
- Ease of collaboration among colleagues in different locations (via email attachments or online services such as Google Docs)
- Professional appearance of the final product

There are, however, other software options for creating posters. Purrington (2012) recommends several text layout programs (including LaTeX and SCRIBUS) and graphics software packages (such as Illustrator and Omnigraffle) for poster preparation. The utility of these packages is often outweighed by the time and effort required to learn how to use them.

When making posters in PowerPoint, the entire poster is contained on one slide, whose size is set in the **Page Setup** dialog box. Text is written in textboxes, which are inserted along with visuals (images and graphs) on the slide as desired. The following sections describe some of the basic tasks you will carry out when making your poster in PowerPoint 2010.

As is the case with all computer files, save your file frequently when making posters in PowerPoint. Consider cloud-based storage as an extra level of backup security.

Page Setup

1. Click **Home** (→Slides) | **Layout** and choose **Blank**.
2. Click **Design** (→Page Setup) | **Page Setup** (see Figure 12.3). Then enter the desired poster dimensions, as shown in Figure 12.4. Portrait or landscape orientation is chosen automatically based on the numbers you enter.

    When entering numbers, keep in mind that posters commonly have the same aspect ratios as other media, such as photographs and iPads (4:3), flat screen TVs and computer screens (16:9), and Android tablets (16:10). Therefore, when deciding what numbers

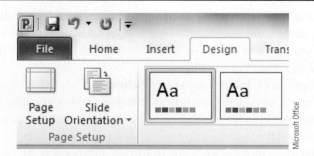

FIGURE 12.3 | Select **Page Setup** from the **Design** tab

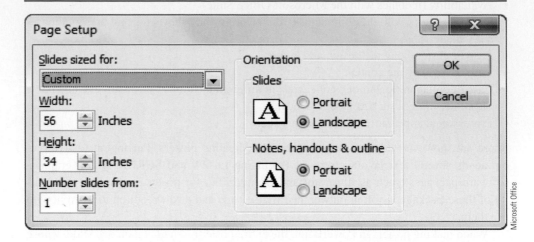

FIGURE 12.4 | Poster dimensions are specified in the **Page Setup** dialog box using the **Design** tab

to enter for the width and height in PowerPoint (see Figure 12.4), make sure the aspect ratio is close to 4:3, 16:9, or 16:10 because that is what people are used to seeing in the media.

Common dimensions for landscape formatted posters are 42 × 32 inches, 48 × 27 inches, 48 × 36 inches, and 56 × 34 inches. If you are using a large-format plotter, the maximum size is usually around 56 × 34 inches, but professional poster-printing services can be engaged to produce even larger format posters, up to the size of billboards.

3. If desired, select a theme or a color for the background on the **Design** tab. Keep any background simple and unobtrusive.

Adding Text, Images, and Graphs

1. First plan a rough layout for the poster (see the section on "Layout"). Then insert textboxes, images, graphs, and other elements at the appropriate locations.
2. To add text to the poster, click **Insert** (→Text) | **Textbox** for each section or block of text. Aim for a consistent look by using the same family of fonts. Adjust the font size for the title, authors' names and affiliations, section headings, and text by making the appropriate selections on **Home** (→Font). To change the properties of the textbox itself, right-click it and select **Format Shape**.
3. To insert images, click **Insert** (→Images) | **Picture**.
4. To insert a graph, copy and paste it from Excel or another graphing program. Right-click the Chart Area to enlarge the font to 24 pt. Right-click the axes and data sets to make the lines thicker.
5. To make your own graphics, see "Line Drawings" in Appendix III.
6. Use the **Zoom slider** in the lower right corner to enlarge the section of the poster on which you are working.

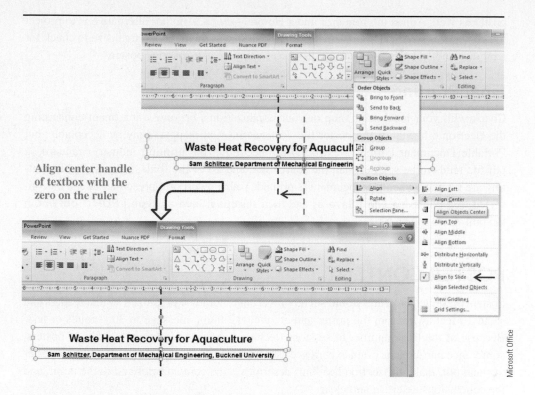

FIGURE 12.5 | Center the title and authors' names and affiliations by clicking **Arrange | Position Objects | Align | Align Center** with **Align to Slide** selected to align the center handle of the textbox with the zero on the horizontal ruler

Aligning Objects

1. To center the title and author name textboxes, hold down the **Shift** key, click each text-box, and click **Arrange | Position objects | Align** and then verify that the **Align to Slide** checkbox is selected. If it is not, click the box and then repeat the command sequence **Arrange | Position objects | Align | Align center** on the **Drawing Tools Format** tab (Figure 12.5).
2. To align the left edges of text boxes and images, hold down the **Shift** key, click each object, and then click **Arrange | Position objects | Align | Align Left** on the **Drawing Tools Format** tab.

Proofread Your Work

Print a rough draft of your poster on an 8½ × 11 inch (or DIN A4) sheet of paper. To do so, click **File | Print** and select the desired printer. Click **Printer Properties | Effects**, and under **Resizing Options | Print Document On**, select **Letter**. The **Scale to Fit** checkbox will be automatically checked. Check the printed page for typographical and grammatical errors and evaluate the layout and overall appearance of your poster. If the print is not large enough to read on the draft, it will probably be hard to read at a distance on the full-size poster as well. Adjust the font size under **Home** (→Font). Another alternative is to project the poster on a screen at approximately the same size as the intended final size to check for legibility and errors before wasting resources printing an erroneous poster.

Final Printing

Check with your local print shop or your organization's printing department concerning the electronic format of the poster file. PowerPoint (ppt or pptx) may be acceptable, but Portable Document Format files (pdfs) are becoming the printing industry standard. A pdf file tends to be smaller than the source file, and all of the fonts, images, and formatting are retained when the document is printed. Your PowerPoint presentation can be converted to a pdf using **File | Save as** and then selecting **Save as type | PDF (*.pdf)** from the pull-down menu.

POSTER CONTENT

Posters presented at **professional society meetings** should be organized so that readers can stand 10 feet away from the poster and get the take-home message in 30 seconds or less. Because of the large number of sessions (lectures) and an even larger number of posters, conference participants often experience "sensory overload." Thus, if you want your poster to stand out, make the section headings descriptive, the content brief and to the point, and the conclusions assertive and clear.

Concise instructions for **undergraduate poster preparation** can be found at the University of Pittsburgh (2011) website, among other sites (type "undergraduate poster preparation" into your browser's search box). Posters for a student audience in the context of an in-house or classroom presentation should follow the same principle of brevity as conference posters, but may retain the sections traditionally found in scientific papers. These include:

- Title banner
- Abstract (optional—if present, it is a summary of the work presented in the poster)
- Introduction
- Apparatus and Procedures
- Results
- Discussion or Conclusions
- References (these should be brief on the poster and provided on a handout)
- Acknowledgments

These sections will be described in detail below.

Title Banner

Use a short, yet descriptive, title since it is the first and most important component for attracting viewers. Try to incorporate the most important conclusion in the title. For example, the title *Waste Heat Recovery for Aquaculture* in Figure 12.5 provides more information than simply *Waste Heat Recovery*. The title is always placed at the top of the poster and is usually centered, but it may sometimes be left aligned. See the "Font" section for tips on selecting a font for the title.

Beneath the title are the authors' names and their institutional affiliation(s). The font size for the authors' names should be smaller than that for the title and slightly larger than that for the affiliations (see Table 12.1).

Introduction

Instead of this conventional heading, consider using a short statement of the topic or introduce the topic as a question. Under the heading, briefly explain the existing state of knowledge of the topic and why the study was undertaken. A bulleted list of objectives may be a good way to present this information.

Apparatus and Procedures

Present the methods used to investigate the problem in enough detail so that someone with training in the field could repeat the measurements. Writing the methods as a series of bulleted statements, and then providing more details in the subsequent text is a recommended format. Be both *brief* and *thorough*. Photographs, flowcharts, and line drawings are effective alternatives to text.

Results

The Results section of a poster consists mostly of visuals (images and graphs) and a minimum of text. Poster viewers do not have time to read the results leisurely, as they do with a journal article. An effective presentation of the results should announce each result with a heading, provide a visual that supports the result, and use text sparingly as a supplement to the visuals.

Tables   Avoid using tables with large amounts of data; if the data are important, prepare a printed table to distribute as a handout. Figures are usually the most effective way of summarizing the data and the take-home message.

Figures   Graphs should summarize the processed data in a manner that allows viewers to appreciate both the general patterns of the data and the degree of variability that they possess. Follow the instructions in Chapter 6 for preparing graphs. Avoid all distractions such as gridlines, colored backgrounds, and 3-D displays when the data are 2-D. The text in which the figure is described should focus on general patterns, trends, and differences in the results, and not on the numbers themselves. For example, the reader can visualize "The price of gasoline increased steeply by about 50% from September 2010 through May 2011" much more easily than "The gas price was about $2.80 in September of 2010 and then rose to about $3.00 in early January 2011, and it rose again to about $3.40 in early March and to $3.60 by early April 2011 and continued to rise to almost $4.00 per gallon by May of 2011."

Figures may contain some statistical information including mean values, standard errors, and minimum and maximum values, where appropriate. Make the data points prominent and use a simple vertical line without crossbars for the error bars. Use the same-sized font for the axis labels and the key as you use for the text on the poster. Do not present the same data in both a table and as a figure.

Images are a great addition to a poster, but only if they are sharp. Picture quality is determined by the size (number of pixels) of the digital image and the printer resolution.

Various sources recommend a printer resolution of at least 240 ppi (pixels per inch) to print quality photos. That means for a $5 \times 7"$ print, the image would have to be at least 1200×1680 pixels to attain the desired resolution on paper. To check the size of an image, right-click it and select **Properties**. Be careful when resizing pictures. Enlarging an image that does not have enough pixels produces a blurry picture.

Visual elements in posters do not necessarily need a caption, although they may have both a number and a title (see examples at the end of this chapter). When there is a caption, use an italic font and place it next to or beneath the figure. When placing the caption next to the figure, right align the caption if it is to the left of the figure and left align the caption if it is to the right of the figure. Alternatively, skip the caption and position the figure just before or after the text in which it is described.

Edit the text ruthlessly to remove nonessential information about the visuals. A sentence such as "The gas price is shown in the following figure" is nothing but deadwood. On the other hand, "Gasoline prices increased rapidly by 50% from September 2010 to May 2011" describes the result in no uncertain terms.

Discussion or Conclusions

Interpret the data in relation to the original objectives and relate these interpretations to the present state of knowledge presented in the introduction. Discuss any surprising results. Discuss the future needs or direction of the research. Where appropriate, identify sources of error and basic inadequacies of the technique. Do not cover up mistakes; instead, suggest ways to improve the experiment if it were to be done again. Speculation on the broader meaning of the conclusions may be included in this section. Use bullets to present this information concisely.

Literature Citations

In technical papers, it is common to cite the work of others, particularly in the Introduction and Discussion sections. Because posters are informal presentations and the author is present to provide supporting details, the References section of a poster can be less comprehensive than that in a paper. Try to keep the reference list to the minimum number of sources needed to provide credibility in the summary of the state of the art. Participants who view the poster are likely to work in the same field and probably are already familiar with most of the literature, so a long list of references would only waste valuable space on the poster. Even visitors who have a general knowledge of the topic, but who work in a different subdiscipline, are not interested in the details. They are interested mainly in how the approach or the results might help them improve their methods or provide insight into their work. Discussions with participants in this second group are valuable to the presenter because they provide a different perspective and may uncover links to other subdisciplines.

Student presentations in an in-house setting are different from poster sessions at large national meetings because students usually do not have the background or the familiarity with the literature that engineers in the workforce have. Researching your topic and presenting your work in the context of the published literature is something you will do throughout your career. Thus, your instructor may ask you to include literature citations on your poster or to provide a list of references on a handout. See the section "Common Engineering Citation Formats" in Chapter 2.

Acknowledgments

In this section, the author acknowledges organization(s) that provided funding and thanks technicians, colleagues, and others who have made significant contributions to the work. Usually, students need not acknowledge the guidance of their faculty mentor, but acknowledging other staff members who provided assistance is appropriate.

White Space

Remember to leave room for white space, which is the blank space around text and figures. White space is an effective way to separate sections, and it also gives the viewer's eyes a rest.

PRESENTING YOUR POSTER

As mentioned at the beginning of this chapter, poster sessions are a great opportunity to showcase the presenter's work and to get immediate feedback from participants. The presenter is expected to stand near the poster during the entire session and be prepared to discuss the work with interested individuals. Analogous to the 3-30-3 rule for white papers (see Chapter 10), we would like to propose a 3-30-5 rule for poster presentation. The presenter has 3 seconds to capture the attention of a passerby with the title, 30 seconds to give an "elevator speech" on the poster (a summary of the work and why it is important), and then, after capturing the interest of the participant, another 3 to 5 minutes to give them the "whole story." Prepare both the 30-second summary of the work and the 3-5 minute detailed explanation beforehand. Anticipate questions and prepare appropriate answers.

Purrington (2012) and Block (1996) provide advice on how to dress and behave when presenting a poster. Their collective advice is summarized as follows:

- Wear attractive professional attire. Keegan and Bannister (2003) found that wearing clothing with a color scheme compatible with the poster colors increases the number of participants stopping to read the poster.
- Do not wear sunglasses.
- Do not wear caps or other headwear (except for cultural or religious reasons).
- Do not chew gum (or tobacco).
- Keep your hands out of your pockets to avoid jingling change or keys.
- Do not stand directly in front of your poster.
- Do not badger passersby.
- Prepare ancillary materials to hand out to interested individuals.

Use the poster session to seek feedback on the presented work. Even if some participants are critical, be receptive and understand that their perspectives may further refine the direction of subsequent research.

Further guidelines for posters can be found through a Web search for "poster preparation guidelines." Among the more useful sources not cited elsewhere in this chapter are Alley (2008), Anderson (2009), CCMR (2013), Erren and Bourne (2007), and Miller (2007). Briscoe (1996), a general reference for preparing visual elements, also contains a section on poster preparation.

EVALUATION FORM FOR POSTER PRESENTATIONS

Good posters are the product of creativity, hard work, and feedback at various stages of the poster preparation process. When you present a poster to your class or at a professional society meeting, visitors may be asked to evaluate your poster using a form such as the "Evaluation Form for Poster Presentations" shown in Table 12.2. This form is also available as a Word document at <www.CENGAGE.com/Knisely>.

TABLE 12.2 | Template for evaluating poster presentations

CRITERIA	COMMENTS
Layout and Appearance	
APPEARANCE: Is the poster neatly constructed? Do the text and the figures stand out against the background? Are colors and fonts used consistently? Is the text large and legible from 3–10 ft away?	
SECTIONS: Does each section begin with a descriptive heading? Is there sufficient space between sections? Do the sections naturally flow from top left to bottom right?	
BALANCE: Is there a nice balance between text and figures? Is there too much text?	
PROOFREADING: Is the text free of typos and grammatical errors?	
Content	
TITLE: Does the title grab your attention?	
AUTHORS: Are the authors' names, affiliations, and contact information provided?	
INTRODUCTION: Were the objectives clearly stated? Do you understand why this study was done? Did you get enough background information to understand the system? Were all abbreviations (if any) defined for the general visitor?	

TABLE 12.2 | Continued

CRITERIA	COMMENTS
Content	
METHODS: Were the methods described clearly and concisely? RESULTS: Were the graphs easy to understand? Were any visual elements distracting? CONCLUSIONS: Do the conclusions match the data? Are reasonable ideas put forth to explain the observed patterns? Is there a clear connection between the conclusions and the original objectives?	

Adapted from Knisely, 2013. Used with permission from the publisher, Sinauer Associates, Inc. and W. H. Freeman & Co.

When you serve as an evaluator, provide comments you would find helpful if you were the presenter. As you know, feedback is most likely to be appreciated when it is constructive, specific, and done in an atmosphere of mutual respect.

SAMPLE POSTERS

Sample posters from a national conference (Figure 12.6) and from a classroom presentation (Figure 12.7) are provided. Each poster is evaluated in Table 12.3 and 12.4, respectively, using the criteria presented in Table 12.2.

Examples of other posters along with reviewer comments are posted on <www .CENGAGE.com/Knisely>. What do you like (or not like) about each poster? Use the evaluation form to determine how well the authors have met the requirements of good poster design.

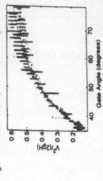

FIGURE 12.6 | Poster presented at Hydrovision 2011 in Sacramento, California by Adam Reich, Natel Energy. The text is legible even on the scale of a single page

Waste Heat Recovery for Aquaculture

Sam Schlitzer, Department of Mechanical Engineering, Bucknell University

Overview

Waste heat recovery

- Efficiently recycles energy
- Improves aquatic habitat
- Serves entrepreneurial ventures

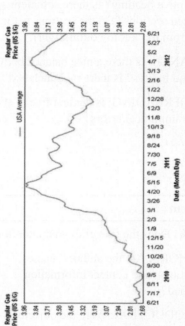

Fig. 1. Overview of a heat recovery system.

Specific Applications

- Waste heat from nuclear power plants heat the surrounding environment for crocodile farm in Civeaux in Provence and sea bass farming in Dunkirk, France [1] (Fig. 2)
- Waste heat from sugar cane mill used to heat water to 40°C to sustain a crocodile farm near Cairns, Australia [2]
- Ice plants utilize waste heat to power refrigeration cycles to produce ice for cooling perishable foods [3] (Fig. 3)

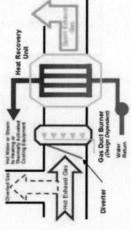

Fig. 2. Measuring a young crocodile raised in a pond heated by waste heat from a nuclear power plant

Facts and Insight

- Waste heat use improves overall energy utilization factor (EUF)
- Waste heat use reduces overall CO_2 emissions

With rising fuel costs (Fig. 4), a foreseeable shortage of fossil fuels, and concerns about global warming, waste heat regeneration is a smart and efficient way to improve the energy utilization factor (EUF) while decreasing the carbon footprint.

Fig. 3. Perishable farmed fish can be maintained with ice from an ice plant powered by waste heat

Conclusion

Waste heat regeneration is an inexpensive and efficient way to raise reptiles and fish species. Waste heat regeneration provides a greener alternative for heating animal habitats or powering ice making plants to produce ice for cooling perishable food products. Waste heat utilization can supply inexpensive heating and cooling for entrepreneurs pursuing unique business ventures.

Acknowledgments

Special thanks to the Bucknell Engineering College computing facilities personnel for their help with printing this poster.

References

1. Christmas cracker: crocs and nukes 2009 downloaded Oct, 28, 2011 from http://environmentalresearchweb.org/blog/2009/12/christmas-cracker-crocs-and-nu.html
2. Ellis. C. 2007 Crocodile Farming, Behind the News (6/27/2007) downloaded Oct, 28, 2011 from http://www.abc.net.au/btn/v3/stories/s2081730.htm
3. Waste heat recovery by triple regenerative effect water–lithium bromide absorption system. 1995 Proceedings of the 1995 ASME International Mechanical Engineering Congress and Exposition, San Francisco, CA.
4. Historical Gas Prices. Web download June 21, 2012 from http://gasbuddy.com/gb_retail_price_chart.aspxm

Fig. 4. Rising cost of gasoline 2010-2012 [4] suggests that waste heat recovery will become more financially lucrative if prices of energy continue rising

FIGURE 12.7 | Sample sophomore-level student poster for classroom presentation based on independent literature reading

TABLE 12.3 | Evaluation of the poster in Figure 12.6

CRITERIA	COMMENTS
Layout and Appearance	
APPEARANCE: Is the poster neatly constructed? Do the text and the figures stand out against the background? Are colors and fonts used consistently? Is the text large and legible from 3–10 ft away?	Poster looks professional. Text and figures stand out against the background. Font is large and legible.
SECTIONS: Does each section begin with a descriptive heading? Is there sufficient space between sections? Do the sections naturally flow from top left to bottom right?	Section headings are familiar to the audience. Two-color scheme for text is attractive. Dark-colored, larger headings make it easy to find information.
BALANCE: Is there a nice balance between text and figures? Is there too much text?	There is good balance between text and figures.
PROOFREADING: Is the text free of typos and grammatical errors?	Photos have high resolution and images look sharp. There are no apparent typos or grammatical errors.
Content	
TITLE: Does the title grab your attention?	**Title** is descriptive
AUTHORS: Are the authors' names, affiliations, and contact information provided?	**Authors and affiliations** are provided.
INTRODUCTION: Were the objectives clearly stated? Do you understand why this study was done? Did you get enough background information to understand the system? Were all abbreviations (if any) defined for the general visitor?	**Objectives** are clearly stated. Relevant details are given in a concise, bulleted list. Photos are very effective in describing the test site and the apparatus in **Site Overview**.
METHODS: Were the methods described clearly and concisely?	The **Measurement Technique** is described clearly with a few pictures and a minimum amount of text.
RESULTS: Were the graphs easy to understand? Were any visual elements distracting?	The **Results and Conclusions** are summarized in a bulleted list. Rather than numbering the figures, the figures are placed near the text in which they are described.
CONCLUSIONS: Do the conclusions match the data? Are reasonable ideas put forth to explain the observed patterns? Is there a clear connection between the conclusions and the original objectives?	There is a clear connection between the original objectives and the results and conclusions.

TABLE 12.4 | Evaluation of the poster in Figure 12.7

Layout and Appearance	
CRITERIA	**COMMENTS**
APPEARANCE: Is the poster neatly constructed? Do the text and the figures stand out against the background? Are colors and fonts used consistently? Is the text large and legible from 3–10 ft away? SECTIONS: Does each section begin with a descriptive heading? Is there sufficient space between sections? Do the sections naturally flow from top left to bottom right? BALANCE: Is there a nice balance between text and figures? Is there too much text? PROOFREADING: Is the text free of typos and grammatical errors?	Poster looks professional. Text and figures stand out against the background. Font is large and legible. Nontraditional section headings are effective and make it easy to find information. Good use is made of white space to separate sections. Two-color scheme for text is attractive. Good balance exists between text and figures. Photos have high resolution and images look sharp. There are no apparent typos or grammatical errors.
Content	
TITLE: Does the title grab your attention? AUTHORS: Are the authors' names, affiliations, and contact information provided? INTRODUCTION: Were the objectives clearly stated? Do you understand why this study was done? Did you get enough background information to understand the system? Were any abbreviations defined for the general visitor? METHODS: Were the methods described clearly and concisely? RESULTS: Were the graphs easy to understand? Were any visual elements distracting? CONCLUSIONS: Do the conclusions match the data? Are reasonable ideas put forth to explain the observed patterns? Is there a clear connection between the conclusions and the original objectives?	The **Title** is appropriate for a classroom presentation, but would need to be more focused for a poster session at a conference. If the author's focus is on using energy more efficiently, then: "Energy savings in Aquaculture" might be a better title. If the author wants to present a case study on a business venture, then: "Aquaculture Prospers Using Waste Heat." **Authors and affiliations** are provided. The **Overview** sends confusing messages. The bulleted list itemizes *uses* of waste heat, but Fig. 1 shows the *process of recovering* waste heat. The author's objectives are ambiguous. Fig. 1 is neither referenced nor explained.

TABLE 12.4 | Continued

	Content
	Under **Specific Applications**, the keywords "waste heat," "nuclear power plants," and "crocodile farm" do not prepare the reader for the tiny crocodile shown in Fig. 2.
	The **Facts and Insights** are not supported with data. Fig. 4 is a generic graph showing rising gas costs. What is the connection between recovering waste heat from power plants and rising gas costs?
	It may be helpful to define *energy utilization factor* (EUF) for nonspecialist readers.
	Fig. 4 is not formatted correctly. The second y-axis is superfluous. The title is descriptive.
	The figures do not have to have a caption when they are placed near the text in which they are mentioned.
	None of the author's **Conclusions** is supported with hard data.
	The original objectives are restated in the conclusions, but no quantitative data has been provided. The poster leaves us with all kinds of unanswered questions like:
	How does the waste heat recovery process work?
	How is energy recycled efficiently in this process?
	In what ways does waste heat "improve" the aquatic habitat?
	What aspects of waste heat recovery lend themselves to entrepreneurial ventures?
	The student's faculty mentor is not acknowledged, as is appropriate.

SUMMARY

Posters are a form of communication that combines elements of written and oral presentations. Similar to slides in an oral presentation, there is a large amount of visual content and a minimum of text on a poster. Poster headings parallel those of written papers, yet the essential information is summarized on a single large-format page. The audience can scan the entire presentation, focusing selectively on sections of interest. The sections of interest provide the entry point for conversations between the audience and the author. Posters are most frequently prepared in PowerPoint, although other options do exist and may be superior, provided the author is familiar with the other software. When preparing your poster, remember that the most important objective is to attract passersby to your poster. Make your title stand out and your poster layout aesthetically pleasing. Once you have "caught" a viewer, reward his or her initial interest in your poster with a well-articulated, 30-second speech that highlights what you did and why your findings are important. If effective, this speech will start a conversation that spurs creativity and may lead to future collaboration.

EXERCISES

1. Prepare a poster presentation on any of the topics given in the exercises in Chapter 2.
2. Prepare a poster on a topic you select with the approval of your instructor.
3. Prepare a poster based on a white paper you read in Chapter 10.
4. Prepare a poster on a laboratory exercise in an engineering course or on a project at work.

REFERENCES

Alley, M. 2008. Design of scientific posters (online). Accessed 3 March 2013 at <http://www.writing.engr.psu.edu/posters.html>.

Anderson, F. 2009. General guidelines for scientific poster presentations (pdf). Accessed 14 May 2013 at <http://www.omnee.net/files/OMNEE%20General%20Guidelines%20for%20Scientific%20Poster%20Presentations.pdf>.

Block, S.M. 1996. Do's and don'ts of poster presentation, Biophysical Journal 71: 3527–3529

Briscoe, M.H. 1996. Preparing Scientific Illustrations: A guide to better posters, presentations, and publications (2nd ed.). New York: Springer.

CCMR. 2013. Scientific poster design (pdf), Cornell Center for Materials Research. Accessed 3 March 2013 at <http://www.cns.cornell.edu/documents/ScientificPosters.pdf>.

Erren, T.C., and Bourne, P.E. 2007. Ten simple rules for a good poster presentation. PLoS Comput Biol 3(5):e102. doi:10.1371/journal.pcbi.0030102. Accessed 3 March 2013.

Keegan, D.A., and Bannister, S.L. 2003. Effect of colour coordination of attire with poster presentation on poster popularity, Canadian Medical Association Journal, 169(12): 1291–1292.

Knisely, K. 2013. A student handbook for writing in biology, 4th ed. Sunderland, MA: Sinauer Associates, Inc.

Miller, J.E. 2007. Preparing and presenting effective research posters, HSR: Health Services Research 42:1, Part I, 311–328, see also Health Research and Educational Trust, doi: 10.1111/j .1475-6773.2006.00588.x 311, accessed 3 March 2013.

Purrington, C. 2013. Designing conference posters (online). Accessed 3 March 2013 at <http://colinpurrington.com/tips/academic/posterdesign>.

University of Pittsburgh. 2013. The natural sciences undergraduate research conference: Guidelines for poster preparation. Accessed 14 May 2013 at <http://www.pitt.edu/~etbell/nsurg/PosterGuide.html>.

Appendix I

INTRODUCTION

This appendix deals exclusively with Microsoft Word 2010 using the Windows 7 operating system. While there are similarities between the Word 2010 and Word for Mac 2011 Ribbon interfaces, some of the commands are accessed differently.

If you have been using Word 2007, then Word 2010 will not seem that different. However, if you are a Mac user, the Ribbon is a new addition to the user interface at the top of the screen. The **Ribbon** is a single strip that displays **commands** in task-oriented **groups** on a series of **tabs** (Figure A1.1). In Word 2010, the new **File** tab replaces the Microsoft **Office Button** that was located in the top left of the screen in Word 2007. Additional commands in some of the groups can be accessed with the **Dialog Box Launcher**, a diagonal arrow in the bottom right corner of the group label. The **Quick Access Toolbar** comes with buttons for saving your file and undoing and redoing commands; you can also add buttons for tasks you perform frequently.

The window size and screen resolution affect what you see on the Ribbon. You may see fewer command buttons or an entire group abbreviated to one button on a smaller screen or if you have set your screen display to a lower resolution. If you want to see the tab names without the command buttons, click **Minimize the Ribbon**.

The **Status Bar** at the bottom of the screen displays information about your document (Figure A1.2). Clicking the **Page position** button opens the **Go To** dialog box, which allows you to reposition the insertion pointer (the blinking vertical bar) on a different page or on a comment, table, graphic, or other landmark in your document. To return to your original position, click the **Undo** button on the **Quick Access Toolbar** at the top of the screen. The **Automatic word count** button updates the number of words as you type. The **Proofing** button shows a green check mark if Word has not found any spelling or grammatical errors; if it has, the button displays a red *X*. To correct the potential error, click the **Proofing** button and select one of the options in the dialog box.

To select which features are displayed on the status bar, right-click it to open the **Customize Status Bar** dialog box. Click the **Language** check box to display the language you are currently working in. If you have no occasion to use languages other than English, it is not necessary to display the **Language** button on the status bar.

The **Layout selector, Zoom level**, and **Zoom slider** buttons on the right side of the status bar are used to adjust how you view your document on the screen. They do not affect formatting.

This chapter has been adapted from *A Student Handbook for Writing in Biology*, Appendix 1. Used with permission from the publishers, Sinauer Associates, Inc. and W. H. Freeman & Co.

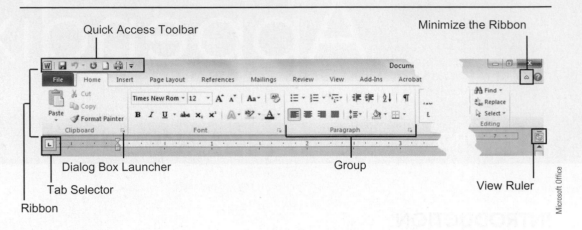

FIGURE A1.1 | Screen display of command area in a Word 2010 document

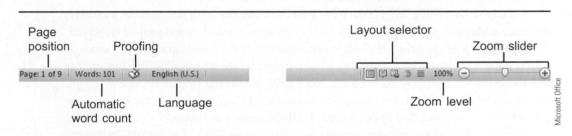

FIGURE A1.2 | **Status bar** at the bottom of the screen. Clicking a button on the left side of the bar opens a dialog box or information window. The right side of the bar is used to change the view.

Word 2010 has many features that make the design more attractive. There are more templates for letters, faxes, resumes, and greeting cards and more clipart and artwork than in earlier versions. The new default typefaces, Calibri and Cambria, were chosen for their improved on-screen appearance and versatility in Web-based and printed media. Although the default font is smaller (11 pt instead of 12 pt), the line spacing is larger (1.15 instead of 1).

Although Word 2010 can make your documents look attractive, it cannot teach you how to write technical papers with proper content and style. That skill is something you can hope to acquire with practice. Consult this appendix to become more efficient at word processing: Look up what you do not know and figure out how to do things faster. Ultimately, you will be able to devote more energy to the content of your technical papers than the format.

INCREASING YOUR WORD PROCESSING EFFICIENCY
Navigation

Use the keyboard shortcuts in Table A1.1 to move around the document quickly.

TABLE A1.1 | Navigate a document using keyboard shortcuts

Where To?	Hot Key*
Beginning of the document	Ctrl + Home
End of the document	Ctrl + End
Beginning of a line	Home
End of a line	End
Up one screen	Page Up
Down one screen	Page Down
Up one page	Ctrl + Page Up
Down one page	Ctrl + Page Down

© Cengage Learning 2015

*The nomenclature **Ctrl + Home** means to hold down **Ctrl** while pressing **Home**.

Navigation Pane

The navigation pane is a convenient way to jump to different sections in longer documents. To display the navigation pane, click the **View** tab and check the **Navigation Pane** box in the **Show** group. The navigation pane has three tabs, which correspond to the **Heading**, **Page**, and **Search results** views (Figure A1.3a). In **Heading view**, click a heading to move to that point in the document. Rearrange entire sections by dragging a heading to a new position on the navigation pane. Similarly, clicking a page thumbnail in **Page view** allows you to move to that page. The third tab lets you search the document not just for words, but also graphics, equations, and comments (click the magnifying glass next to the search box to display options). The search result(s) are displayed in boxes that, if clicked, move you to that location in the document (Figure A1.3b).

Text Selection

Word offers you multiple ways to select text with the mouse or the keyboard or a combination of both. Word processing experts tend to prefer the keyboard because taking their hands off the keyboard to edit or format text with the mouse takes time. On the other hand, clicking command buttons with the mouse is more intuitive and does not require memorizing keyboard shortcuts. Table A1.2 shows you how to select text with both keyboard and mouse so that you can copy, move, format, or delete it. Before you follow these instructions, position the insertion pointer (the blinking vertical bar) next to the character, word, or block of text you want to select.

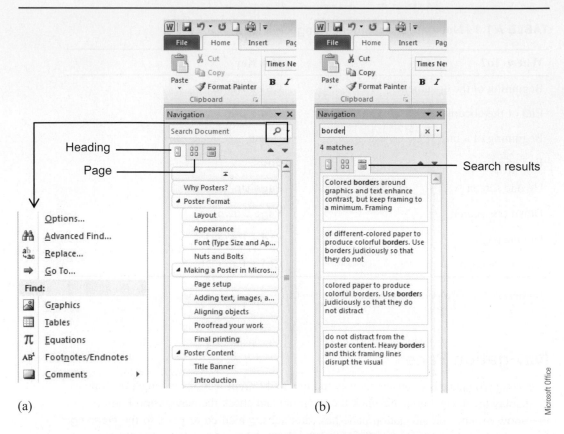

(a)   (b)

FIGURE A1.3 | The Navigation pane facilitates rapid navigation through long documents. **(a)** Click Heading view to jump to a particular section or Page view to look at a thumbnail of a particular page. **(b)** Click Search results view to go to each location where a word or another document element was found. Clicking the magnifying glass next to the search box provides more options.

TABLE A1.2 | Select text with the keyboard and mouse after positioning the insertion pointer

Text	Keyboard*	Mouse
Character	**Shift + →** or **Shift + ←**	Press and hold the left mouse button and drag over the character.
Word	**Ctrl + Shift + →** or **Ctrl + Shift + ←**	**Double-click** the word.
Sentence	Repeat sequence for a word	Hold down the **Ctrl** key and **click**.

TABLE A1.2 | Continued

Text	Keyboard*	Mouse
Line	**Shift + ↓** or **Shift + ↑**	Position the insertion pointer in the left margin so that it changes to an arrow. **Click.**
Block of text	**Shift + arrow keys, Home, End, Page Up,** or **Page Down**	Position the insertion pointer at the beginning of the block. Hold down the **Shift** key. **Click** at the end of the block.
Paragraph	**Ctrl + Shift + ↓** or **Ctrl + Shift + ↑**	**Triple-click** anywhere in the paragraph.
Entire document	**Ctrl + A**	Position the insertion pointer in the left margin so that it changes to an arrow. **Triple-click.**

*The nomenclature **Shift +** → means to hold down **Shift** while pressing →; similarly, **Ctrl + Shift +** → means to hold down both **Ctrl and Shift** while pressing →.

© Cengage Learning 2015

COMMANDS IN WORD 2010

The best way to get up to speed with Word 2010 is to learn where the commands you used most frequently in earlier versions of Word are located. Table A1.3 is designed to help you with this task. Frequently used commands are listed in alphabetical order in the first column, and the Word 2010 command sequence is given in the second column. The PC keyboard shortcuts for most of the commands are shown in the third column. When using the **Alt** commands, hold down the **Alt** key while pressing the first letter. Release the **Alt** key and type the next letter(s) in the sequence. If you are a Mac user,

Ctrl = Command (⌘)
Alt = Option (however, some **Alt** key sequences do not have a Mac equivalent)

In this book, the nomenclature for the command sequence is as follows:

   Ribbon tab (→Group) **| Command button | Additional Commands** (if available).

For example, **Home** (→Editing) **| Find | Find** means "Select the **Home** tab on the Ribbon," and in the **Editing** group (the unbolded text with (→Group) tells you in which group to find the next button, but there is no need to click), click the down arrow on the **Find** button to select **Find** from the menu."

For an overview of what is new in Word 2010, go to Microsoft's homepage (http://office .microsoft.com) and type "what's new in word 2010" into the search box. The link for Word for Mac 2011 is <http://www.microsoft.com/mac/word/whats-new>.

Microsoft Office also has an online interactive guide to help you transition from Word 2003 to Word 2010: Go to Microsoft's Home Page (<http://office.microsoft.com>) and type "office 2010 interactive" into the search box. Select "Learn where menu and toolbar commands are in Office 2010 and related products" and follow the instructions on the website.

TABLE A1.3 | Common commands (listed alphabetically) and how to carry them out in Word 2010. Many of the Word 2010 commands can also be accessed by selecting text or an object and right-clicking. For **Alt** commands, hold down **Alt** while pressing the first letter key, which corresponds to a tab name

| File F | Home H | Insert N | Page Layout P | References S | Mailings M | Review R | View W | Add-Ins X | Acrobat B |

. Then release **Alt** and type the next letter(s) in the sequence. **F1**, **F4**, **F7**, and **F12** are function keys. The nomenclature **Ctrl+C** means to hold down **Ctrl** while pressing **C**; similarly, **Ctrl+Shift+E** means to hold down both **Ctrl** and **Shift** while pressing **E**.

Command	Word 2010 Command Sequence	PC Hot Key
Editing		
AutoCorrect	**File | Options | Proofing | AutoCorrect Options**	Alt+FT
Comments	**Review** (→Comments)	Alt+R
Copy	**Home** (→Clipboard) **| Copy**	Ctrl+C or Alt+HC
Cut	**Home** (→Clipboard) **| Cut**	Ctrl+X or Alt+HX
Find	**Home** (→Editing) **| Find**	Ctrl+F or Alt+HFDF
Paste	**Home** (→Clipboard) **| Paste**	Ctrl+V or Alt+HV
Paste Special	**Home** (→Clipboard) **| Paste | Paste Special**	Ctrl+Alt+V or Alt+HVS
Redo last action	**Quick Access Toolbar | Redo button** ↻	Ctrl+Y
Replace	**Home** (→Editing) **| Replace**	Ctrl+H orAlt+HR
Spelling & Grammar Check	**Review** (→Proofing) **| Spelling & Grammar**	F7 or Alt+RS
Track changes	**Review** (→Tracking) **| Track Changes**	Ctrl+Shift+E or Alt+RG
Undo		
Last action	**Quick Access Toolbar | Undo button** ↺	Ctrl+Z
Several previous actions	**Undo button dropdown menu**	
File		
Close	**X on right side of Title bar**	Ctrl+W or Alt+F4 or Alt+FX **(Mac: Cmd+Q)**
Document properties	**File | Info**	Alt+FIQ
Open existing	**File | Open**	Ctrl+O or Alt+FO
Open new	**File | New**	Ctrl+N or Alt+FN
Print	**File | Print** **Quick Access Toolbar** 🖨	Ctrl+P or Alt+FP
Save	**File | Save** **Quick Access Toolbar** 💾	Ctrl+S or Alt+FS
Save as	**File | Save As**	F12 or Alt+FA

TABLE A1.3 | Continued

Command	Word 2010 Command Sequence	PC Hot Key
Formatting		
Align text		
Left/center/right/ justified	**Home** (→Paragraph)	Ctrl+L/E/R/J or Alt+ HAL/HAC/HAR/HAJ
Font face, size, style, color	**Home** (→Font)	Ctrl+D or Ctrl+Shift+P or Alt+HFN
Format Painter		
Copy formatting	**Home** (→Clipboard) | **Format Painter**	Ctrl+Shift+C or Alt+HFP
Apply formatting	**Drag cursor (Paintbrush I-beam) over text to be formatted**	Ctrl+Shift+V
Lists		
Bulleted	**Home** (→Paragraph) | **Bullets**	Alt+HU
Numbered	**Home** (→Paragraph) | **Numbering**	Alt+HN
Show/Hide		Ctrl+Shift+* or Alt+H8
Formatting symbols	**Home** (→Paragraph) | **Show/Hide¶**	
Gridlines	**View** (→Show) | **Gridlines**	Alt+WG
Ruler	**View** (→Show) | **Ruler**	Alt+WR
Spacing		
Lines	**Home** (→Paragraph) | **Line and Paragraph Spacing**	Alt+HK
Paragraphs		
Subscript	**Home** (→Font) | x_2	Ctrl+ = or Alt+H5
Superscript	**Home** (→Font) | x^2	Ctrl+Shift = or Alt+H6
Symbols	**Insert** (→Symbols) | **Symbol**	Alt+NU
Tabs	**Home** (→Paragraph) | **Tabs** or **Page Layout** (→Paragraph) | **Tabs**	Alt+HPG
Page Layout		
Headers & Footers	**Insert** (→Header & Footer) | **Header/Footer**	Alt+NH/NO
Margins	**Page Layout** (→Page Setup) | **Margins**	Alt+PM
Orientation	**Page Layout** (→Page Setup) | **Orientation**	Alt+PO
Columns	**Page Layout** (→Page Setup) | **Columns**	Alt+PJ
Page break	**Page Layout** (→Page Setup) | **Breaks** | **Page**	Ctrl+Enter or Alt+PBP
Page numbers	**Insert** (→Header & Footer) | **Page Number**	Alt+NNU

TABLE A1.3 | Continued

Command	Word 2010 Command Sequence	PC Hot Key
Paragraphs		
Dialog box launcher	**Home** (→Paragraph) or **Page Layout** (→Paragraph)	Alt+PPG
Hanging indent	**Dialog box launcher	Indentation	Special: Hanging**	Ctrl+T
Indent first line	**Dialog box launcher	Indentation	Special: First line**	Ctrl+M, then Ctrl+Shift+T
Indent whole paragraph	**Page Layout** (→Paragraph) **| Indent**	
Keep with next (prevents unwanted page breaks)	**Dialog box launcher	Line and Page Breaks tab	Pagination	Keep with next**	Ctrl+M or Alt+PIL/PIR
Shapes		
Align	**Shift+Click** individual shapes. **Drawing Tools Format** (→Arrange) **| Align**	Alt+JAA
Display gridlines	Click a shape. **Drawing Tools Format** (→Arrange) **	Align	View Gridlines**	Alt+JAAS
Format	**Right-click** the shape. **Format AutoShape.**	Alt+JO
Group	**Shift+Click** individual shapes. **Drawing Tools Format** (→Arrange) **	Group	Group**	Alt+JAGG
Insert	**Insert** (→Illustrations) **| Shapes**	Alt+NSH
Nudge to position precisely		Click shape, then Ctrl+arrow keys.
Position freely	Click a shape. **Drawing Tools Format** (→Arrange) **	Align	Grid Settings	Uncheck Snap objects to grid when the gridlines are not displayed**	Alt+JAAG
Special Characters		
Equations	**Insert** (→Symbols) **| Equation**	Alt+NE
Hyphen, nonbreaking		Ctrl+Shift+-
Space, nonbreaking		Ctrl+Shift+spacebar
Symbols	**Insert** (→Symbols) **| Symbol**	Alt+NU
Tab stops	**Home** (→Paragraph) **dialog box launcher	Tabs**	Alt+HPG

TABLE A1.3 | Continued

Command	Word 2010 Command Sequence	PC Hot Key
Table of Contents		
Generate	**References** (→Table of Contents) | **TOC**	Alt+ST
Mark headings to use as entries	**Home** (→Styles) | **Heading 1 (2 or 3)**	Alt+HL or Ctrl+Alt+1/2/3
Update	**References** (→Table of Contents) | **Update Table**	Alt+SU
Tables		
Insert	**Insert** (→Tables) | **Table**	Alt+NT
Format	Click inside table, then **Table Tools Design and Table Tools Layout** tabs	
Repeat table headings on multiple-page tables	Click anywhere in the first row, then **Table Tools Layout** (→Data) | **Repeat Header Rows.**	Alt+JLJ
View		
Document views	**View** (→Document Views) or Status bar icons lower right corner of screen	
Multiple open documents side by side	**View** (→Window) | **View Side by Side**	Alt+WB
Multiple open documents horizontally	**View** (→Window) | **Arrange All**	Alt+WA
Print preview	**File | Print**	Alt+FP
Split current window (keep one part of document fixed while scrolling through the rest)	**View** (→Window) | **Split**	Alt+WS

UNFURLING THE RIBBON

If you prefer a more visual approach to the commands, the following sections show you the buttons on each tab. If you want more choices, click the **dialog box launcher** (the arrow on the right side of some groups) to open a dialog box for that particular task. The dialog boxes are mostly the same as those in Word 2003 and 2007.

The File Tab

Clicking the **File** tab opens **Backstage View** (Figure A1.4). The left panel lists the file-related commands, the center panel provides the options for that command, and the right panel shows a preview or other options. Document commands such as **Save, Save As, Open, Close, Recent, New**, and **Print**, which were located on the **Office Button** in Word 2007 and on the **File** menu in Word 2003, are located on the **File** tab. Other commands on the **File** tab are

- **Info**, which provides new file management tools for setting permissions, inspecting the document for hidden properties and personal information, and recovering earlier autosaved or unsaved versions of a file.
- **Save & Send** makes it easier to collaborate with others on files. All collaborators must use Office 2010 and possess SharePoint 2010 software or a SkyDrive account to access and save the files online. Similar to Google Docs, this feature allows two or more people to work on the same document at the same time.

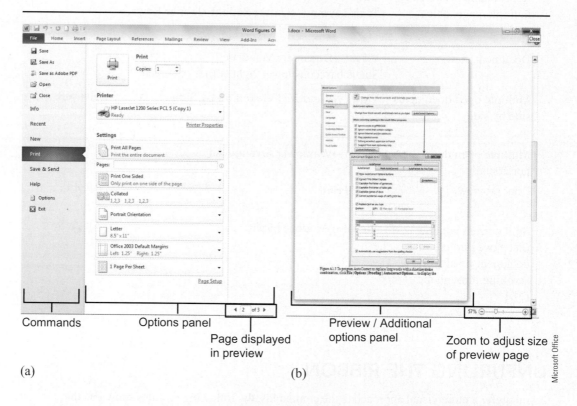

Commands   Options panel   Preview / Additional
                                                                    options panel
                        Page displayed   Zoom to adjust size
                        in preview   of preview page

(a)   (b)

Microsoft Office

FIGURE A1.4 | Backstage View is displayed when clicking the File tab. **(a)** The left panel shows file-related commands such as **Save, Save As, Open, Close,** and **Print.** The options panel shows additional commands related to the command selected on the left panel, in this case **Print. (b)** The preview panel appears on the right in Backstage View and here shows page 2 as it will be printed. The zoom slider affects the view, but not the page format.

- **Help**. The **Help** menu can also be accessed by pressing the **F1** function key.
- **Options** allows you to change program settings such as user name and initials, rules for checking spelling and grammar, AutoCorrect options, how often documents are saved automatically, and so on. The **Options** button is also the gateway to commands for customizing the Ribbon and the Quick Access Toolbar and for managing Microsoft Office add-ins and security settings.

AutoCorrect   AutoCorrect is useful for more than just correcting spelling mistakes. Increase your efficiency by programming AutoCorrect to replace an expression that takes a long time to type with a simple keystroke combination. You must choose the simple keystroke combination judiciously, however, because every time you type these keystrokes, Word will replace them with what you programmed in AutoCorrect. One "trick" is to precede the keystroke combination with either a comma or semicolon (, or ;). Since punctuation marks are typically followed by a space and not a letter, adding one before your "code letter(s)" creates a unique combination.

For example, assume "flow-induced noise and vibration" is a phrase you have to type frequently. You choose ";fin" as the abbreviated keystroke combination for this phrase. To program AutoCorrect for this entry:

1. Select **File | Options | Proofing | AutoCorrect Options**. This opens the **AutoCorrect** tab in the AutoCorrect dialog box (Figure A1.5).
2. Type ";fin" in the **Replace** text box.
3. Type "flow-induced noise and vibration" in the **With** text box.
4. Click the **Add** button.
5. Click **OK** to exit the **AutoCorrect** dialog box.
6. Click **OK** to exit the **Word Options** box.

AutoCorrect can also take the work out of formatting the mathematical expressions of variable names, such as freestream velocity, V_∞. Notice that this expression contains a symbol (∞) and a subscript and, because it is a variable, the whole expression is italicized. To program AutoCorrect to automate formatting, follow these steps:

1. First type the expression without format (e.g., V∞). The infinity symbol is inserted by clicking **Insert** (→Symbols) **| Symbol** and selecting ∞ from the dialog box.
2. Select the infinity symbol and subscript it by clicking the **Subscript** button on **Home** (→Font) (or use **Ctrl+=**). The expression then becomes V_∞.
3. To italicize the entire expression, select it and click **Home** (→Font) **| Italic** (or use **Ctrl+I**).
4. Now select the entire formatted expression and click **File | Options | Proofing | Auto-Correct Options** (see Figure A1.5).
5. On the **AutoCorrect** tab, you will notice "V∞" already entered in the **With** text box. Click the **Formatted text** option button to add the subscripting and italics.
6. Type a mnemonic abbreviation of your choosing in the **Replace** textbox. For example, let ";vi" stand for "v infinity."
7. Click the **Add** button.
8. Click **OK** to exit the **AutoCorrect** dialog box.
9. Click **OK** to exit the **Word Options** box.

After AutoCorrect has been programmed in this way, every time you type ";vi" and hit the space bar, Word automatically changes the abbreviation to V_∞.

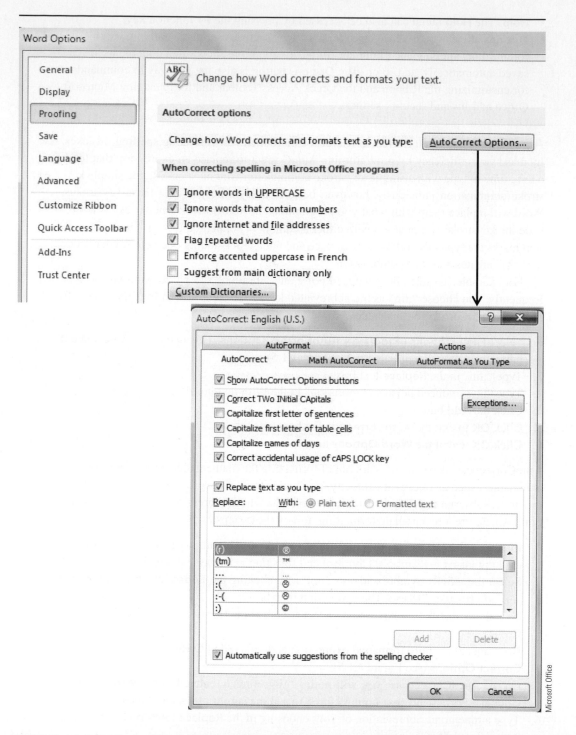

FIGURE A1.5 | To simplify the formatting of italicized expressions or those with mathematical symbols, sub- and superscripts, program AutoCorrect. Click **File | Proofing | AutoCorrect Options** to open the dialog box.

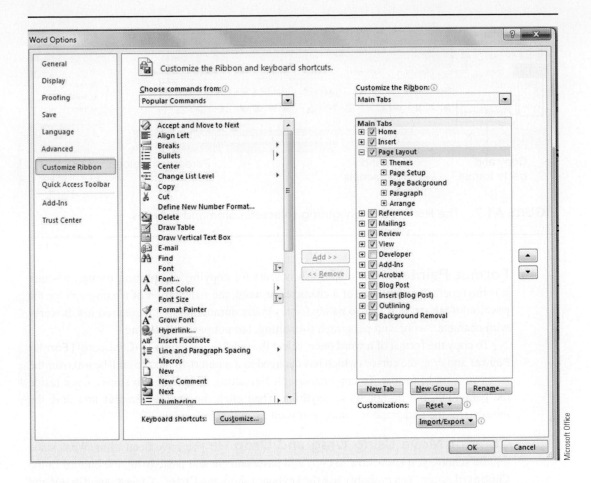

FIGURE A1.6 | To add more tabs with custom commands to the Ribbon, click **File | Options | Customize Ribbon** and then click **New Tab**.

Customize the Ribbon If you frequently use commands that are buried deep within menus and submenus, then this feature is for you. Make custom tabs on the Ribbon, and add the hard-to-find commands for easy access. In Word 2010, click **File | Options | Customize Ribbon**. After clicking the **New Tab** button, add groups and commands to the new tab (Figure A1.6). It is not possible to customize the Ribbon in Word for Mac 2011, but the menu bar and the standard toolbar can be customized.

The Home Tab

The **Home** tab contains commands for formatting text and paragraphs (Figure A1.7). The familiar **Find** and **Replace** buttons are also located here.

Paste Special Use **Paste Special** rather than **Paste** when you want to copy Web page content without the HTML (Hypertext Markup Language) formatting. Click the (→Clipboard) | **Paste** down arrow and select **Paste Special | Unformatted Text** and then **OK**.

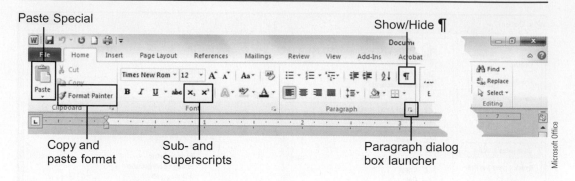

FIGURE A1.7 | The **Home** tab highlighting selected command buttons.

Format Painter   This paintbrush is great for copying and pasting format, whether it is the typeface or font size of a character or word, the indentation of a paragraph, or the position of tab stops. It is also handy for resuming numbering in a numbered list. It works with character, word, and paragraph formatting, but not page formatting.

To copy the format of a word once, select the word, click **Home** (→Clipboard) | **Format Painter** and drag the cursor (which has changed to a paintbrush next to an I-beam) over the text you want to format. To copy paragraph formatting (including tab stops), click inside the paragraph, but do not select anything. Then click the **Format Painter** and drag the mouse over the paragraph to which you want to apply the format.

Copy or Move Using Drag and Drop   If you have been using Word since middle school, you know how to copy and paste and cut and paste using the buttons in the **Clipboard** group. You probably use the keyboard shortcuts **Ctrl+C**, **Ctrl+X**, and **Ctrl+V** and the right mouse button as well. A fourth way to copy or move text is Drag and Drop. This method is best for short distances within view of your screen; it works within and between documents (e.g., when you have two or more documents open side by side).

To *move* text with Drag and Drop, select the block (see Table A1.2) and hold down the left mouse button. The arrow changes to an arrow with a box. Continue to hold down the mouse button and drag the selected text to a new insertion point (location).

To *copy* text with Drag and Drop, select the block and hold down the **Ctrl** key while holding down the left mouse button. The arrow changes to an arrow with a plus sign in the box. Drag the selected text to a new insertion point.

If you expect to paste the same block more than once, use the **Clipboard** buttons or keyboard shortcuts rather than Drag and Drop. Drag and Drop does not copy to **Clipboard**.

Sub- and Superscripts   Writers and journalists may wonder why these buttons have been placed in such a prominent location in Word 2010. Having them there, however, saves engineers quite a few keystrokes! To superscript or subscript text, first type the expression without formatting. Then highlight the character(s) to be sub- or superscripted and click the appropriate button in the **Font** group. Alternatively, you can apply and remove sub- and super-scripting using keyboard commands. Select the character(s), and then hold down **Ctrl** while pressing = for subscript or hold down both **Ctrl** and **Shift** while pressing = for superscript.

Expressions with sub- or superscripting are common in engineering, and their formatting makes them readily identifiable to members of the technical community. Therefore, it would be incorrect to write sub- or superscripted characters on the same line as the rest of the text. Similarly, when using scientific notation, exponents are always superscripted. It is not acceptable to write exponents preceded by a caret (^) or by an uppercase E to designate superscript.

RIGHT: 2×10^{-3} (exponent is superscripted)

WRONG: $2 \times 10{-}3$ or $2 \times 10{\wedge}{-}3$ or $2 \times 10\mathrm{E}{-}3$

If you frequently have to type expressions with sub- and superscripts, save time by programming them in AutoCorrect.

Show/Hide   This button allows you to view formatting marks such as ¶ (paragraph), → (tab), and · (space). Normally you would not display these marks because they are distracting. You might *want* to see them, however, when you are troubleshooting formatting problems.

Paragraph Dialog Box   The **Paragraph** dialog box can be accessed from both the **Home** and **Page Layout** tabs. Clicking the diagonal arrow on the **Paragraph** group label launches a dialog box (Figure A1.8) where you can set

- **First line indent** to denote the beginning of a new paragraph. Next to **Indentation | Special**, select **First line** from the dropdown menu and then specify **By: 0.5"** (or your preference).

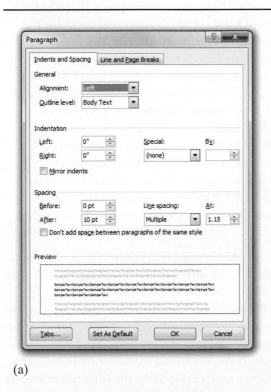
(a)

(b)

Microsoft Office

FIGURE A1.8 | The **Paragraph** dialog box is launched by clicking the arrow on the **Paragraph** group label. **(a)** Click the **Indents and Spacing** tab to format first line and hanging indents (under "Special"); to adjust spacing between lines and between paragraphs; and to set tab stops. **(b)** Click the **Line and Page Breaks** tab and check **Keep with next** to prevent section headings from separating from the body.

- **Hanging indent** for listing full references. The first line of each reference begins on the left margin, and the subsequent lines are indented. First type each reference so that it is aligned on the left margin and ends with a hard return ¶. When you are finished, select all of the references (see Table A1.2, Block of text) and click **Home** (→Paragraph) **dialog box launcher**. Next to **Indentation | Special**, select **Hanging** from the dropdown menu and then specify **By: 0.25"** (or your preference).
- **Line spacing** for all lines of text. The default spacing in Word 2010 is **Multiple at 1.15 lines**; your instructor may ask you to change the spacing to **Double** in order to have a little extra space to write comments. There is also a **Line spacing** command button in the **Paragraph** group on the Ribbon where you can make this change.
- **Paragraph spacing.** Besides indenting the first line, paragraphs can be separated with a blank line. To automatically add extra space after each paragraph, thereby saving you the trouble of pressing the **Enter** key twice, put the insertion pointer anywhere in the paragraph. In the **Paragraph** dialog box, select **12 pt** in the **Spacing | After** box. 12 pt paragraph spacing corresponds to one blank line when using a 12 pt font size for your paper.
- **Line and page breaks.** This feature is handy for preventing section headings from being separated from the body due to a natural page break. Select the section heading (for example, "Apparatus and Procedures"). Then open the **Paragraph** dialog box and click the second tab called **Line and Page Breaks**. Under **Pagination**, check the box next to **Keep with next** to keep the heading together with the body. To keep all the lines of a paragraph together on the same page, check **Keep lines together**.
- **Tabs.** Whenever you want to align text in columns, use tab stops or make a table without borders. Never use the space bar for more than one space. Even the old custom of putting two spaces after periods and colons has become obsolete as a result of proportionally spaced computer fonts. The **Tabs** button is located in the bottom left corner of the **Paragraph dialog box**. To add a new tab stop, open the **Tabs** dialog box and enter a **Tab stop position**. Choose an alignment (left, center, right, or decimal). Click the Set button. Repeat this procedure to add more tab stops. When you are done, click **OK**. The new tab will appear on the ruler at the top of the page. If the ruler is not displayed, click **View Rule**r at the top of the scroll bar on the right side of the page (see Figure A1.1).

A faster, but slightly less precise, way to add tab stops is to select an alignment by repeatedly clicking the **Tab selector** button on the left side of the ruler (see Figure A1.1) and then simply clicking the ruler where you want to add the new tab stop.

To remove the tab stop, point to it, hold down the left mouse button, and drag it off the ruler. To move the tab stop, simply drag it to another position on the ruler. You can also carry out these tasks in the **Tabs** dialog box.

The default units of the ruler are inches. If you would like to use metric units instead, click **File | Options | Advanced**. In the Display area, next to **Show measurements in units of**, select the desired units from the dropdown menu.

The Insert Tab

The **Insert** tab has buttons for inserting pages and page breaks, tables, illustrations, links, headers and footers, textboxes, equations, and symbols (Figure A1.9).

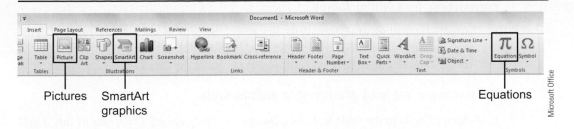

Pictures   SmartArt   Equations
            graphics

FIGURE A1.9 | The Insert tab highlighting selected commands.

Tables   By convention, tables in technical reports do not have vertical lines to separate the columns, and horizontal lines are used only to separate the table caption from the column headings, the headings from the data, and the data from any footnotes (notice the format of the tables in this book).

Creating a Table

1. Position the insertion pointer where you want to insert the table.
2. Click **Insert** (→Tables) | **Table**. Highlight the desired number of columns and rows.
    A blank table appears with the cursor in the first cell. To change the table format, use the **Table Tools Design** and **Layout** tabs (Figure A1.10). These tabs only appear when the insertion pointer is inside the table.
3. To insert or delete columns or rows or to merge or split cells, use the command buttons on the **Table Tools Layout** tab.

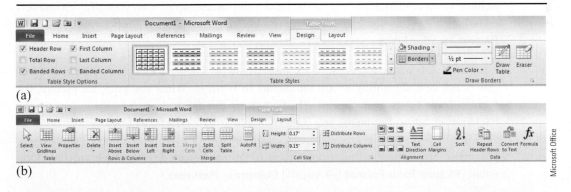

(a)

(b)

FIGURE A1.10 | Table formatting tabs appear when you click inside a table. **(a)** The **Design** tab is used to modify shading and borders. **(b)** The **Layout** tab is used to insert or delete rows and columns, split or merge cells, adjust cell size, and sort data.

Formatting Text within Tables

1. To apply formatting to adjacent cells, first select the cells by clicking the first cell in the range, holding down the **Shift** key, and then clicking the last cell.
2. To apply formatting to selected cells, repeat (1) using the **Ctrl** key.

Formatting a Table by Modifying a Built-in Style

1. Select all cells in the table and click **Design** (→Table Styles). Click one of the "Light Shading" styles. These table styles have the appropriate horizontal lines and no vertical lines preferred in technical papers.
2. With the whole table still selected, click the down arrow next to the **Shading** button and select **No Color**.

Navigating in Tables

- To jump from one cell to an adjacent one, use the arrow keys.
- To move forward across the row, use the **Tab** key. *Note:* If you press **Tab** when you are in the last cell of the table, Word adds another row to the table.
- To align text on a tab stop in a table cell, press **Ctrl+Tab**.

Viewing Gridlines It is easier to enter data in a table when the gridlines are displayed. Position the insertion pointer inside the table and click **Layout** (→Table) | **View Gridlines**. Gridlines are not printed.

Anchoring Images within Tables Getting images to line up horizontally and vertically on a page can be tricky. It is possible to make multiple columns on the **Page Layout** tab or to adjust the position of each picture on the **Picture Tools Format** tab, but quite often the images become misaligned when the document is edited. To make images stay put, create a table. Select the desired layout (for example, 3 columns × 1 row), size the images so they fit into the cells, and then paste the images into the table.

Repeating Header Rows When a table extends over multiple pages, it is convenient to have Word automatically repeat the header row at the top of each page. To activate this command, click anywhere in the header row and then select **Table Tools Layout** (→Data) | **Repeat Header Rows**.

Picture
To insert a picture, click **Insert** (→Illustrations) | **Picture**. While that function is not new, the ability to edit the picture in Word is. When you click the picture, the **Picture Tools Format** tab appears. The buttons on this tab make it possible to remove the background, adjust brightness and contrast, apply artistic effects, and add various borders. Because high-resolution images can increase the size of a Word document significantly, compress the pictures if the document will be printed, posted to the Web, or shared via email (**Picture Tools Format** (→Adjust) | **Compress Pictures**).

Shapes
To make line drawings or add simple graphics to your document, click **Insert** (→Illustrations) | **Shapes**. To insert a line, for example, click a line to display crosshairs. Position the mouse pointer on the page where you want the line to start, hold down the left mouse button, drag to where you want the line to end, and release the mouse button. To make perfectly horizontal or vertical lines, hold down the **Shift** key while holding down the left mouse button and drag. In general, when it is important that a shape retain perfect

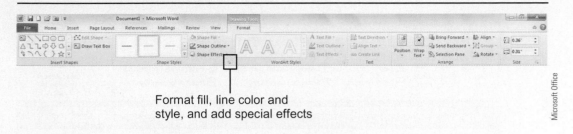

Format fill, line color and
style, and add special effects

FIGURE A1.11 | To format a shape, click it to open the **Drawing Tools Format** tab. Right-clicking the shape or selecting the dialog box launcher opens a dialog box with additional options.

proportions (e.g., circle not oval; square not rectangle), hold down the **Shift** key while drawing or resizing the shape with the mouse.

To change the angle or length of the line, click the line and then position the mouse pointer over one of the end points of the line to display a two-headed arrow. Hold down the left mouse button, drag to where you want the line to end, and release the mouse button.

To change the location of the line, click the line and then position the mouse pointer over the line to display crossed arrows. Hold down the left mouse button, drag the line to the desired location, and release the mouse button. For more precise positioning, right-click the line and select **Format AutoShape** from the dropdown menu. On the **Layout** tab, click the **Advanced** button. Change the **Picture Position** to place the line exactly where you want it.

To format the line, right-click it and select **Format AutoShape.** This dialog box gives you options for changing the appearance of the line, resizing it precisely, and laying it out in relation to other objects. Many of the same formatting commands are located on the **Drawing Tools Format** tab that appears when you select the line (Figure A1.11).

If several objects make up a graphic, then it may make sense to group them as a unit. Grouping allows you to copy, move, or format all the objects in the graphic at one time. To group individual objects into one unit, click each object while holding down the **Shift** key. Release the **Shift** key and then click **Group** under **Drawing Tools Format** (→Arrange). One set of selection handles surrounds the entire unit when the individual objects are grouped. To ungroup them, simply select the group and click **Group | Ungroup**.

The **Align** button is handy for precisely lining up objects (shapes as well as text boxes). Click each object that you want to line up while holding down the **Shift** key. Release the **Shift** key, click **Align** under **Drawing Tools Format** (→Arrange), and select one of the alignment options. Another useful feature for arranging objects on a page is **Align | View Gridlines**. This command puts a non-printing grid on your page. Click **Align | Grid Settings** to adjust the spacing between the gridlines or to snap objects to the grid. Check the **Snap Objects to grid when the gridlines are not displayed** box if you want to position objects on gridlines. Do not check this box if you prefer to position the objects freely. Finally, to move objects just a fraction of a millimeter from their current position, click the object, hold down the **Ctrl** key, and use the arrow keys to nudge the object exactly where you want it on the page.

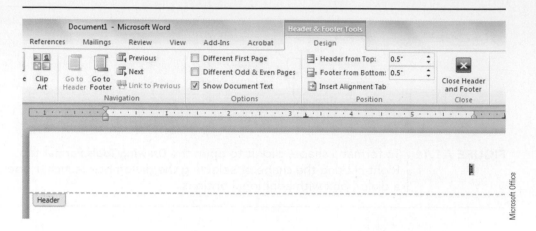

FIGURE A1.12 | Page numbers are part of headers and footers. To remove the page number from the top of the first page, select **Different First Page** from the **Header & Footer Tools Design** tab.

SmartArt   SmartArt is a collection of graphics templates within MS Word for lists, processes, cycles, and relationships. Before you go to the trouble of drawing your own graphics, see if one of the SmartArt templates might suit your needs. If you have to draw more complicated graphics, try Microsoft Visio. For highly technical drawings, a computer-aided design program like AutoCAD®, SolidWorks, or Pro/ENGINEER, among others, may be required.

Page Numbers   Page numbers make it easier for you to assemble the pages of your document in the correct order. To insert a page number, click **Insert** (→Header & Footer) | **Page Number** (see Figure A1.9) and then select a position for the numbers. As shown in Figure A1.12, after the page number is inserted at the top right of the page, the **Header & Footer Tools Design** tab appears. In the **Options** group, you have the opportunity to check **Different First Page** to remove the page number from the first page. To remove all page numbers, click **Insert** (→Header & Footer) | **Page Number** | **Remove Page Numbers**. This command is not available in documents saved in earlier versions of Word. For such documents, double-click the page number to open the header or footer and then double-click the page number again and press **Delete**. When you are finished inserting and formatting the page numbers, click **Close Header and Footer**.

Equations   To insert a common mathematical equation or make your own, click **Insert** (→Symbols) | **Equation** to display the **Equation Tools Design** tab (Figure A1.13a). Peruse the built-in equations in the dropdown menu for the **Equation** button on the left side of the tab. If this menu does not contain the equation you would like to write, choose a form to modify from the **Structures** group. Fill in numbers or select Greek letters, arrows, and mathematical operators from the eight symbol sets that appear after clicking the **More** arrow (Figure A1.13b).

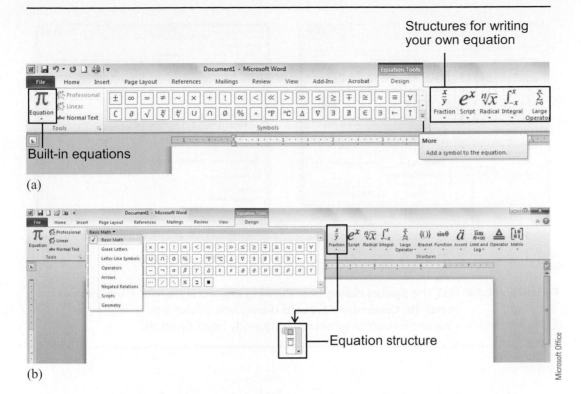

Structures for writing your own equation

Built-in equations

(a)

Equation structure

(b)

FIGURE A1.13 (a) The **Equation Tools Design** tab lets you choose built-in equations or select a structure to modify. (b) Insert symbols, Greek letters, and operators for the equation by clicking the **More** arrow in the **Symbols** group and selecting from one of the symbols sets.

Symbols

Greek letters and mathematical symbols can also be inserted in running text. To do so, click **Insert** | (→Symbols) | **More Symbols** (Figure A1.14a). The characters that are displayed in the **Symbol** dialog box depend on which font is selected. Under **Font: (normal text)**, you will find commonly used symbols such as degrees (°), plus or minus (±), micro (μ), parts per thousand (‰), infinity (∞), is less than or equal to (≤), is greater than or equal to (≥), middle dot (·), and arrows for chemical reactions (→ and ←). Normal text font also includes Greek letters and characters with diacritical marks used in European languages other than English (for example, à, â, ã, ä, and ç).

If you use a particular Greek letter or mathematical symbol frequently, you can make a shortcut key to save time. Consider the degree sign, for example (see Figure A1.14a). After clicking this symbol in the **Symbol** dialog box, click the **Shortcut Key** button at the bottom of the box. In the **Customize Keyboard** dialog box (Figure A1.14b), define a combination of keystrokes using **Ctrl**, **Alt**, **Ctrl+Shift**, or **Ctrl+Alt** plus some other character. Because the degree sign looks like a lowercase *o*, we will use **Alt+o**. To define **Alt+o** as the shortcut key for the degree sign, type **Alt+o** in the **Press new shortcut key** box. Word notifies you that this combination is unassigned (as in Figure A1.14b) or that it has already been assigned to another command. (*Note*: Even though you typed Alt and lowercase o, the shortcut appears as Alt and uppercase O in the box. If you had typed Alt and uppercase O, this would have appeared as Alt+Shift+O.)

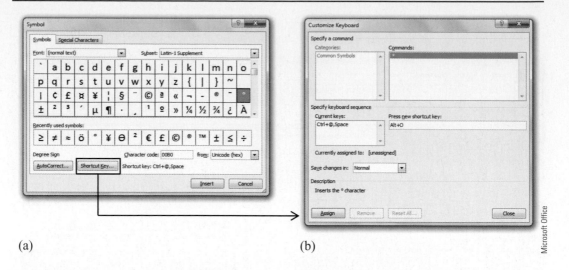

(a)   (b)

FIGURE A1.14 | **(a)** The **Symbol** dialog box showing characters under **Font: (normal text)**. **(b)** **Customize Keyboard** dialog box allows you to define a keystroke combination for frequently used symbols.

If the shortcut is already assigned to a different command, determine if you use that command very often. If you do not, you can remove the shortcut from the seldom-used command and assign it to the symbol you want. To do this:

1. Select **Assign** from the buttons at the bottom of the dialog box (see Figure A1.14b). **Alt+o** will then be listed under **Current keys**. If Word assigned a different shortcut key to the degree sign, this will also be displayed under **Current keys** (in this example, **Ctrl+@, Space**).
2. Click the previously assigned shortcut key, and then select **Remove** from the buttons at the bottom of the dialog box.
3. Close the **Customize Keyboard** dialog box.
4. Close the **Symbol** dialog box.

Next time you have to write the symbol for "degrees Celsius," simply type **Alt+o** followed by uppercase C to get: °C.

Another way to simplify inserting symbols is to program AutoCorrect. Once again using the degree sign as an example, open the **Symbol** dialog box, click the degree sign, and then click the **AutoCorrect** . . . button at the bottom of the box (see Figure A1.14a). The ° symbol will already be entered in the **With** textbox, and the **Formatted text** option box will be selected. Type a unique combination of characters in the **Replace** textbox. Because you may not want to replace the word "degrees" with the degree symbol (°) every time, use the trick of preceding the word with either a comma or semicolon (, or ;) (see the "AutoCorrect" section). Type ";degrees" in the **Replace** box. Click the **Add** button, click **OK** to exit the **AutoCorrect** dialog box, and finally click **OK** to exit the **Word Options** box.

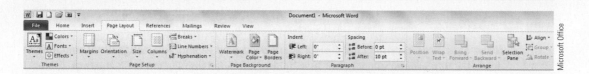

FIGURE A1.15 | The Page Layout tab is used to format pages.

The Page Layout Tab

The **Page Layout** tab has buttons for formatting whole pages (Figure A1.15). The **Page Setup** and **Paragraph** groups are the ones you are likely to use for your technical reports. The dialog box launchers for these groups open the dialog boxes that you are familiar with from using previous editions of Word. The commands in the **Themes** and **Page Background** groups affect the overall design of the page and are useful for designing your own stationery, invitations, and so on.

Document Orientation Most lab reports are printed in Portrait Orientation. Occasionally, however, it may be necessary to insert a figure or large table in Landscape Orientation. To do so, insert a section break (*not* a page break) at the end of the previous page by clicking **Page Layout** (→Page Setup) | **Breaks** | **Section Breaks** | **Next Page**. Then click **Page Layout** (→Page Setup) | **Orientation** | **Landscape** to create a page turned sideways. After you are finished typing the table or pasting a figure created in Excel, insert another section break and choose Portrait to return to the usual orientation.

Sometimes section breaks throw off the page numbering. If this happens to your document, link the header or footer that contains the page number to the previous section. To do so, click **Insert** (→Header & Footer) and select either **Header** or **Footer**, depending on where the page number is located. Then click **Edit Header** (or **Edit Footer**) and scroll to the section in which the page number is messed up. Under **Header & Footer Tools Design** (→Navigation), click **Link to Previous** to continue the numbering from the previous section. Click **Close Header and Footer** to exit (see Figure A1.12).

The References Tab

Insert Endnotes The most useful feature on the References tab, at least with regard to technical papers, is the Insert Endnote (*not* footnote!) command for in-text references in the citation-sequence (C-S) system (Figure A1.16). To insert superscripted endnotes, type the sentence containing information that requires a reference. End the sentence with a period. Then click **References** (→Footnotes) | **Insert Endnote**. Type the full reference in proper format (see "The Citation-Sequence System" in Chapter 2). Click **References** (→Footnotes) | **Show Notes** to exit the endnote and return to the text. To change the endnote number style from Roman to Arabic, click the **Footnotes dialog box launcher** and select Arabic numeral format from the dropdown menu.

Word updates endnotes automatically so that they appear sequentially. To see an endnote, point the mouse at the superscripted number. To edit an endnote, double-click the

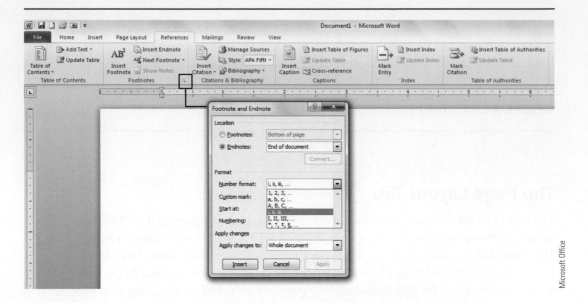

Microsoft Office

FIGURE A1.16 | The **Footnote and Endnote** dialog box opens after clicking the **Footnotes dialog box launcher** on the **References** tab. Number format is selected from the dropdown menu.

number. To delete an endnote, highlight the superscripted number and press **Delete**. The other endnote numbers will be renumbered automatically.

Table of Contents

For longer papers, such as dissertations and honors theses, you might like to provide a table of contents. The easiest way to generate a table of contents is to use the built-in heading styles on the **Home** tab. First decide which headings (and the level of each) in your document will be used as entries in the table of contents. Second, generate the table.

Mark the Entries as Headings in Your Document   It does not matter if you define the heading level before or after you type the heading. To define the heading level first, click **Home** (→Styles) and then select **Heading 1 (2, 3,** or **4)**. Then type the heading and press **Enter**. Alternatively, to mark the text as a heading after you have typed it, highlight it, and then click **Home** (→Styles) and select the appropriate heading level.

Generate the Table of Contents   Place the insertion pointer at the beginning of the document. Click **References** (→Table of Contents) **| Table of Contents** and select a style for the table. Word automatically generates the table from the built-in headings. If you add or delete headings while revising your document, simply click **References** (→Table of Contents) **| Update Table**.

Citations & Bibliography

The **Citations & Bibliography** group promises to cite sources in one of the common styles, but this feature has major deficiencies. First, the Council of Science Editors (CSE) style is not among those listed. Second, APA style

(American Psychological Association) has similarities to CSE style, but the in-text citation comes out wrong. Whether the publication has one, two, or more authors, Word only lists one author in the in-text reference. The *correct* way to cite references in the name-year (N-Y) system is to give both authors' names when there are two authors and the first author's name plus et al. when there are three or more. Until Word's software developers are able to correct these mistakes, use another references management tool or type in-text citations and full references as described in Chapter 2.

The Review Tab

The **Review** tab is where you will find commands for checking spelling and grammar, looking up synonyms and antonyms, adding comments, tracking changes made by others, and other functions related to editing (Figure A1.17).

Proofing   The **Proofing** group contains a spelling and grammar checker and buttons that connect to online dictionaries and other references. Spelling and grammar issues are discussed more fully in the next major section, "Proofreading Your Documents."

Comments   Comments are frequently used to exchange ideas when collaborating on a paper. You are most likely to add a comment, read someone else's comment and respond with a comment of your own, or delete one or more comments. You may also like to change how the comments are displayed on your screen.

Adding a Comment

1. Select the text you want to comment on.
2. Click **Review** (→Comments) | **New Comment**.
3. Type your comment in the box. Click outside of the comment area to return to the document.
4. To change the initials used in comments, go to **File | Options | General | Personalize your copy of Microsoft Office**.

Reviewing Comments   When there are no comments in a document, the **Delete**, **Previous**, and **Next** buttons in the **Comments** group are grayed out (as in Figure A1.17). When there are comments, these commands become available. After you have read and, if appropriate, taken action on all of the comments, delete them with **Review** (→Comments) | **Delete | Delete All Comments in Document**. To delete comments one at a time, right-click the balloon or the inline comment and select **Delete Comment**.

FIGURE A1.17 | The **Review** tab provides options for revising your document, including spelling and grammar check, making comments, and tracking changes.

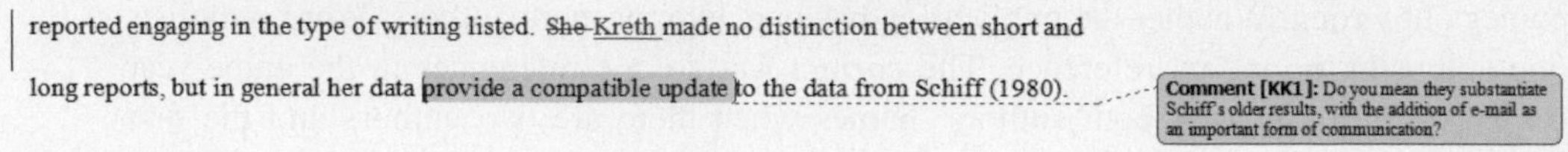

reported engaging in the type of writing listed. She Kreth made no distinction between short and

long reports, but in general her data provide a compatible update to the data from Schiff (1980). ⌐ **Comment [KK1]:** Do you mean they substantiate Schiff's older results, with the addition of e-mail as an important form of communication?

Microsoft Office

FIGURE A1.18 | Comments are displayed in numerical order in balloons (as in this screenshot) or on a **Reviewing Pane**. Command buttons related to viewing, adding, and deleting comments are located on the **Review** tab.

Displaying Comments Comments can be displayed as balloons in the margin (as in Figure A1.18) or within the document itself (inline). To see the difference, click **Review** (→Tracking) | **Show Markup** | **Balloons** and click one of the three options: **Show Revisions in Balloons**, **Show All Revisions Inline**, or **Show Only Comments and Formatting in Balloons**. When the inline option is selected, revisions appear in a **Reviewing Pane** at the bottom or left-hand side of the screen (depending on the layout selected under (→Tracking) | **Reviewing Pane**).

Tracking Changes Made by Others

It may not always be possible for you and your collaborator (or reviewer) to find a common time to meet to go over your paper. Email makes the review process more convenient. You can send your paper to a reviewer in electronic format as an attached file, and your reviewer can send it back to you after making comments or revisions directly in the document.

It is important for you to be able to distinguish your original text from the comments and changes suggested by your reviewer. After all, you are the author, who has the right to accept or decline the reviewer's suggestions.

Before you send your lab report to a reviewer, click **Review** (→Tracking) | **Track Changes** to turn on tracking (the button will turn orange). As long as this command is activated, anything someone types in the document will be colored and underlined. Anything someone deletes will be colored and crossed out. Each reviewer gets his or her own color, so you can identify the changes made by multiple reviewers. In addition, a black vertical line in the left margin alerts you to changes in your document.

When you get your document back, open it, and turn off tracking by clicking the **Track Changes** button again. Then click (→Changes) | **Next** to go to the first location where a comment or revision was made. Use the dropdown menus for **Accept or Reject** or simply click **Next** to read the suggestion without taking any action.

To make sure you have addressed all tracked changes, save your document and then click **File** | **Info** | **Check for Issues** | **Inspect Document**. The **Document Inspector** box will notify you if any comments, revision marks, personal information, or other issues were found. Rather than clicking **Remove All**, close the box and look for the changes you missed. To do so, click **Review** (→Changes) | **Next**. Delete the stray revision mark after deciding whether any action is necessary.

The View Tab

Print Layout view is the one you see when you open a Word 2010 document. This view shows you how your page will look when it is printed, including headers and footers and

FIGURE A1.19 | The **View** tab contains commands that were present on the **View** and **Window** buttons on the menu bar in Word 2003.

images. Other views can be selected from the **Document Views** group on the **View** tab (Figure A1.19) and on the status bar at the bottom of the screen (see Figure A1.2, Layout selector).

PROOFREADING YOUR DOCUMENTS

Before you waste reams of paper printing out drafts of your document, have Word check spelling and grammar and inspect the document for stray revision marks and comments. Use your eyes to look over format on screen before proofreading the printed document.

Spelling and Grammar

Word gives you visual indicators to alert you to possible spelling and grammar mistakes:

- The **Proofing** button on the status bar at the bottom of the screen shows a red X instead of a green ✓.
- Words that are possibly misspelled are underlined with a wavy red line.
- Phrases that may contain a grammatical error are underlined with a wavy green line.

Do not ignore these visual cues! Click the **Proofing** button to see what potential problem(s) Word has identified. Take action by choosing one of the options in the dialog box. Alternatively, to deal with a word underlined in red, right-click it. A pop-up menu appears with commands and suggestions for replacements (Figure A1.20). **Ignore All** applies only to the current document. **Add to Dictionary** applies to all future documents. It makes sense to add technical terminology to Word's dictionary after you consult your textbook or laboratory manual to confirm the correct spelling. After making a selection on the pop-up menu, the wavy red underline is deleted and the word is ignored in the manual, systematic spelling, and grammar check. Similarly, to deal with a possible grammatical error underlined in green (including extra spaces between words), right-click it to accept or ignore Word's suggestions.

The **AutoCorrect** function corrects common types of spelling mistakes as you type. To see the list of commonly misspelled words that Word corrects automatically, click **File | Options | Proofing | AutoCorrect Options**, and scroll through the words on the **AutoCorrect** tab. If you do not want AutoCorrect to change a particular keystroke combination (e.g., do not change (c) to ©), then select this combination on the list, and click the **Delete** button. You can also add to the list words that you know you misspell frequently.

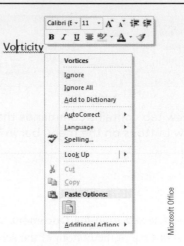

Microsoft Office

FIGURE A1.20 | Dialog box for spelling suggestions and other ways to handle words not in Word's dictionary. When you right-click a word in Word 2010, a mini-toolbar also pops up with options for changing typeface, character size, color, and so on.

If AutoCorrect changes text that you do not want changed while you are typing, click the **Undo** button on the **Quick Access Toolbar**.

Format

Use **Print Layout** view to proofread your document on screen. This view closely resembles what your printed page will look like with headers, footers, blank areas, figures, and page breaks. **Print Layout** is the view you see when you first open a Word 2010 document.

When proofreading for format, pay attention to the following items:

Margins   1.25" left and right and 1" top and bottom margins give your instructor room to make comments. Margins can be adjusted by clicking **Page Layout** (→Page Setup) | **Margins** or **Page Layout** (→Page Setup) **dialog box launcher**.

Paragraph Spacing   Paragraphs end with a ¶ symbol. The next paragraph starts either with the first line indented or preceded by an empty line. Word 2010 automatically adds 10 pt spacing after each paragraph to simplify formatting. If you like the automatic space, then do not use the first line indent method for starting a new paragraph. If you prefer the first line indent method, go to **Home** (→Paragraph) **dialog box launcher** and change **Indentation: Special** to **First line** and **Spacing: After** to **0**.

Document Orientation   Most lab reports are printed in Portrait Orientation. If, however, you have one or more pages in Landscape Orientation to accommodate a figure or large table, for example, check the format of the margins, page numbers, and orientation of *all* the pages in the document. If you find any orientation-related formatting problems, correct them with commands on the **Page Layout** tab.

Page Numbers Page numbers make it easier for you to assemble the pages of your document in the correct order. Check that the pages are numbered consecutively, especially the page after a section break. See the "Page Numbers" section to insert or remove page numbers. See "The Page Layout Tab" section for troubleshooting tips.

Section Headings Must Not Be Separated from the Body Technical papers are divided into sections that have headings such as Abstract, Introduction, Apparatus and Procedures, Results, Discussion, References, and Acknowledgments. The heading is typed on a separate line, with the section body starting on the next line. When you check the format of your paper, make sure the heading is not cut off from the body.

To prevent heading–body separation problems, use one of these options:

- Select the heading. Click **Home** (→Paragraph) **dialog box launcher | Line and Page Breaks** tab. Under **Pagination**, check **Keep with next**.
- Insert a hard page break (**Ctrl+Enter**) to the left of the heading to force the heading onto the next page with the body.

Figures and Tables Must Not Be Separated from Their Captions If necessary, insert a hard page break (**Ctrl+Enter**) to the left of the table caption (which, by convention, is typed above the table) or the figure itself (the figure caption goes below the figure).

Errant Blank Pages To delete blank pages in the middle of a document, go to the blank page, click it, and display the hidden symbols (**Home** (→Paragraph) **| Show/Hide ¶**). Delete spaces and paragraph symbols until the page is gone. Similarly, if the blank page is at the end of the document, press **Ctrl+End** and remove the hidden symbols to delete the extra page.

Document Inspector

To check that you have removed all comments and tracked changes, click **File | Info | Check for Issues | Inspect Document**. Rather than selecting **Remove All**, click **Close**. Find the stray revision marks by clicking **Review** (→Changes) **| Next** and take action, if necessary, and delete or reject them.

Finally, Print a Hard Copy

When you are confident that you have found and corrected all mistakes on screen, print a hard copy and proofread your paper again. Some mistakes are more easily identified on paper, and it is always better for you, rather than your instructor or supervisor, to find them.

HOUSEKEEPING

Organizing Your Files in Folders

You can expect to type at least one major paper and perhaps several minor writing assignments in every college course each semester. This amounts to a fair number of documents on your computer. One way to organize your files is to make individual course folders that contain subfolders for lecture notes, homework, and lab reports, for example. To create a new folder in Word when you are saving a file:

1. Click **File | Save As**. The **Save As** dialog box appears.
2. Click a link in the breadcrumb trail in the box at the top or use the navigation pane on the left to locate the folder in which you want to create a new folder.
3. Click the **New Folder** button. The **New Folder** dialog box appears.
4. Give the folder a short, descriptive name. Click **OK**.

To create a new folder in Windows Explorer:

1. Click **Start | Documents**. The **Documents library** dialog box appears.
2. Click **New Folder**. Type a name for the new folder and press **Enter**.

Accessing Files and Folders Quickly

To make frequently used files or folders readily accessible, consider pinning them either to the **Recent** list in Word or to the **Favorites** list in Windows Explorer.

To pin files or folders to **Recent** in Word:

1. Click **File | Recent** to display recent documents and recent folders.
2. Right-click a file or folder and select **Pin to list** from the menu.
3. The file (folder) will be pinned to the top of the **Recent Documents (Recent Places)** list.
4. When you are finished working on the file (folder), click **File | Recent**, right-click the pinned file (folder), and select **Unpin from list**.

To pin files or folders to **Favorites** in Windows Explorer:

1. Open a **Windows Explorer** dialog box by clicking **Start | Documents**.
2. Browse your documents library until you find the frequently used file (folder).
3. Hold down the left mouse button and drag the file (folder) to **Favorites** in the navigation pane on the left (Figure A1.21).

Naming Your Files

File names can include letters, numbers, underlines, and spaces as well as certain punctuation marks such as periods, commas, and hyphens. The following characters are not allowed: \, /, :, *, ?, ", <, >, and |.

In general, file names should be short and descriptive so that you can easily find the file on your computer. If you intend to share your file, however, consider the person with whom you are sharing it. When you send your lab report draft to your instructor for feedback, put your name and the topic in the file name (for example, Miller_TransientCond); while "HTlab1" may seem unambiguous to you, it is not exactly informative for your heat transfer professor!

If you have forgotten the exact file name and location, you can search for it using Windows. Click **Start | Computer | Local Disk (C:)** (or **Documents** if you are able to narrow down the location) and search for files and folders.

Saving Your Documents

When you write the first draft of a paper by hand, you have tangible evidence that you have done the work. When you type something on the computer, however, your work is

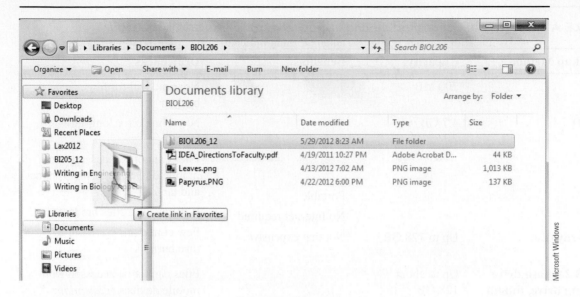

FIGURE A1.21 | Pin frequently used files and folders to **Favorites** in Windows Explorer by clicking the file or folder and dragging it to the navigation pane on the left.

unprotected until you save it. That means that if the power goes off or the computer crashes before you save the file, you have to start again from scratch. Hopefully, it will not take the loss of a night's work to convince you to **save your work early and often**. Do not wait until you have typed a whole page; save your file after the first sentence. Then continue to save it often, especially when the content and format are complicated. Think about how long you would need to retype the text if it were lost and if you can afford to spend that much time redoing it.

Word automatically saves your file at certain intervals. You can adjust the settings by clicking **File | Options | Save | Save AutoRecover information every __ minutes**. It is also a good idea to save the file manually from time to time by clicking 🖫 on the **Quick Access Toolbar**.

Backing Up Your Files

It should be obvious that your backup files should not be saved on your computer's hard drive. Table A1.4 lists some offline and online backup options along with their advantages and disadvantages. In terms of offline options, USB flash drives are probably the best option for compactness and convenience, but external hard drives come with software that lets you schedule automatic backups. With an account and an Internet connection, you can take advantage of any number of online options. These include Google Drive and other cloud services that store your files virtually, saving files to your organization's server, and even emailing files to yourself.

TABLE A1.4 | Possible options for backing up your electronic files

Backup Method	Capacity	Benefits	Drawbacks
CD	700 MB		Not practical for large-scale backup
DVD	4.7 GB		Not reliable for long-term storage
			Files cannot be accessed from mobile devices like smartphones and tablets
		Portable	If disc is lost, files are lost
		No Internet required	
Blu-ray disc	Up to 128 GB	Not that expensive	Few computers have blu-ray disc burners
USB 2.0 flash drive (jump drive, thumb drive)	Up to about 128 GB		Files cannot be accessed from mobile devices (*Exception*: Some mobile devices support SD cards and MicroSD cards)
MicroSD card	Up to 64 GB		
SD card	Up to 128 GB		If disc, drive, or card is lost, files are lost
External hard drive	Up to about 3 TB		
Internal server	Depends on organization	Free	In-house Ethernet required
		Files can be shared or kept private	Internet and private network software required to access off-campus
			Limited storage capacity
Send emails to yourself	Depends on the email provider, but typically in the GB range	Free	Internet required
		Potentially infinite storage capacity	Limit to size of file that can be sent as attachment (typically 10–25 MB)
		Can be accessed from other computers and mobile devices	
Cloud storage (remote server)	Infinite	All services offer a free storage option (2–7 GB)	Internet required
Box		Can be accessed from other computers and mobile devices	Fees increase as storage needs increase
Dropbox			
Google Drive[*]			
Microsoft SkyDrive		Files can be shared or kept private	Limit to file size that can be uploaded
SugarSync			

[*]Google Drive provides up to 5 GB of free storage for non-Google formats (ppt, xls, jpg, mov, etc.) and unlimited for Google formats like Google Docs (D. Hiller, personal communication, 20 July 2010).

The most important thing about backing up your files is to have a plan. The odds are in your favor, unfortunately, that you will have a computer malfunction at least once every three years. For that reason alone,

- Schedule regular, automatic backups for your hard drive (that way you will not forget).
- Have a second backup option for important files that you are currently working on (especially projects with tight deadlines).
- Make sure your backup options do not run out of storage space. As a rough guide, a 10-page text file has a size of about 200 KB. In comparison, a picture taken with your digital camera is typically between 1000 KB and 3000 KB, and each of your music files is around 4000 KB.

The bottom line is: **Protect your valuable files with a reliable backup system.**

Working with Previous Versions of MS Word

Starting with Word 2007, Microsoft introduced a new file format that decreases file size and increases security and reliability. The file extension of this new format is **.docx**; earlier versions of Word end with **.doc**. If you have an earlier version of Word, you will not be able to open **.docx** files unless you have installed the Compatibility Pack. To download this pack, type "office 2010 compatibility" into your browser's search box and download the Compatibility Pack from Microsoft's home page.

If Office 2007 or Office for Mac 2008 or later is installed on your computer, you will be able to open documents created in earlier versions of Word. After you open a **.doc** file, the title will appear with "Compatibility Mode" in parentheses on the title bar. To convert an older Word document to the new file format, click **File | Info | Convert**. A message warns you that the layout may change, but converting to the new file format allows you to take advantage of new features such as:

- Modifying imported Excel graphs directly in your Word document. In earlier versions, all corrections and modifications to graphs had to be made in Excel. In Word 2007 and later versions for both PCs and Macs, you *can* make changes to graphs after they have been imported into Word.
- Inserting equations by selecting a common equation from Word's list or typing your own equation into one of the built-in equation structures (see "The Insert Tab, Equations").
- Checking for personal information or hidden content before sharing the document with others (see "The File Tab"). This feature is handy for removing all comments and tracked changes on final versions of lab reports.
- Protected View to view files from potentially untrustworthy sources. You can manage the Trust Center settings by clicking **File | Options | Trust Center | Trust Center Settings**.

Finally, if you would like to share a Word 2007 document with someone who has an earlier version of Word and who has not installed the Compatibility Pack, click the **File** tab and select **Save As | Word 97-2003 Document**.

REFERENCES

Meece, M. 2012. A User's Guide to Finding Storage Space in the Cloud. [Internet] The New York Times. [publ. 16 May 2012]. Accessed 17 July 2012 at <http://www.nytimes.com/2012/05/17 /technology/personaltech/a-computer-users-guide-to-cloud-storage.html?pagewanted=all>

Wynn, E.S. The Best Ways To Backup Your Files And Photos—Advice And Tips. HubPages, Inc. [updated 5 Jan 2012]. Accessed 17 July 2012 at <http://earlswynn.hubpages.com/hub/What-is-the-best-way-to-backup-your-files-and—photos——advice-and-tips-_>

Appendix II

MAKING GRAPHS IN MICROSOFT EXCEL 2010 AND EXCEL FOR MAC 2011

INTRODUCTION

The focus of this appendix is on Microsoft Excel 2010 for Windows. The commands for making graphs in Microsoft Excel 2010 for PCs and Excel for Mac 2011 are quite similar, but the user interface to access those commands is a little different. In this appendix, screenshots and instructions for both systems will be given unless the command sequence is the same, in which case Excel 2010 will be the default. Screenshots for Excel 2010 were taken on the Windows 7 operating system; those for Excel for Mac 2011 were taken on the OS X operating system.

If you are used to using Excel 2007, then Excel 2010 will not seem that different. However, if you are a Mac user, the Ribbon is a new addition to the user interface at the top of the screen. The **Ribbon** is a single strip that displays **commands** in task-oriented **groups** on a series of **tabs** (Figure A2.1). In Excel 2010, the new **File** tab replaces the Microsoft **Office Button** that was located in the top left of the screen in Excel 2007. Additional commands in some of the groups can be accessed with the **Dialog Box Launcher**, an inset diagonal arrow in the lower right corner of the group label (circled in red in Figure A2.1a). The **Quick Access Toolbar** comes with buttons for saving your file and undoing and redoing commands; you can also add buttons for tasks you perform frequently. The commands on the highlighted tab will always be displayed when there is an upward pointing "hat" symbol (^) next to the blue question mark in the upper right corner. If the arrow points down ($_\vee$), the tab closes when you click on any other area of the screen.

The window size and screen resolution affect what you see on the Ribbon. On a smaller screen or with lower screen resolution, you may see fewer command buttons or an entire group incorporated into one button. Furthermore, in Excel 2010 you can create custom tabs on the Ribbon. The beauty of this function is that if you frequently use commands that are buried deep within menus and submenus, you can add these commands to your custom tabs for easy access. See the section "Accessing Commands in Excel" for instructions on creating custom tabs. It is not possible to customize the Ribbon in Excel for Mac 2011, but the **menu bar** and the **standard toolbar** can be customized.

This chapter has been adapted from *A Student Handbook for Writing in Biology*, Appendix 2. Used with permission from the publishers, Sinauer Associates, Inc. and W. H. Freeman & Co.

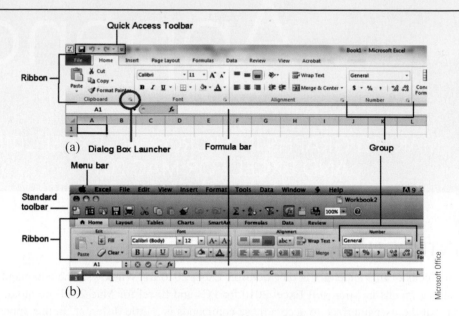

FIGURE A2.1 | Screen display of part of the command area in **(a)** an Excel 2010 worksheet and **(b)** an Excel for Mac 2011 worksheet.

HANDLING COMPUTER FILES

The "Housekeeping" section in Appendix I applies equally to Word documents and Excel workbooks. Read over this section to develop good habits for naming, organizing, and backing up computer files. In addition, if you are making the transition from Microsoft Office 2003 (or earlier) to 2010, there are some things you should know about file compatibility, which is also covered in the "Housekeeping" section.

ACCESSING COMMANDS IN EXCEL

The best way to gain proficiency with Excel 2010 is to learn where the commands you used most frequently in previous versions of Excel are located. Table A2.1 is designed to help you with this task. Frequently used commands are listed in alphabetical order in the first column, and the Excel 2010 command sequence is given in the second column. The PC keyboard shortcuts for most of the commands are shown in the third column. If you are a Mac user,

> Ctrl = Command ($\mathcal{H}$)
> Alt = Option (however, some **Alt** key sequences do not have a Mac equivalent)

In this appendix, the nomenclature for the command sequence is as follows:

> **Ribbon tab** (→Group) **| Command button | Additional Commands** (if available).

For example, the command **Insert** (→Charts) **| Scatter | Scatter with only Markers** means "Select the **Insert** tab on the Ribbon, then in the Charts group click **Scatter**, then select **Scatter with only Markers** from the drop down menu. We will use this command sequence to generate an *x*–*y* plot in a subsequent example.

TABLE A2.1 | Common commands (listed alphabetically) and how to carry them out in Excel 2010. Many of the Word 2010 commands can also be accessed by selecting text or an object and right-clicking. For **Alt** commands, hold down **Alt** while pressing the first letter key, which corresponds to a tab name

Then release **Alt** and type the next letter(s) in the sequence. **F11** is a function key.

Command	Excel 2010 Command Sequence	PC Hot Key
Editing	See Table A1.3	
File	See Table A1.3	
Formatting	Also see Table A1.3	
Cells	**Home** (→Cells) | **Format** | **Format Cells**	Ctrl+1 or Alt+HOE
Column width	**Home** (→Cells) | **Format** | **Column Width**	Alt+HOW
Numbers		
General, currency, date, etc.	**Home** (→Number)	Alt+HN or Alt+HFM
Increase or decrease decimal	**Home** (→Number) | **Increase (Decrease) Decimal**	Alt+H0 (increase) or Alt+H9 (decrease)
Paste Special	**Home** (→Clipboard) | **Paste** | **Paste Special**	Ctrl+Alt+V or Alt+HVS
Row height	**Home** (→Cells) | **Format** | **Row Height**	Alt+HOH
Sort data	**Data** (→Sort & Filter) | **Sort** or **Home** (→Editing) | **Sort & Filter** | **Sort**	Alt+AS or Alt+HS
Sub- and superscript	**Home** | (→Font) **dialog box launcher** | **Effects** | **Sub- or Superscript**	Alt+HFN
Symbols	**Insert** (→Editing) | **Symbol**	Alt+NU
Page Layout		
Headers & Footers	**Insert** (→Text) | **Header & Footer**	Alt+NH
Insert cells, columns, rows, worksheets	**Home** (→Cells) | **Insert**	Alt+HI
Cells		Ctrl+Shift+=
Worksheet		Shift+F11
Insert charts (graphs)	**Insert** | (→Charts)	Alt+N
Margins	**Page Layout** (→Page Setup) | **Margins**	Alt+PM
Orientation	**Page Layout** (→Page Setup) | **Orientation**	Alt+PO
Page break	**Page Layout** (→Page Setup) | **Breaks**	Alt+PB
Page numbers	**Insert** (→Text) | **Header & Footer** | (→Header & Footer Elements) | **Page Number**	Alt+NH

TABLE A2.1 | Continued

Command	Excel 2010 Command Sequence	PC Hot Key
Page Layout		
Scaling	**Page Layout** (→Page Setup)	**Size**	**More Paper Sizes**	Alt+PSZM
Set print area	**Page Layout** (→Page Setup)	**Print Area**	**Set Print Area**	Alt+PRS
Print gridlines	**Page Layout** (→Page Setup)	**Print Titles**	Alt+PI
Repeat row or column titles on multiple-page worksheets	**Page Layout** (→Page Setup)	**Print Titles**	Alt+I
View		
Headers and Footers	**View** (→Workbook Views)	**Page Layout**	Alt+WP
Page breaks	**View** (→Workbook Views)	**Page Break Preview**	Alt+WI
Print preview	**File**	**Print**	Alt+FP
Freeze (keep row(s) or column(s) fixed while scrolling through the rest of the worksheet)	**View** (→Window)	**Freeze**	Alt+WF

© Cengage Learning 2015

For an overview of what is new in Excel 2010, go to Microsoft's homepage (http://office.microsoft.com) and type "what's new in excel 2010" into the search box. The link for Excel for Mac 2011 is <http://www.microsoft.com/mac/excel>.

FORMULAS IN EXCEL

Excel is a popular spreadsheet program in the business world, but its "number crunching" capabilities make it a powerful tool for data reduction and analysis as well. You can use Excel like a calculator by typing numbers and mathematical operators into a cell and then pressing **Enter**. Most likely, however, you will write your own formulas or choose common functions from Excel's collection and apply them to cell references instead of numbers. Excel is great for doing repetitive calculations quickly.

In this section you will learn how to write formulas. Even after you have become proficient at writing formulas in Excel, it is still a good idea to *do a sample calculation by hand (using your calculator) to make sure the formula you entered in Excel gives you the same result*. If you find a discrepancy, check the formula and check your math. Make sure the answer makes sense.

Formulas in Excel always start with an equal sign (=) followed by the commands, cell references, numbers, and operators that make up the formula. Some commonly used operators are shown in Table A2.2. Excel performs the calculations in order from left to right according to the same order of operations used in algebra:

- First negation (−)
- Then all percentages (%)

TABLE A2.2 | Operators commonly used for calculations in Excel

Operator	Meaning
+	Addition
−	Subtraction or negation
*	Multiplication
/	Division
%	Percentage
^	Exponentiation
:	Range of adjacent cells
,	Multiple, nonadjacent cells

- Then all exponentiations (^)
- Then all multiplications and divisions (* or /)
- Finally, all subtractions and additions (− or +)

For example, the formula "**= 100−50/10**" would result in "**95**", because Excel performs division before subtraction. To change the order of operations, enclose in parentheses the part of the formula you wish to have priority evaluation. For example, "**= (100−50)/10**" would force the evaluation of the subtraction to occur before the division, and thus result in the value "**5**".

Writing Formulas

Tasks you will encounter when writing formulas include entering a formula in the correct format or selecting a function from Excel's collection and applying the function to a range of cells.

Type a Formula in the Active Cell

1. Click a cell in which you want the result of the formula to be displayed (the so-called **active cell**). The selected cell will have a dark, black border, as in cell **O4** in Figure A2.2a. The **Formulas** tab in Excel for Mac 2011 is shown in Figure A2.2b.
2. Type "**=**" (equal sign).
3. Type the constants, operators, cell references, and functions that you want to use in the calculation. See Table A2.3 for examples.
4. Press **Enter**. The result of the calculation is displayed in the active cell. You can view and edit the formula on the formula bar.

Apply a Function to Values in Adjacent Cells

1. Click a cell in which you want the result of the formula to be displayed.
2. Type "**=**", the name of the function (e.g., **AVERAGE**), and then "**(**".
3. Hold down the left mouse button and select the cells you want to average. If there are a lot of cells to average, click the first cell, hold down the **Shift** key, and then click the last cell in the range.

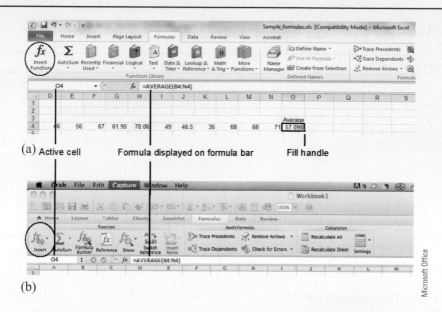

FIGURE A2.2 | The **Formulas** tab has an **Insert Function** button and other commands for doing calculations in **(a)** Excel 2010 and **(b)** Excel for Mac 2011.

TABLE A2.3 | Examples of formulas written in Excel

Formula	Meaning
=16*31−42	Assigns the product of 16 times 31 minus 42 to the active cell
=A1	Assigns the value in cell A1 to the active cell
=A1+B1	Assigns the sum of the values in cells A1 and B1 to the active cell
=(O4/H4)^(10/(I4−B4))	Value in cell O4 is divided by the value in cell H4. This number is raised to the power of 10 divided by the difference between the values in cells I4 and B4.
=((J5−K5)*60*L5)/(M5*N5)	Value in cell J5 minus the value in cell K5 is multiplied by 60 and multiplied by the value in cell L5. This result is divided by the product of the values in cells M5 and N5.
=SUM(A1:A26)	Assigns the sum of the values in cells A1 through A26 inclusive to the active cell
=SUM(B4,J4,L4,T4)	Assigns the sum of the values in cells B4, J4, L4, and T4 to the active cell
=AVERAGE(B4:U4)	Assigns the average of the values in cells B4 through U4 inclusive to the active cell
=AVERAGE(B4,J4,L4,T4)	Assigns the average of the values in cells B4, J4, L4, and T4 to the active cell
=log(B1)	Assigns the base-ten log of the value in cell B1 to the active cell
=ln(B1)	Assigns the natural log of the value in cell B1 to the active cell

Regardless of the selection method, Excel automatically inserts the first cell and last cell of the range, separated by a colon, in the formula bar (see Figure A2.2a).

4. Type ")" and press **Enter**.

Apply a Function to Values not in Adjacent Cells

1. Click a cell in which you want the result of the formula to be displayed.
2. Type "=", the name of the function (e.g., **AVERAGE**) and then "**(**".
3. Hold down the **Ctrl** key and click the cells you want to average. Excel automatically inserts the selected cells, separated by commas, in the formula bar.
4. Type ")" and press **Enter**.

Select an Excel Function with the Insert Function Button

1. Click a cell in which you want the result of the formula to be displayed (the active cell).
2. Click **Formulas** (→Function Library) | **Insert Function** (Figure A2.3a). Or, from any tab, click the **Insert Function** button (f_x) to the left of the formula bar.
3. If the function you are looking for is not shown in the **Recently Used** category, click the down arrow to display another category.
4. When you locate the function you want to use, click it in the **Select a function** list box and then click **OK**.

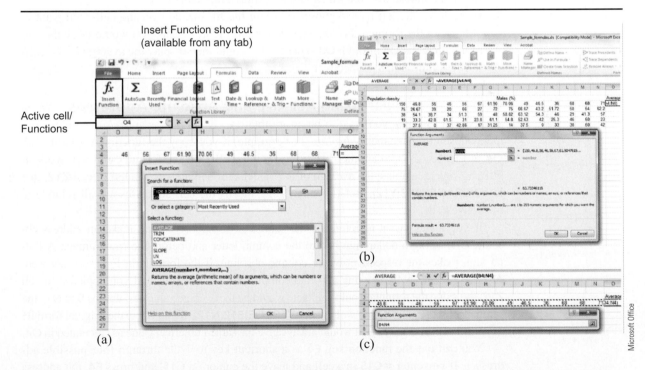

FIGURE A2.3 | **(a)** Selecting a function from Excel's **Function Library** and **(b)** Applying it to a range of cells by typing the cell references separated by a colon. **(c)** Instead of typing the cell references (and potentially introducing typos), click the first cell, hold down the **Shift** key, click the last cell, and then hit **Enter**.

5. In the **Function Arguments** dialog box (Figure A2.3b), Excel tries to guess the range of cells to which you want to apply the function. If the range is incorrect, use the **Shift** key or the **Ctrl** key as explained previously to select the correct range (Figure A2.3c).
6. Click **OK**.

Select an Excel Function by Typing an Equal Sign in an Active Cell

1. Click a cell in which you want the result of the formula to be displayed (the active cell).
2. Type "=" in the active cell. The most recently used function is displayed in the **Active Cell/Functions** box to the left of the formula bar (see Figure A2.3b).
3. Click the down arrow to select a function from the dropdown menu. The **Function Arguments** dialog box appears (Figure A2.3b).
4. Follow steps 5 and 6 above.

Copying Formulas Using the Fill Handle

Quite frequently, you may want to perform the same calculation on data contained in cells of neighboring rows or columns. Instead of copying and pasting the formula, you can simply drag the formula into adjacent cells. To do so, follow these steps:

1. Click the cell containing the formula you wish to copy.
2. Locate the fill handle, which is a small black square on the bottom right corner of the cell (see Figure A2.2a).
3. Move the mouse over the fill handle to display cross hairs (+).
4. Hold down the left mouse button and drag the fill handle over the cells you want to "fill" with the formula. For the sample data in Figure A2.3b, you would copy the formula into cells **O5** through **O8**. These cells would then display the average of the cells in rows 5 through 8, respectively.

Copying Formulas Using Relative and Absolute Addresses

When you copy a formula, you need to be aware that the cell addresses are *relative* unless you have chosen to make them *absolute*. For example, if you copy the function located in cell **O4** in Figure A2.2a to another location, say into cell **Z12**, the formula will still work. It will, however, calculate the average of the 12 cells immediately to the left of cell **Z12**, that is, **AVERAGE(M12:Y12)**. Relative addressing means the cell addresses are all relative to the active cell.

If you always want to reference the same cell location, you can make an address absolute by placing a dollar sign before the column letter and before the row number. A dollar sign makes the respective row or column absolute. If the formula in Figure A2.2a were **AVERAGE($B4:$N4)**, denoting absolute column and variable row, and you copied it to cell **Z12**, the value the formula would calculate would be the average value of columns **B** to **N** in the same row as **Z12**, that is, you would get **AVERAGE(B12:N12)**. If you wrote the original formula as **AVERAGE($B$4:$N$4)**, it would always give the same numerical value as calculated in **O4**.

You can use the function key **F4** as a shortcut key to cycle through four possible addresses. If you enter **=C15** in a cell and leave the cursor on C15, and press **F4**, the address changes to $C$15; pressing **F4** again changes the address to C$15; a third toggle of **F4** changes the address to $C15, and finally a fourth toggle of **F4** brings you back to the relative address you started with. Absolute addresses are one way to include constants such as density and gravitational acceleration in formulas.

Copying Cell Values, but not the Formula When you select **Copy**, Excel copies everything in the cell—the formula, the number, and the text. When you select **Paste**, however, you may get a "#REF!" error instead of the entry you expected. This error typically occurs when the connection between the cell references and the formula is lost. To paste the value without the formula, click **Home** (→Clipboard) | **Paste** | **Paste Values**.

UNFURLING THE RIBBON

If you prefer a more visual approach to the commands, the following sections show you the tabs you are likely to use for entering data, plotting graphs, formatting a spreadsheet before you print it, and selecting the way your document appears on the screen. In addition to the buttons on the tabs, some groups have a diagonal arrow that launches a dialog box for those particular tasks (Excel 2010 but not Excel for Mac 2011). The dialog boxes are similar (if not identical) to those in Excel 2003 and 2007.

The File Tab

Clicking the **File** tab opens a new dialog box called the **Backstage View** (Figure A2.4). The left panel lists the file-related commands, the center panel provides the options for the selected command, and the right panel shows a preview or other options. Document

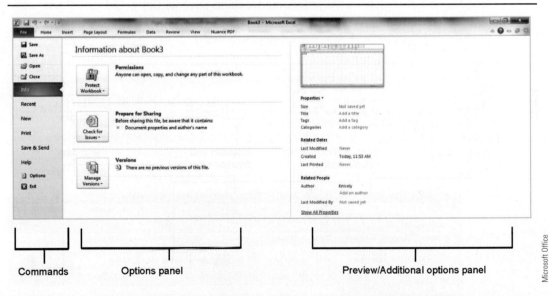

Commands   Options panel   Preview/Additional options panel

Microsoft Office

FIGURE A2.4 | This **Backstage View** of the **File** tab consists of three panels: commands, options, and preview. Workbook-related commands such as **Save, Save As, Open, Close,** and **Print** are shown on the left panel. The options panel shows additional commands related to the command selected on the left panel. The preview panel displays the page as it will be printed. For commands other than **Print**, additional options are shown on the right panel.

commands such as **Save, Save As, Open, Close, Recent, New**, and **Print**, which were located on the **Office Button** in Word 2007 and on the **File** menu in Word 2003, are located on the **File** tab. Other commands on the **File** tab are

- **Info**, which provides new file management tools for setting permissions, inspecting the document for hidden properties and personal information, and recovering earlier autosaved or unsaved versions of a file.
- **Save & Send** makes it easier to collaborate with others on files. Similar to Google docs, this feature allows two or more people to work on the same document at the same time and to store the document online.
- **Help**. The **Help** menu can also be accessed by pressing the **F1** function key.
- **Options** allows you to change program settings such as user name, how formulas are calculated and how errors are checked, AutoCorrect options, how documents are saved and printed, and so on. The **Options** button is also the gateway to commands for customizing the Ribbon and the Quick Access Toolbar and for managing Microsoft Office add-ins and security settings.

Customize the Ribbon   This feature lets you create a new tab on the Ribbon, to which you can add hard-to-find commands that you use frequently. Click **File | Options | Customize Ribbon**. After clicking the **New Tab** button, add the desired groups and commands to the new tab (see Figure A1.6).

The Home Tab

The **Home** tab contains word processing functions such as **Copy, Cut**, and **Paste**; commands for formatting text; and the familiar **Find** and **Replace** buttons (Figure A2.5).

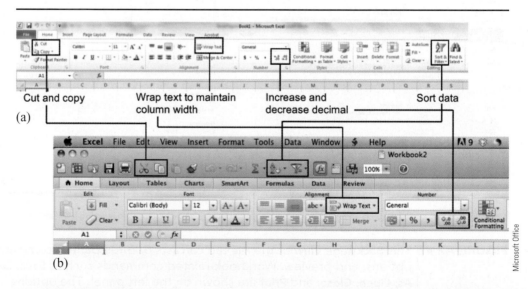

FIGURE A2.5 | Useful commands found on the **Home** tab of **(a)** an Excel 2010 worksheet and **(b)** an Excel for Mac 2011 worksheet. The **Sort** and **Filter** commands are also on the **Data** tab in Excel for Mac 2011.

In addition, there are commands for formatting and sorting data. From left to right in Figure A2.5, these command buttons are **Wrap Text, Increase / Decrease Decimal, Format Cells,** and **Sort & Filter.**

Wrap Text   When you type a long text in a cell, it runs into the adjacent cell to the right. If there is text or a numerical entry in the adjacent cell, the long text is hidden (it appears to be cut off). To make the long text visible, you could widen the columns, but then fewer columns will be visible on the screen. To be able to maintain column width and still view the entire text, click the desired cell and then **Home** (→Alignment) **| Wrap Text.** Excel expands the row height to accommodate the contents of the cell.

Increase or Decrease Decimal   When a calculation results in a value that has more decimal places than the measurements from which it originated, round off. In other words, display the same number of significant figures as the smallest number of significant figures in any measurement (see the "Significant Figures" section in Chapter 6). To round off, click the desired cell and then **Home** (→Number) **| Decrease Decimal.**

Format Cells   **Home** (→Cells) **| Format Cells** opens a dialog box with six tabs: **Number, Alignment, Font, Border, Fill**, and **Protection**. The options selected can be applied to one or many cells.

Sort Data   If you would like to alphabetize a list of names or arrange numerical values in ascending or descending order, use the **Home** (→Editing) **| Sort & Filter** command. First, select *all of the columns that contain data* that you want to sort. This is important, because if you select only one column, then only the data in that column will be sorted. Data in the adjacent columns will then no longer correspond to the correct data in the sorted column. After selecting all of the data, click **Custom Sort** on the **Sort & Filter** button's dropdown list. Select the criteria according to which you want to sort: by column, by values, and in which order. Then click **OK**.

The Insert Tab

In Excel 2010, the **Insert** tab has commands for inserting tables, illustrations, charts, links, text, and symbols (Figure A2.6a). **Insert** (→Charts) is where you begin the process of making graphs for your lab reports. In Excel for Mac 2011, the commands for inserting tables and charts are on separate tabs. Detailed instructions for making *x–y* graphs, vertical and horizontal bar graphs, and pie charts are given in the following sections.

The Page Layout Tab

The **Page Layout** tab is where you format your spreadsheet after you have entered data (Figure A2.6b). The buttons on this tab allow you to

- Change the margins
- Add headers, footers, or page numbers
- Scale the content to fit on one sheet of paper
- Adjust page breaks
- Set the print area
- Repeat row or column titles on multiple-page worksheets

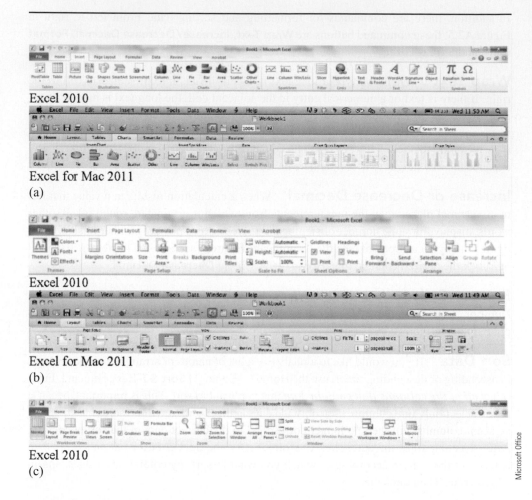

Excel 2010

Excel for Mac 2011

(a)

Excel 2010

Excel for Mac 2011

(b)

Excel 2010

(c)

FIGURE A2.6 | **(a)** The **Insert** tab (Excel 2010) and the **Charts** tab (Excel for Mac 2011) have commands for graphing data. **(b)** **Page Layout** buttons are used for formatting the spreadsheet. **(c)** The **View** tab affects how you see your worksheet on the screen. Some of the buttons on the Excel 2010 **View** tab are found on the **Layout** tab in Excel for Mac 2011.

The View Tab

The **View** tab in Excel 2010 gives you options for displaying your workbook(s) on the screen (Figure A2.6c). In Excel for Mac 2011, some of these buttons are found on the **Layout** tab. There are three workbook views that can also be accessed from the status bar (Figure A2.7).

- **Normal** view is the default view and is typically used when entering data.
- **Page Layout** view displays margins; headers, footers, and page numbers; and page breaks.
- **Page Break Preview** is handy for adjusting page breaks to keep blocks of text or data together.

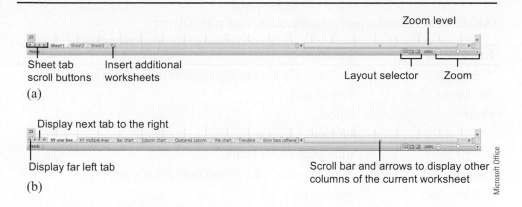

FIGURE A2.7 | Customizing, inserting, and navigating between and within worksheets. **(a)** Excel 2010 default workbook with three worksheets. **(b)** Excel 2010 workbook with extra, customized worksheet tabs.

When a spreadsheet contains a lot of data, the row and column titles disappear when you scroll down or to the right. To be able to view the titles as you scroll through the worksheet, click **View** (→Window) | **Freeze Panes** | **Freeze Top Row** (or **Freeze First Column**). **Split** view splits the worksheet into two or four sections (based on the cell that is selected), each of which can be scrolled independently. Use **Split** view to view different sections of the worksheet simultaneously. Use **Freeze Panes** specifically to lock row and column titles while scrolling through the rest of the worksheet.

More Tabs Below

Each Excel 2010 workbook initially contains three worksheets, named **Sheet1**, **Sheet2**, and **Sheet3** (Figure A2.7a). Each sheet has its own tab, located in the bottom left corner of the screen, just above the status bar. You can customize the tabs for each sheet by double-clicking the name on the tab and typing a better descriptor (see Figure A2.7b). You can also rearrange the worksheets by dragging a tab to a different position. Finally, if you would like to add more worksheets, click the **Insert Worksheet** button, located to the right of the last sheet tab. Some good reasons for keeping multiple worksheets in one workbook are:

- To collect replicate data, yet keep individual trials separate
- To collect data from multiple lab sections for the same experiment, to pool or keep separate as needed
- To simplify file organization

To the left of the sheet tabs, you will see four sheet tab scroll buttons. The scroll buttons are only needed if the workbook contains so many worksheets that their tabs cannot all be displayed at once, as in Figure A2.7b. In that case, the first scroll button causes the far left worksheet tab to be displayed, the second button displays the next tab to the left, the third button displays the next tab to the right, and the fourth button displays the far right worksheet tab.

TABLE A2.4 | Excel-specific terminology and translation

Excel Term	Description
Workbook	An Excel file that initially consists of three worksheets
Worksheet	Spreadsheet
Chart	Graph
Charting	Plotting
Scatter chart	Line graph or x–y graph
Line chart	Not a line graph; do not use this type of chart
Column chart	Vertical bar graph
Bar chart	Horizontal bar graph
Data series	Set of related data points
Plot area	Area of the graph inside the axes
Chart area	Area outside the axes but inside the frame
Legend	Legend or key
Marker	Data symbol

© Cengage Learning 2015

The bottom right corner of the workbook has three features related to the view: Layout Selector, Zoom Slider, and Scroll bar and arrows. The **Layout Selector** is similar to the one in Excel 2003, in which you can select among three layouts: **Normal, Page Layout View**, and **Page Break Preview**. The **Zoom Slider** lets you zoom in or out on certain cells in the worksheet. The horizontal scroll bar and arrows let you scroll across columns; the vertical scroll bar and arrows allow you to move across rows. Alternatively, if your mouse has a wheel, rotate the wheel to navigate vertically or hold down the wheel and move the mouse horizontally until the desired row or column is displayed.

EXCEL TERMINOLOGY

Every discipline has its own terminology, and the Microsoft-dominated computer world is no exception (Table A2.4). In the sections that follow, Excel terminology is used to make it easier for you to find the buttons for making and formatting graphs. Keep in mind, however, that this is not really the language of engineers!

GOODBYE CHART WIZARD, HELLO INSERT CHARTS

In Excel 97 through 2003, Chart Wizard prompts you to select a chart type, confirm the arrangement of the source data, input chart, and axis titles, and select where in the workbook the graph should be inserted. In Excel 2007 and 2010, Chart Wizard has been replaced by chart-specific command buttons on the **Insert** tab (see Figure A2.6a).

Plotting x–y Graphs (Scatter Charts)

Engineers may use the phrase *line graphs* to refer to x–y graphs, but you should not confuse line graphs with Excel's *line charts*. Line charts do not space data proportionally on the x-axis. For example, intervals of 5, 20, and 50 units would be spaced equally, when in fact there should be 5 units in the first interval, 15 in the second, and 30 in the third. The bottom line is: When you want to make an x–y line graph in Excel, choose **Insert** ($\rightarrow$Charts) | **Scatter**.

Entering Data in Worksheet   Before you enter data in an Excel worksheet, you must first have a clear understanding of plotting conventions that dictate the appearance of your x–y graph. Which parameter should be plotted on the x-axis and which one on the y-axis? By convention, the x-axis of the graph shows the independent variable, the one that was varied during the experiment. The y-axis of the graph shows the dependent variable or system response, the variable that changes in response to changes in the independent variable.

On the worksheet, the leftmost of two columns, here Column A, is used to enter the data for the x-axis, whereas the next column to the right is used for data for the y-axis. Subsequent cells further to the right might contain additional y-axis data to be plotted as separate data sets, if all y values depended on the same x values.

Assume you carried out a hypothetical experiment in which you measured the acoustic intensity (in decibels) from a point source (a small speaker) as a function of the horizontal distance (in meters) away from the speaker (Figure A2.8). Because distance is the parameter you varied, enter the distance data in Column A. Enter the corresponding acoustic intensity data in Column B. For a simple graph like this with only one data set, it is not necessary to provide column headings, although for future reference it is always good to identify the variable name and the units on the data. To gain the most from this explanation, open Excel 2010 on your computer and reproduce the steps as they are presented.

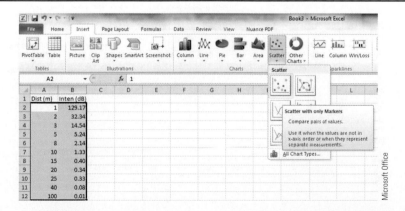

FIGURE A2.8 | Enter the data for the independent variable (x-axis) in Column A. Enter the data for the dependent variable (y-axis) in Column B. Select **Insert** ($\rightarrow$Charts) | **Scatter with only Markers**, not Line, to make an x–y graph in which the data points (markers) are displayed. The chart type is selected with **Charts** in Excel for Mac 2011.

Creating the Scatter Chart

1. Hold down the left mouse button and select the cells containing the values you want to plot. If there are a lot of cells, click the first cell, hold down the **Shift** key, and then click the last cell in the range. *Note*: If the data to be plotted are not in adjacent columns, select the first column, hold down the **Ctrl** key, and then select the data in any other column(s). In Excel 2010, click **Insert** (→Charts) | **Scatter** | **Scatter with only Markers** (see Figure A2.8).

   Holding the mouse over each type of scatter chart opens a pop-up window with a description of that type of graph. Never use **Line Chart**, as this option spaces the *x*-axis values at equal intervals, instead of according to the intervals of the data. In Excel for Mac 2011, click **Charts** | **Scatter** | **Marked Scatter**.

2. Clicking the plot area (the area inside the axes) or the chart area (the area outside the axes) of the newly created graph activates the **Chart Tools** contextual tab, which has three tabs of its own. Click the **Chart Tools Design** tab to view the buttons for modifying the fundamental design of the graph (Figure A2.9). Then in the **Data** group, click **Select Data** (Figure A2.10a) to bring up the **Select Data Source** dialog box. In this menu, select the data set of interest and click the **Edit** button. Three data boxes will appear, one for the name of the data set, one for the range of *x*-values, and the third for the range of the *y*-values. Provide the desired name for the data set (for example, "Measured Intensity") and click **OK** twice. If Excel adds a chart title, left-click it and then right-click to bring up the menu to delete the title. Similarly, place the cursor on a horizontal grid line, right-click it, and select delete.

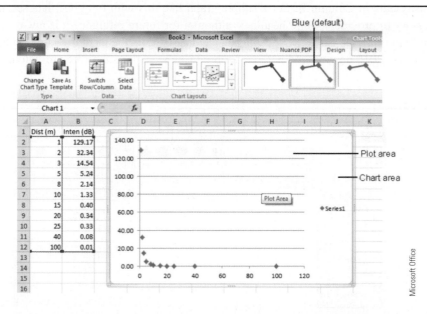

FIGURE A2.9 The **Chart Tools** tab is displayed in Excel 2010 when clicking the plot area or chart area of the graph. One of the three tabs associated with the **Chart Tools** tab is the **Chart Tools Design** tab, which has commands for changing the chart type, saving the chart format as a template, and changing the color of the data markers and lines.

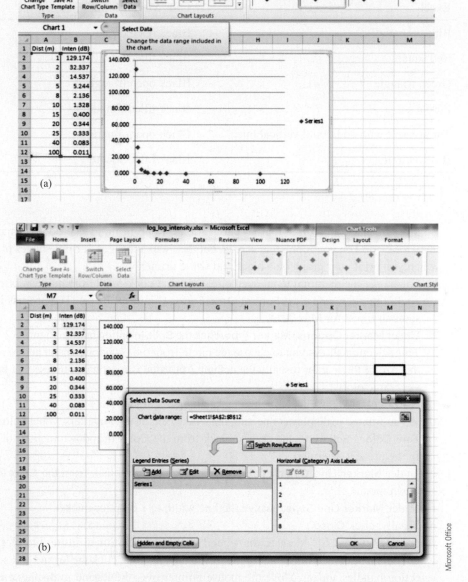

FIGURE A2.10 | **(a)** On the **Chart Tools Design** tab, click the **Select Data** button. **(b)** In the **Select Data Source** dialog box, click **Edit** to change the name of the data series. Alternatively, right-clicking a data marker gets to the same dialog box shown in (b).

3. The type of marker chosen by Excel—the diamond—is not among the top choices recommended in the Council of Science Editors Manual (2006) for research articles. The symbol hierarchies of Excel and the CSE Manual are compared in Table A2.5. The science editors' recommendation for symbols is based on ease of recognition and good

TABLE A2.5 | Comparison of Excel and CSE Manual symbol hierarchy for line graphs

Excel	CSE Manual
Blue diamond	Black open circle
Red square	Black filled circle
Green triangle	Black open triangle
Purple x	Black filled triangle
Turquoise x with additional vertical line	Black open square
Brown circle	Black filled square

© Cengage Learning 2015

contrast in black and white journal publications. Keep in mind that colored markers may look good on your computer screen, but they will appear as various shades of gray when you print your graph on a black and white printer.

a. To change the diamond to an open circle, double-click one of the markers or right-click a marker and select **Format Data Series** (Figure A2.11a). The process is the same in Excel for Mac 2011.

b. Under **Marker Options|Marker Type**, click the **Built-in** option button (Figure A2.11b). This selection allows you to choose the circle from the dropdown menu and to adjust the size of the marker (usually the default 7 pt value is reasonable unless you have a very large plot).

c. Now click **Marker Fill | No Fill** to make an open circle.

d. If you wanted to display connecting lines between the data markers, you would click **Line Color | Solid line** and choose black from the **Color** dropdown menu (not used in the present example).

e. Under **Marker Line Color**, click **Solid line** and choose black from the **Color** drop-down menu.

f. Under **Marker Line Style**, change the line width to 1.00 pt or thicker.

g. Finally, click **Close**.

4. According to the CSE Manual, the legend should be positioned inside the plot area when there is room. To move the legend inside the plot area, click the legend box so that selection handles appear. Move the mouse pointer over the legend to display crossed double-headed arrows and drag the legend to an open space inside the plot area. Alternatively, if there is no space for the legend on the plot area, move the legend to the chart area below the x-axis, but above the figure caption.

    After moving the legend to the plot area and stretching the plot area to fill the chart area, your plot should resemble that in Figure A2.11c.

5. Double-click or right-click the number scale along the x-axis and open the **Format Axis** dialog box (Figure A2.11d). In acoustics, it is advantageous to plot these data on log-log coordinates. Click the **Logarithmic scale** box and then select the **Inside** option for the major and minor tick marks. Repeat the formatting for the y-axis. You will now have a log-log plot of acoustic intensity from a point source.

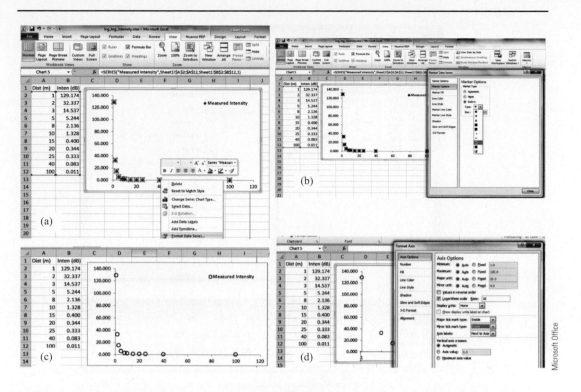

FIGURE A2.11 | (a) Double-click a data marker to open the **Format Data Series** dialog box. (b) Under **Marker Options** select **Built-In** and click the circle. Adjust marker fill, marker line color, and marker line style as described in the text. (c) Move the legend box into the plot area and stretch the plot area to fill the chart area. (d) Click each axis and adjust format in the **Format Axis** dialog box.

6. Now click on the plot area and select **Chart Tools Layout | Trendline | More Trendline Options** to open the **Format Trendline** dialog box. Alternatively, select any data marker and right-click to arrive at the same dialog box. Because we know that intensity follows a $1/r^2$ law, we will select a **Power Law** fit. Add the phrase "Power Law Curve Fit" to the **Custom Trendline Name** box. Finally, check the **Display Equation on Chart** and the **Display R-squared Value on Chart** boxes (Figure A2.12a). After you click **Close**, drag the equation box to an empty space on the plot area. The resulting graph should look like the one in Figure A2.12b.

7. To add a textbox for the *x*-axis title, click **Chart Tools Layout | Axis Titles | Primary Horizontal Axis Title | Title Below Axis** (Figure A2.13a). In Excel for Mac 2011, the same commands are found under **Chart Layout | Axis Titles** (Figure A2.13b). In the present example, the *x*-axis title is "Distance from source, x (m)." If the title were to include a Greek or math symbol, type the usual text, switch to the **Insert** tab and click (→Symbols) | **Symbol**. Find and click the desired symbol, click **Insert**, and then click **Close**. Unfortunately, it is not possible to make shortcut keys or to program symbols in AutoCorrect in Excel 2010. In Excel for Mac 2011, click the **Media Browser** icon on the standard toolbar and then **Symbols** (Figure A2.13b) or **Insert | Symbol** on the menu bar.

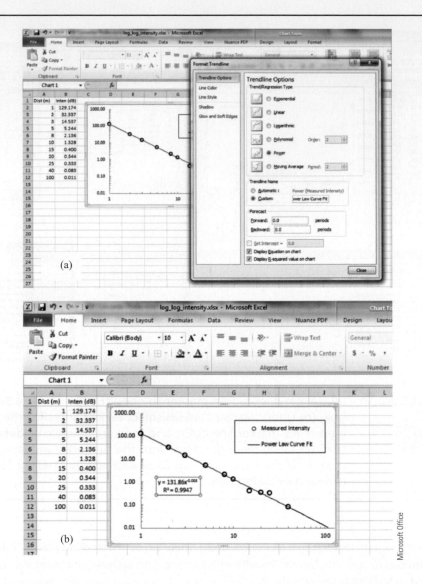

FIGURE A2.12 | **(a)** The **Format Trendline** dialog box is used to fit a power law trendline to the data and to display the equation and the R-squared value on the chart. **(b)** The resulting graph shows the trendline and equation of the curve fit.

8. To add a textbox for the *y*-axis title, click **Chart Tools Layout | Axis Titles | Primary Vertical Axis Title | Rotated Title** (Figure A2.14). In Excel for Mac 2011, the same commands are found under **Chart Layout | Axis Titles** (see Figure A2.13b).

The *y*-axis title in the present example is "Acoustic Intensity, I_A (dB)." This title contains the subscript A. In Excel 2010, type "Acoustic Intensity, IA (dB)." Press **Enter** and the *y*-axis title appears on the graph. Now select the letter to be subscripted, right-click, and select **Font**. Then check the **Subscript** box and **OK** (Figure A2.15).

In Excel for Mac 2011, select the *A* to be subscripted, right-click to open the **Format Text** dialog box, and then select **Subscript**.

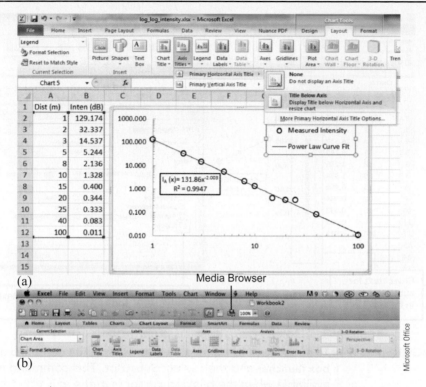

FIGURE A2.13 | **(a)** To add a textbox for the x-axis title in Excel 2010, click **Axis Titles | Primary Horizontal Axis Title | Title Below Axis** on the **Chart Tools Layout** tab. **(b)** In Excel for Mac 2011, the same commands are found under **Chart Layout | Axis Titles**.

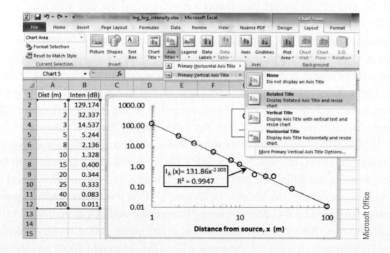

FIGURE A2.14 | To add a textbox for the *y*-axis title, click **Axis Titles | Primary Vertical Axis Title | Rotated Title** on the **Chart Tools Layout** tab. In Excel for Mac 2011, the same commands are found under **Chart Layout | Axis Titles**.

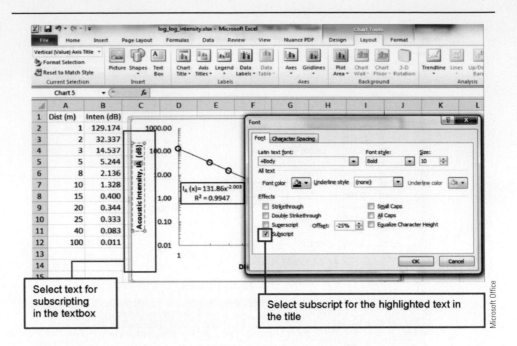

FIGURE A2.15 | To subscript part of an axis title, click **Home** (→Font) **dialog box launcher** and then select **Subscript**. This command is only available when the blinking cursor is in the axis title textbox, not in the formula bar. In Excel for Mac 2011, select the text to be subscripted, right-click to open the **Format Text** dialog box, and then select **Subscript**.

If the axis title is longer than the axis itself, part of the title will wrap onto the next line. To avoid wrapping, simply enlarge the chart: Position the mouse pointer on the lower right corner of the graph so that it changes to a double-headed arrow. Drag the corner to enlarge the graph.

9. Because there is only one data set on this graph, no legend is required. However, we added a fitted trend line, so the legend is retained in this example to distinguish the experimental data from the fitted trend line. If you wanted to delete the legend, on the **Chart Tools Layout** tab, click the **Legend** button and select **None, Turn off Legend**. Alternatively, click on the textbox containing "Measured Intensity" and "Power Law Curve Fit" and hit the **Delete** key, or right-click the highlighted legend, and select **Delete** from the menu.

10. If your data are not filling the plot area, you should adjust the minimum and maximum axis values so your graph has no "dead space." Use the **Chart Tools Layout | Axes | Primary Horizontal Axis | More Primary Horizontal Axis Options** to open the **Format Axis** dialog box. Alternatively, you can get to the same dialog box by double-clicking or right-clicking a number on the *x*-axis. The CSE Manual (2006) recommends using multiples of 1, 2, or 5 for the intervals on the axes. If the intervals have changed as a result of adjusting the minimum or maximum axis values, be sure to adjust **Format Axis | Axis Options | Major Unit** accordingly. Then adjust the *y*-axis scaling appropriately so there are no large regions along the *y*-axis with no data.

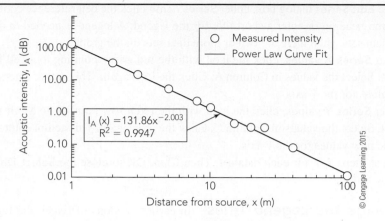

FIGURE A2.16 | Final product of the graphing process in Excel

11. To remove the border around the graph, double-click anywhere in the chart area or right-click and select **Format Chart Area** from the menu. In Excel 2010, inside the **Format Chart Area** dialog box, click **Border Color | No Line** and then **Close**. In Excel for Mac 2011, inside the dialog box click **Line | No Line**.

The final graph, formatted according to CSE Manual recommendations, is shown in Figure A2.16. This graph was copied from Excel 2010 and pasted into a Word 2010 document. Further changes can be made to the graph in Word 2010. Simply double-click on the element and make the appropriate revisions in the dialog box.

Adding Data after Completing Graph

If you have gone to all this trouble to format a graph, and realize after the fact that you need to add another data point, the last thing you want to do is to start from scratch. Fortunately, Excel 2010 makes it possible to add data after the graph has been made.

To Incorporate Additional Data Points in the Same Series
Insert a new row in the worksheet for the new data points. To do this, click a cell in the row below where the new row is to be inserted. Then click **Home** (→Cells) | **Insert | Insert Sheet Rows**. Enter the value for the x-axis in Column A and that for the y-axis in Column B. The graph is automatically updated to include these new values.

To Incorporate Additional Data Sets on the Same Graph
These instructions assume that the x-axis values remain the same and that you would like to insert additional y-axis values that represent data for a different treatment or condition. The y-axis values for the existing line are already in Column B. Type the y-axis values for the second data set in Column C. For each additional data set, enter data in the next column.

Now right-click anywhere in the chart area or the plot area of the existing graph and click **Select Data**.

1. In the **Select Data Source** dialog box, click the **Add** button in the **Legend Entries (Series)** list box (see Figure A2.10).

2. In the **Edit Series** dialog box, under **Series name** click the cell reference containing the column heading or enter a short title for the legend. A legend is needed to distinguish the data sets when there are two or more data sets on the graph.
3. Under **Series X values**, click the icon with the red arrow pointing to a cell in a worksheet. Select the values in Column A. Click the icon again. This action enters the range of values for the x-axis.
4. Under **Series Y values**, click the icon with the red arrow pointing to a cell in a worksheet. Select the values in Column C. Click the icon again. This action enters the range of the new values for the y-axis.
5. Repeat Steps 2–4 for each data set. Then Click **OK** to close the **Select Data Source** dialog box.

To Change the Legend Titles　In earlier versions of Excel, the legend titles could be changed by simply double-clicking the textbox containing the title. In Excel 2010, this method no longer works. Here are the new instructions:

1. Right-click a marker of the data series whose title you want to change. Click **Select Data** from the dropdown menu.
2. Click **Legend Entries (Series) | Edit.**
3. In the **Edit Series** dialog box, enter the new name for the series. Click **OK** twice.

Save Chart Formatting　x–y graphs are very common in engineering. While they may have different coordinate axes—Cartesian, semi-log on the y-axis, semi-log on the x-axis, and the log-log format we just described—each type of graph has a consistent format. It is possible to save graph formats as a chart template (*crtx) in Excel. Next time you need to make a similar graph, you can select the chart template and apply it to a new data set. Details of creating templates will be explained in the following section on multiple data sets plotted on a single graph.

Multiple Data Sets on an x–y Graph

Plotting multiple data sets on one graph is often the most efficient way to compare the results from several different treatments. How many lines should you put on one set of axes? The CSE Manual (2006) recommends no more than eight, but many other sources recommend no more than six data sets on one plot. Use common sense. You should be able to follow each line individually, and the graph should not look cluttered.

Once you create a graph with up to eight data sets, you can save the format as a template so you will not have to re-format the graph every time you create a new graph.

Entering Data in Worksheet　Type the data for the x-axis in Column A and the data for the dependent variable for each line in Columns B, C, and so on. In the first row of each column, enter a short title that will be used in the legend to identify the different treatments or conditions. To keep the example simple, assume you want to plot a simple family of functions $y_N = N\sqrt{x}$ with N = 1, 2 . . . 8. In Column A we will enter our x-value, as the square of integers from 1 to 12 (Figure A2.17).

Creating the Scatter Chart

1. Select the data you just entered in Columns A through F (including the headings) and click **Insert** ($\rightarrow$Charts) | **Scatter | Scatter with Straight Lines and Markers** (Excel 2010) or **Charts | Scatter | Straight Lined Scatter** (Excel for Mac 2011).

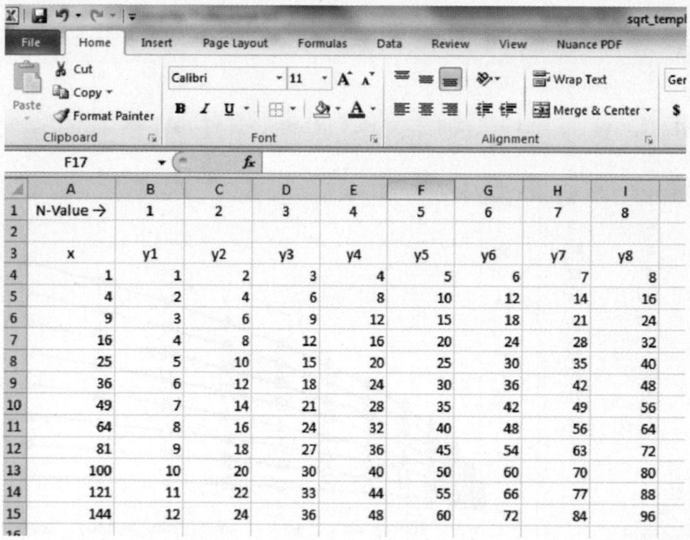

	A	B	C	D	E	F	G	H	I
1	N-Value →	1	2	3	4	5	6	7	8
2									
3	x	y1	y2	y3	y4	y5	y6	y7	y8
4	1	1	2	3	4	5	6	7	8
5	4	2	4	6	8	10	12	14	16
6	9	3	6	9	12	15	18	21	24
7	16	4	8	12	16	20	24	28	32
8	25	5	10	15	20	25	30	35	40
9	36	6	12	18	24	30	36	42	48
10	49	7	14	21	28	35	42	49	56
11	64	8	16	24	32	40	48	56	64
12	81	9	18	27	36	45	54	63	72
13	100	10	20	30	40	50	60	70	80
14	121	11	22	33	44	55	66	77	88
15	144	12	24	36	48	60	72	84	96

Microsoft Office

FIGURE A2.17 | Data generated to plot eight data sets on a single graph.

2. Begin with the data set for N = 1, and change the symbol to an open black circle using **Right-click Data Series y1 | Format Data Series | Marker Options | Built-in | Select circle**. Change to **Marker Fill | No Fill**. Then click **Marker Line Color | Solid line | Color = Black. Marker Line Type | Width = 1.25**. Then change **Line Color | Solid Line** and the **Line Type | Width = 1.5.** Follow similar procedures for data series *y*2 but make the marker a filled circle. Then follow similar procedures for each successive data series, assigning markers as given in Table A2.6. After converting all data symbols to the CSE hierarchy and removing gridlines and the chart border, the graph will look like Figure A2.18.

TABLE A2.6 | Assignment of symbols (markers) to data series in CSE hierarchy

Data Series	Marker
y1	open circle
y2	filled circle
y3	open triangle
y4	filled triangle
y5	open square
y6	filled square
y7	open diamond
y8	filled diamond

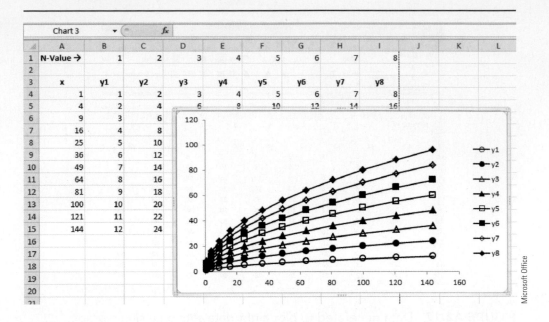

FIGURE A2.18 | Eight data series plotted on a single graph using CSE hierarchy of symbols.

3. Now you have a plot that has up to eight data sets on Cartesian coordinates. Save this plot as a template by clicking **Chart Tools Design** (→Type) **| Save As Template | my_scatter_line** in Excel 2010. Notice that the **Save as type** is "Chart Template Files (*crtx)". In Excel for Mac 2011, the command sequence is **Charts | Other | Save as Template**.

4. After saving your template, test it out on your square root data in various combinations. Select the data you want to plot and click **Insert** (→Charts) **dialog box launcher | Templates** and select **my_scatter_line**.

5. Now systematically delete all lines between the data points by right-clicking each series and then **Format Data Series | Line Color | No Line**. Save the template as **my_marker_scatter**.

6. Test out your new template by selecting just the data in Columns A, D, and F, as shown in Figure A2.19a. Then click **Insert** (→Charts) **dialog box launcher | Templates.** Select **my_marker_scatter**. To get the desired plot you will need to click the plot area or chart area and then **Chart Tools Design** (→Data) **| Switch Row/Column** once and then again. In Excel for Mac 2011, on the **Charts** tab, click **Other | Templates** and select the desired template. Then click **Charts | Switch Plot** and click **Plot series by row** followed by **Plot series by column**.

   Your plot should contain two data sets, the first with open circles (y3) and the second with filled circles (y5), as in Figure A2.19b. The template will always assign open circles to the first data set, filled circles to the second data set, and so on for up to eight data sets.

7. We can make similar scatter chart templates with semi-log and log-log coordinates. To make a linear y-axis and log x-axis, change the x-axis to a logarithmic axis in the **Format Axis** dialog box as we did previously and save the template as **my_semilog_x**. Then uncheck the **Logarithmic scale** x-axis and make the y-axis scale logarithmic. Save the template as **my_semilog_y**. Re-check the **Logarithmic scale** x-axis and save the template as **my_log_log**. You could also change the major tick marks to be inside instead of outside and save this format in the template (not done in this example).

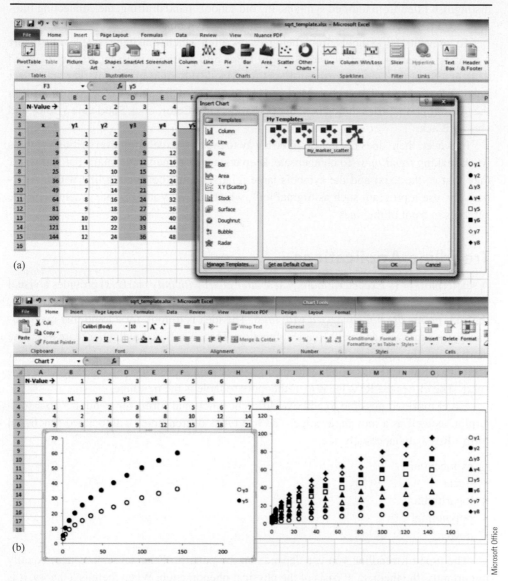

Microsoft Office

FIGURE A2.19 | (a) To load a template to plot y3 and y5 as a function of x, click **Insert** (→Charts) **dialog box launcher | Template | my_marker_scatter**. (b) The template-generated plot (lower left) follows the CSE hierarchy, assigning open circles to y3 and closed circles to y5.

Once the basic format of the graph is set, the remaining work is to annotate the graph with labels on the *x*- and *y*-axis and to modify, as needed, the legend box. By convention, the legend should be positioned inside the plot area when there is room.

Overwritten Symbols
As shown in Figure A2.18, when lines are used to connect data points, Excel overwrites the symbols with the lines. Especially when the graph is relatively small, lines that are too wide relative to the symbol size make it difficult to

distinguish filled from unfilled symbols. In "old school" hand drafting, the connecting lines are drawn between the symbols and do not cover the symbols.

To prevent the symbols from being overwritten when making *scatterplots* (**Insert** (→Charts) | **Scatter** | **Scatter with Straight Lines and Markers** in Excel 2010 or **Charts** | **Scatter** | **Straight Lined Scatter** in Excel for Mac 2011, right-click a marker and select **Format Data Series**. Then click **Marker Fill** | **Solid Fill** | **Fill color** | **Color: White**. Still in the **Format Data Series** dialog box, under **Marker Line Color**, click **Solid Line** | **Color** and choose **Black**.

However, there does not appear to be a way to prevent Excel from overwriting symbols when making *trendlines*. To compensate, keep connecting line width small (but still wider than that of the axis) and the symbols large relative to the connecting line width. Alternatively, use a program such as SigmaPlot™, which allows you to create graphs with the symbols in front of the lines.

Trendline Applications

A scatterplot (or in Excel terminology, *a scatter chart with only markers*) provides a visual method of assessing the relationship between two quantitative variables. On a scatterplot, *the data points are not connected with lines*. The idea is to look for a trend that may ultimately be expressed as a mathematical function. This so-called trendline or regression line or "best-fit" line can then be used to make predictions about one variable when the other is known. How well the trendline fits the measured data is described by the R-squared (R^2) value. The closer the R^2 value is to 1, the more reliable is the trendline.

Excel makes it possible to add five different functional forms for trendlines to a scatterplot, as well as a moving average. The five types of trendlines with defined functional forms, listed alphabetically, are

- exponential
- linear
- logarithmic
- polynomial
- power

The type of trendline you use depends on the pattern of the data points and, equally importantly, the theoretical basis of the physical phenomenon. When there is a theory, it is fairly clear what type of curve fit we should seek when fitting a trendline in Excel. Examples of such theoretical relationships are given in Table A2.7. When there is no theory, the choice of trendline that best fits the data (i.e., the trendline that gives the highest R-squared value) may aid in the development of an empirical correlation or perhaps a new theory.

To add a trendline to a data set, follow the instructions in the section "Plotting *x–y* Graphs (Scatter Charts)." In the following sections, we give examples of engineering applications for different kinds of trendlines. For more information on standard math functions, see Chapter 6.

Exponential Trendline
As an example of fitting an exponential curve to a data set, consider the heat transfer from a small conductive body. Often the body can be assumed to have a constant temperature throughout if the convective resistance controls the heat transfer. The example is representative of first-order systems. The problem is that proper nondimensionalization requires knowledge of the time constant. There are several ways to find the time

TABLE A2.7 | Examples of theoretical relationships that predict behavior of experimental data

Independent Parameter	Dependent Parameter	Expected Theoretical Relationship
Voltage in a circuit (V)	Current (I)	Linear: $I = V/R$ where R = constant resistance
Force on an object (F)	Acceleration (a)	Linear: $a = F/m$ where m = constant mass
Specific volume (v)	Temperature (T) during constant pressure heating	Linear: $T = pv/R$ where p = constant pressure, and R = gas constant
Pressure (p) during isentropic compression	Specific volume (v)	Power law: $v = (Const/p)^{(1/k)}$ where k = constant
Distance from source (r)	Acoustic intensity (I_A)	Power law: $I_A = Const/r^2$
Time (t)	Temperature (T) of lumped mass with convective cooling	Exponential: $T = T_i - (T_i - T_f)\left(1 - exp\left(\frac{-t}{\tau}\right)\right)$ where T_i and T_f are the initial and final temperatures, respectively, and τ is the time constant

constant. The first is to plot the dimensional data and determine the approximate time for the temperature difference to decrease to 63.2% of the total difference. The second is to interpolate the experimental data numerically. The final determination of the correct time constant value is when the coefficient in the fitted exponential expression takes on a value of 1.0.

From theory, we find that the temperature T for a cooling process of a lumped mass system starting at T_i and going to T_f can be represented nondimensionally as $\theta = (T - T_f) / (T_i - T_f)$. This nondimensional temperature is a decaying exponential, given as

$$\theta = \exp\left(-t/\tau\right) \tag{A2.1}$$

From the measured data, an estimate for τ is calculated by interpolation (Figure A2.20) since we know the time constant is defined as the time it takes for the system to undergo 63.2% of the temperature change. The measured dimensional time is nondimensionalized using τ. An exponential trendline is fit to the data, yielding a coefficient of 0.984 in the exponent. Dividing our first estimate of τ by 0.984 and renormalizing the time data provides an exponential relationship with a coefficient of 1.0 in the form of Equation A2.1. The best fit for the time constant is 1.746 min.

Linear Trendline

A linear trendline would be appropriate for any of the cases in Table A2.7, in which the theoretical relationship is expected to be linear. Excel provides the linear least-squares values for the slope and intercept for the equation of the best-fit line. The user has the option of specifying a zero intercept value.

As an example of a linear curve fit, Figure A2.22 shows deflection data for a coiled tension spring for a series of loads. As the load increases, the spring elongates linearly up to a point. The traditional way of plotting the force-displacement diagram for a spring is with the deflection along the x-axis and the force (i.e., the load) along the y-axis, even though the data are usually obtained by varying the load. The reason for plotting the

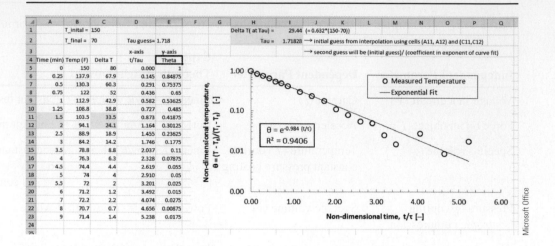

FIGURE A2.20 | Data for lumped mass cooling of an object with time in Column A and temperature in Column B. Interpolation of the data points highlighted in yellow provided an estimate of the time constant, τ, from which the first estimated curve fit of the nondimensional temperature versus nondimensional time was obtained. Columns D and E were plotted in the above graph. From the curve fit, an improved estimate of the time constant was obtained by dividing the initial guess for the time constant τ by 0.984 to yield the final plot given in Figure A2.21.

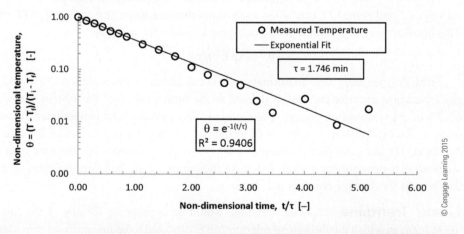

FIGURE A2.21 | Example of an exponential curve fit for a first-order system plotted on a semi-log axis showing the limited resolution of the temperature measurements as the temperature difference became less than one degree. The exponential coefficient of -1.0 and the R-squared value of 94% suggest that the curve fit represents the data well and the estimate of the time constant at $\tau = 1.746$ min is accurate.

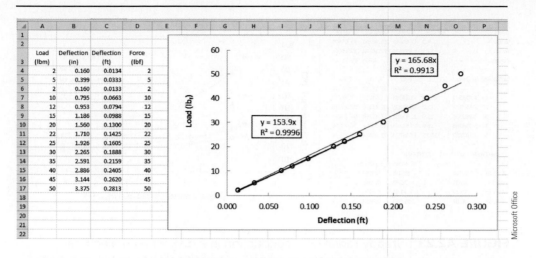

	Load (lbm)	Deflection (in)	Deflection (ft)	Force (lbf)
4	2	0.160	0.0134	2
5	5	0.399	0.0333	5
6	2	0.160	0.0133	2
7	10	0.795	0.0663	10
8	12	0.953	0.0794	12
9	15	1.186	0.0988	15
10	20	1.560	0.1300	20
11	22	1.710	0.1425	22
12	25	1.926	0.1605	25
13	30	2.265	0.1888	30
14	35	2.591	0.2159	35
15	40	2.886	0.2405	40
16	45	3.144	0.2620	45
17	50	3.375	0.2813	50

FIGURE A2.22 | Linear least-squares curve fit for coiled spring force-displacement data. The slope of the line is the value of the spring constant. The smaller spring constant ($K_s = 153.9$ lb_f/ft) is appropriate for smaller loads, while the larger value ($K_s = 165.7$ lb_f/ft) would be more appropriate for larger loads.

force-displacement diagram this way is that springs are characterized by their spring constants, as used in the calculation of the force needed for a given deflection:

$$F = K_s \Delta x \tag{A2.2}$$

Equation A2.2 defines the spring constant as $K_s = F/\Delta x$, which is the slope of the force-displacement graph.

The force-displacement diagram in Figure A2.22 has two linear curve fits, one for loads less than 25 lb_f (C4:D12) and the other including all loads (C4:D17). The displacement becomes increasingly nonlinear at higher loads. Depending on the application (small loads or large loads), either value of the spring constant might be appropriate.

To generate the two curve fits, plot the small load data (up to 25 lb_f) as one data set and add a linear trendline using **Right-click Data Series | Add Trendline | Linear**. Then turn off the data markers using **Right-click Data Series | Format Data Series | Marker Option | None.** Plot the full range of data as the second data set, and then fit the linear trendline using **Right-click Data Series | Add Trendline | Linear**.

Logarithmic Trendline   Radial conduction heat transfer has the interesting feature that due to increased cross-sectional area (normal to the axis) with increasing radius, the rate of heat transfer per unit area (the heat flux) decreases while obviously the total radial heat flow must be conserved. In an experiment, thermocouples were inserted into a disk that had a 7 mm-radius electrical heater press fit into a bored hole along its centerline and a cooling water flow around its outer periphery. Temperature was measured at seven specified radii for four values of electrical power (VI) into the heater. Data from up to seven trials were averaged for each of the heater settings. The average temperature data were then plotted against the nondimensional radial distance r/r_{inner}. Figure A2.23 shows these data with fitted logarithmic trendlines, with R-squared values in excess of 0.999.

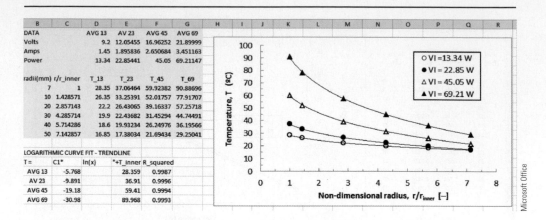

Microsoft Office

FIGURE A2.23 | Steady radial heat conduction in a disk with an electrical heating element along the axis of the disk. The temperature rises with increased energy (VI) input, but in all cases decreases in proportion to the logarithm of the radius. The logarithmic trendline fits the data almost perfectly with R-squared values in excess of 0.99.

How would you know to use the logarithmic trendline if you knew nothing about the theory of radial conduction heat transfer? Depending on your tenacity, you might simply try all possible trendlines in Excel and choose the trendline that gave the highest R^2 value. The results of this endeavor are shown in Table A2.8. The logarithmic trendline has the highest R^2 value for each data set but one (Avg 13). If you chose to use the third-order polynomial curve fit for this data set, the resulting mathematical description of the system would contradict the established theory. For this reason, it is essential to understand the theory that underpins an experiment.

Polynomial Trendline   As an example of a polynomial trendline, consider the pump data shown in Figure A2.24. The head-flowrate was measured at three pump speeds (RPM), and second-order polynomial trendlines were fit to each data set. Because the *x*-axis values

TABLE A2.8 | Measures of fitness for various trendlines for the data sets plotted in Figure A2.23

Data Set	R^2 Value for Applied Trendline					
	LOG	POLY 3	POLY 2	POWER	EXP	LIN
AVG 13	0.9987	0.9999	0.9908	0.9934	0.9669	0.9392
AVG 23	0.9996	0.9992	0.9918	0.9908	0.9757	0.9361
AVG 45	0.9994	0.9982	0.9913	0.9841	0.9846	0.9344
AVG 69	0.9993	0.9979	0.9914	0.9811	0.9873	0.9326

© Cengage Learning 2015

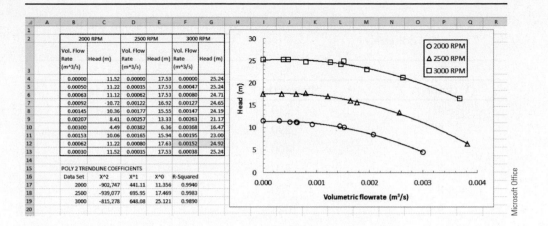

	2000 RPM		2500 RPM		3000 RPM	
	Vol. Flow Rate (m^3/s)	Head (m)	Vol. Flow Rate (m^3/s)	Head (m)	Vol. Flow Rate (m^3/s)	Head (m)
	0.00000	11.52	0.00000	17.53	0.00000	25.24
	0.00050	11.22	0.00035	17.53	0.00047	25.24
	0.00063	11.12	0.00062	17.53	0.00080	24.71
	0.00092	10.72	0.00122	16.92	0.00127	24.65
	0.00145	10.36	0.00177	15.55	0.00147	24.19
	0.00207	8.41	0.00257	13.33	0.00263	21.17
	0.00300	4.49	0.00382	6.36	0.00368	16.47
	0.00153	10.06	0.00165	15.94	0.00195	23.00
	0.00062	11.22	0.00080	17.63	0.00152	24.92
	0.00030	11.52	0.00035	17.53	0.00038	25.24

POLY 2 TRENDLINE COEFFICIENTS				
Data Set	X^2	X^1	X^0	R-Squared
2000	-902,747	441.11	11.356	0.9940
2500	-939,077	695.95	17.469	0.9983
3000	-815,278	648.08	25.121	0.9890

FIGURE A2.24 | Centrifugal pump head-flowrate data for three pump speeds (RPMs), with corresponding graph. Second-order polynomial trendlines provided a good fit to the data. The slightly lower R^2 value for the 3000 RPM data is due to the one data point in the center of the plot (highlighted on the spreadsheet) that was most likely misread or improperly transcribed. With the data point removed, the R^2 value becomes 0.9976.

are different for each data set, the graph cannot be made using the procedure described in the section "Multiple Data Sets on an x–y Graph." Instead, to plot these data, follow these steps:

1. Select the data you entered in the two columns for any pump speed (for example, Columns B and C) and click **Insert** (→Charts) | **Scatter** | **Scatter with only markers**.
2. Click the plot area or chart area and then click **Chart Tools Design** (→Data) | **Select Data**. In the **Select Data Source** dialog box, click the **Add** button in the **Legend Entries (Series)** list box (see Figure A2.10).
3. In the **Edit Series** dialog box, under **Series name** click the cell reference containing the column heading or enter a short title for the legend (for example, "2500 RPM").
4. Under **Series X values**, click the icon with the red arrow pointing to a cell in a worksheet. Select the range of values for the x-axis for 2500 RPM (Column D). Click the icon again.
5. Under **Series Y values**, click the icon with the red arrow pointing to a cell in a worksheet. Select the range of values for the y-axis for 2500 RPM (Column E). Click the icon again.
6. Repeat Steps 2–5 for each data set. If necessary, modify the names of the data sets in the legend by clicking **Edit**. Then Click **OK** to close the **Select Data Source** dialog box.

Again, use theory to select the appropriate trendline. The input energy to the pump is added either to the pressure head or to the kinetic energy of the water. Thus the sum of the pressure head and the kinetic energy must be constant. Kinetic energy is proportional to velocity squared, which in turn, is proportional to the volumetric flowrate for an incompressible fluid (water). Thus, we expect the pressure to decrease with the square of the flowrate, as is shown by the second-order polynomial trendline.

TABLE A 2.9 | Ship velocity and power data with adjusted data in the third column.

B	C	D
V (knots)	Power (hp)	Adjusted hp-129
4	210	81
5	290	161
6	400	271
7	560	431
8	780	651
9	1084	955
11	1810	1681
12	2300	2171
16	5380	5251

© Cengage Learning 2015

Power Law   A power law is any data set that can be represented as $y = Ax^B$ in which A and B are constants. Table A2.9 shows the data set that will be plotted. The measured data include only the first two columns of data. The third column was generated by the procedures given in Chapter 6 for finding the appropriate value for a constant offset that reduces the precision of a power law curve fit. With the data given in the first two columns of Table A2.9, the coefficient was found to have a value of 129.

Initially, the raw data (Columns B and C) were plotted and a power law trendline was added (Figure A2.25a). The trendline gave an equation that did not fit as well as expected

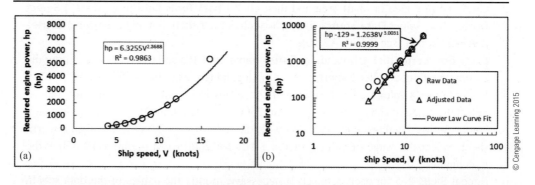

© Cengage Learning 2015

FIGURE A2.25 | Ship power requirements for a given velocity **(a)** on linear axes, and **(b)** as a log-log plot of both the raw data, showing deviation from the power law at the low-power end of the plot, and the adjusted data, showing almost perfect agreement with the expected V^3 characteristic of the power requirement.

with the knowledge that the power required to propel a ship should be approximately proportional to the ship's velocity to the third power,

$$hp = AV^3 \qquad\qquad (A2.3)$$

The linear plot in Figure A2.25a might suggest that the highest velocity data point is the point that is deviating from the curve fit. However, by looking at the data on a log-log plot, as given by the open circles in Figure A2.25b, it is apparent that the deviation from the power law behavior was at the low-speed range of the data. Data exhibiting power law characteristics should fall on a straight line on log-log coordinates. In Figure A2.25b, the adjusted data (the original power data minus the constant value of 129) are plotted on the log-log plot. The constant is small relative to the high-end power magnitudes and hardly changes the power value in this range, but becomes a substantial adjustment in the low-power range.

Notice that the adjusted power data show the expected behavior with the ship's velocity, and the fit of the trendline to the data has become almost perfect.

PLOTTING BAR GRAPHS

Bar graphs are used to compare individual data sets when one of the variables is categorical (not quantitative). The most common types of bar graphs used in engineering applications are described below. However, Excel offers additional options for column and bar graph types should the need arise.

Column Charts

Column charts are bar graphs with vertical bars.

Entering Data in Worksheet   Before you enter data in an Excel worksheet, you must have a clear idea of what your column chart should look like. By convention, the feature that all the columns have in common (the variable that was measured) lies on the axis parallel to the columns. Enter the values for this feature in a column (Column D in Figure A2.26a) and the labels for the categories in the column immediately to the left (in Column C).

If there is no particular order to the categories, arranging the bars from shortest to longest (or vice versa) makes the results easier to comprehend. When deciding what order to enter the categories in the worksheet, remember that the *lowest* row number contains the category label for the *leftmost* column.

Creating the Column Chart

1. Select the data you just entered in Columns C and D. In Excel 2010, click **Insert** (→Charts) **| Column | 2-D Column, Clustered Column** (see Figure A2.26a). In Excel for Mac 2011, the command sequence starts on the **Charts** tab.
2. Format the graph as described in the instructions for *x–y* graphs: Delete the gridlines, add axis titles, but not a chart title (unless you are preparing the graph for an oral presentation), and remove the border.
3. All of the columns in the graph should be the same width, and the columns should always be wider than the space between them. To adjust the width of the columns,

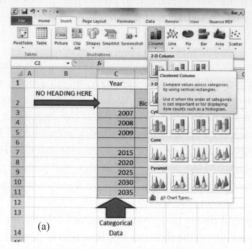

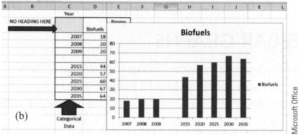

FIGURE A2.26 | **(a)** Enter the categorical data headings in the leftmost column of data, followed by the quantitative data in the adjacent column immediately to the right of the categorical headings. Then click **Insert** (→Charts) | **Column** | **2-D Column, Clustered Column** to create the column chart shown in **(b)**.

double-click or right-click one of them. In Excel 2010, in the **Format Data Series** dialog box, drag the slider for **Gap Width** to the left to decrease the gap and simultaneously increase the width of the columns. In Excel for Mac 2011, click the down arrow to decrease the **Gap Width** number.

4. Still in the **Format Data Series** dialog box, click **Fill | Solid fill | Fill Color: Black** to make the columns black against a white background for best contrast in black and white publications. The procedure is the same for Excel 2010 and Excel for Mac 2011. If you use colored fill, make sure it is a dark color for good contrast when printing in black and white.

Older versions of Excel allowed you to apply a black-and-white pattern fill to the columns. Although this function has been removed from Excel 2010, you can install an add-in called PatternUI. Type "excel patterns" into the Google search box and click the "Microsoft Excel: Chart Pattern Fills" link to access this add-in (http://blogs.msdn.com /excel/archive/2007/11/16/chart-pattern-fills.aspx, accessed 2013 March 22). Once you have downloaded the add-in to your computer, install it by clicking **File | Options | Add-ins**. Click the **Go...** button at the bottom of the dialog box. Then click **Browse...**

and navigate to the location where you saved the PatternUI file. Click it and then select **OK**. Make sure there is a checkmark next to this file in the **Add-ins** dialog box and click **OK**. When you open a new or existing Excel 2010 workbook, a new **Patterns** button will be available on the **Chart Tools Format** tab. To apply a pattern, click a column and then make a selection from the **Patterns** dropdown menu.

5. Depending on the discipline (or your instructor's preference), the baseline may or may not be visible, but all the columns must be aligned as if there were a baseline. To remove the baseline, double-click a category title on the *x*-axis to open the **Format Axis** dialog box. Click **Line Color | No line** and then click **Close**. The procedure is the same for Excel 2010 and Excel for Mac 2011.

Further changes to the graph can be made in Word 2010 after copying it from Excel 2010. Because column charts are common in some disciplines, consider saving the format in a chart template, as described earlier in the section "Multiple Data Sets on an *x–y* Graph."

Clustered Column Charts

Clustered column charts may represent different data sets for each entry in the categorical heading column. Each column in the cluster must be easy to distinguish from its neighbor. Colorful columns may look good on your computer screen, but they may turn out to be the same shade of gray when the graph is printed on a black and white printer. If the person who evaluates your work receives a black and white copy of your paper, be sure to proofread the hard copy and check that it is clear which data sets the columns represent.

Entering Data in Worksheet   Consider the global governmental support for the development of biofuels with that for other renewable energy sources in different years. As shown in Figure A2.27, the categorical data are entered in Column C. The data for the two types of energy are entered in the adjacent columns to the right. The titles entered for each energy source in the first row of Columns D and E are used by Excel to generate the legend. *A title should not be entered for the categorical data in Column C, because category labels are not part of the legend.*

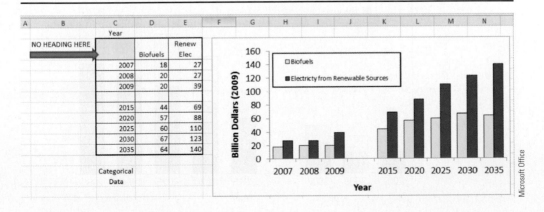

FIGURE A2.27 | Creating a clustered column graph from multiple data sets for each categorical heading.

Creating the Clustered Column Chart

1. Select the data you just entered in Columns C, D, and E, including the titles in the first row. In Excel 2010, click **Insert** (→ Charts) | **Column | 2-D Column, Clustered Column** (see Figure A2.27). In Excel for Mac 2011, the command sequence starts on the **Charts** tab.
2. Format the graph as described in the instructions for *x–y* graphs: Delete the gridlines, add axis titles, but not a chart title (unless you are preparing the graph for an oral presentation), modify the legend, and remove the chart border.

The final graph is shown in Figure A2.27. Further changes to the graph can be made in Word 2010 after copying it from Excel 2010. If you expect to use clustered column charts frequently, consider saving the format in a chart template, as described earlier in the section "Multiple Data Sets on an *x–y* Graph."

Stacked Column Charts

Stacked column charts are a way to display the sum of multiple data sets for a given categorical variable. For example, instead of displaying the two types of energy side by side in a clustered column chart (see Figure A2.27), we can show the combined support for these forms of renewable energy in a stacked column chart (Figure A2.28). Each component in a column stack is entered as a column in the worksheet.

Entering Data in Worksheet   Data entry is the same for stacked and clustered column charts: The categorical data are entered in the first column. The data for the two energy sources are entered in the adjacent columns to the right. The titles entered for each energy source in the first row of the respective columns are used by Excel to generate the legend. *A title should not be entered for the categorical data in the first column because category labels are not part of the legend.*

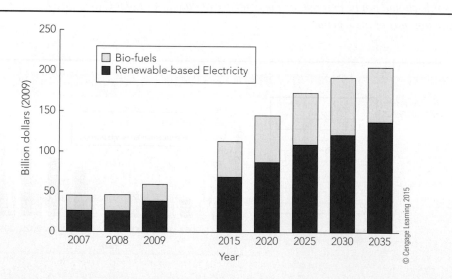

FIGURE A2.28 | Stacked column charts show how the individual components change and how the sum of all components changes with the categorical variable, in this case, time.

Creating the Stacked Column Chart

1. Select the data previously entered in Columns C, D, and E, including the titles in the first row. In Excel 2010, click **Insert** (→Charts) | **Column** | **2-D Column, Stacked Column**. In Excel for Mac 2011, the command sequence starts on the **Charts** tab.
2. Format the graph as described in the instructions for *x*–*y* graphs: Delete the gridlines, add axis titles, but not a chart title (unless you are preparing the graph for an oral presentation), modify the legend, and remove the chart border.

The final graph is shown in Figure A2.28. Further changes to the graph can be made in Word 2010 after copying it from Excel 2010. If you expect to use stacked column charts frequently, consider saving the format in a chart template, as described earlier in the section "Multiple Data Sets on an *x*–*y* Graph."

The same data sets were used to construct the clustered column chart (Figure 2.27) and the stacked column (Figure A2.28). However, the two graphs emphasize different things. The clustered column chart makes it easier to compare how much money is being spent on each quantitative category (energy source) over time, while the stacked column chart provides a better overview of the total money spent on these energy sources over time.

Bar Charts

Bar charts are bar graphs with horizontal bars. Bar charts are more practical than column charts when the category labels are long.

Entering Data in Worksheet   By convention, the feature that all the bars have in common (the variable that was measured) lies on the axis parallel to the bars. Enter the values for this feature in the second column and the labels for the category in the first column (Figure A2.29). If there is no particular order to the categories, arranging the bars from

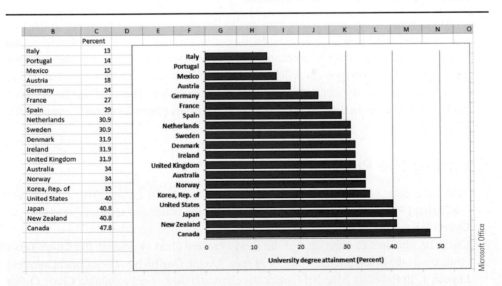

FIGURE A2.29 | A bar graph displays quantitative data relative to a categorical parameter. Notice how the arrangement of short bars to long bars from top to bottom makes relative ranking easier. Vertical gridlines are included to facilitate a quantitative comparison between countries.

shortest to longest (or vice versa) makes the results easier to comprehend. When deciding in what order to enter the categories in the worksheet, remember that the *lowest* row number contains the category label for the *shortest* bar.

Creating the Bar Chart

1. Select the data you just entered (Columns B and C in Figure A2.29). It is not necessary to select the column headings. In Excel 2010, click **Insert** (→Charts) | **Bar** | **2-D Column, Clustered Bar**. In Excel for Mac 2011, the command sequence starts on the **Charts** tab.
2. Format the graph as described in the instructions for *x–y* graphs: Delete the gridlines, add axis titles, but not a chart title (unless you are preparing the graph for an oral presentation), and remove the chart border.

The final graph is shown in Figure A2.29. Further changes to the graph can be made in Word 2010 after copying it from Excel 2010. If you expect to use bar charts frequently, consider saving the format in a chart template, as described earlier in the section "Multiple Data Sets on an *x–y* Graph."

PIE GRAPHS

Pie graphs or **pie charts** are commonly used to show the percentage of constituents out of the whole. Financial data are often displayed in a pie graph. As an example of using a pie chart to display technical data, we plot the consumption of different types of energy in 2006 (information found by searching the Internet).

Entering Data in Worksheet

The name of each pie segment (corresponding to the energy source) is entered in the first column, and the corresponding percentages are entered in the column immediately to the right. Sort the data according to percentage, with the highest percentage in the first row (Figure A2.30). This arrangement will result in the largest segment of the pie beginning at 12 o'clock, with the segments decreasing in size clockwise. There should be between two and eight segments in the pie. Combine small segments (those less than 5%) under the heading "Other."

Creating the Pie Chart

1. Select the data you just entered in Columns A and B. In Excel 2010, click **Insert** (→Charts) | **Pie** | **2-D Pie**. In Excel for Mac 2011, the command sequence starts on the **Charts** tab.
2. Click the plot area or chart area of the newly created chart to activate the **Chart Tools** contextual tab, which has three tabs of its own. On the **Design** tab, click **Chart Layouts | Layout 1**. In Excel for Mac 2011, click the chart and then select a suitable **Chart Quick Layout** and **Chart Style** on the **Charts** tab.
3. Delete the **Chart Title** textbox unless you are preparing this graph for an oral presentation.

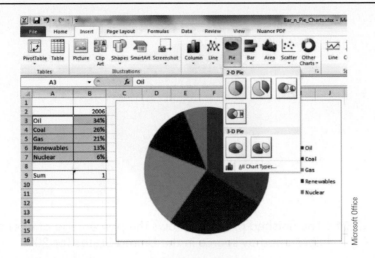

FIGURE A2.30 | Data for a pie graph are sorted from highest to lowest percentage, so that the segments are arranged in order of size in the clockwise direction starting at 12 o'clock. Pie graphs are created using the **Insert** (→Charts) | **Pie** | **2-D Pie** command sequence.

4. How the segments should be labeled depends on the publisher or professional society. One option is to put the label on the pie piece if it fits completely within the piece and outside the pie if it does not fit. To put labels inside the segments, click **Chart Tools Layout** (→Labels) | **Data Labels** | **Best Fit**. To put labels outside of the pie, click **Chart Tools Layout** (→Labels) | **Data Labels** | **Outside End**. The procedure is the same for Excel 2010 and Excel for Mac 2011. Then right-click the pie to open the **Format Data Labels** dialog box and click the **Category Name** and **Percentage** checkboxes. To change how the name and percentage are separated, select an option from the pulldown menu next to **Separator** in the same dialog box.

5. Right-click and delete the legend if you have added labels.

6. To outline the pie pieces to make them easier to distinguish, right-click the pie chart and select **Format Data Series**. Select **Border Color: Solid White** and **Border Styles | Width: 2 pt**.

7. To remove the border around the graph, double-click anywhere in the chart area or right-click and select **Format Chart Area**. Inside the **Format Chart Area** dialog box, click **Border Color | No line** and then **Close**. In Excel for Mac 2011, inside the dialog box, click **Line | No Line**.

The final graph is shown in Figure A2.31. Further changes to the graph can be made in Word 2010 after copying it from Excel 2010. For example, if you notice a mistake in the percentages, right-click the plot area or chart area and select **Edit Data**. This command links you directly to the Excel worksheet, and any changes you make in the worksheet are automatically transferred to the graph in the Word 2010 document.

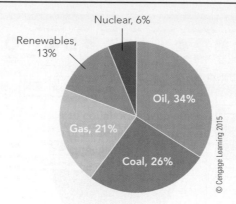

FIGURE A2.31 | The finished pie chart shows the percentage of energy consumed by source in the United States for 2006.

ERROR BARS AND VARIABILITY

The reliability of engineering data depends on good experimental design, the skill and experience of the person collecting the data, the reliability of the equipment, and the proven effectiveness of the methods and procedures, among other things. Even when all of these factors have been optimized, variability in the measurement data is still likely. For example, different students measuring the same sample with the same equipment may come up with different results. Experienced engineers measuring the same variable in replicate experiments are likely to find slight variations in their data. Small electromagnetic disturbances from unrelated machinery switching on and off may add noise to electrical signals. How can we be confident that our results accurately represent the phenomena we are trying to understand when there is variability in the measurement data?

One way to depict variability is to show all of the measured data on a scatterplot. This approach, however, makes it hard to see if there is a trend or relationship between the dependent and independent variables. To reduce the amount of data and begin to make sense of the values, we can take the average of multiple measurements as our best estimate of the true value. In statistics, the average is called the arithmetic mean, and it is calculated by dividing the sum of all the values by the number of values. The formula for calculating the mean value in Excel is

$$=\text{AVERAGE }(\ldots),$$

where "…" is the range of cells to be averaged.

A graph of the mean values is less cluttered, but potentially important information about the variability has been lost. There are two common statistical methods for describing variability: standard deviation and standard error of the mean. Both measures are based on a statistic called variance, which describes how far each measurement value is from the mean. Standard deviation is the square root of the variance, and standard error is the standard deviation divided by the square root of the number of measurements. Your handheld calculator and Excel both make it easy to calculate standard deviation and standard error, but "plugging and chugging" is not the same as understanding what you are doing. It is worth your while to take a statistics course to learn how these formulas are derived and how to use and interpret statistics appropriately.

Adding Error Bars about the Means

Standard deviation and standard error can be depicted graphically in the form of error bars around the means. To add error bars to the mean values, follow these steps:

1. Select an empty cell to enter the formula for standard deviation, which is

   =STDEV(…)

   where "…" is the range of cells that were averaged to calculate the mean
2. Select another empty cell to enter the formula for standard error, which is

   STDEV/$\sqrt{N}$

   or, in Excel format, = Cell reference for STDEV/SQRT(N)
   where SQRT is the formula for square root and N is the number of measurement values that were averaged to calculate the mean.
3. Now click the chart area to display the **Chart Tools Layout** (→Analysis) | **Error Bars** | **More Error Bars Options**. This opens the **Format Error Bars** dialog box. If you would like the vertical error bars to extend above and below the mean value, keep the default **Direction: Both**. Under **Error Amount**, click **Custom** and then **Specify Value** to open a second small **Custom Error Bars** dialog box. Click the red arrow for **Positive Error Value** and select the cells in your worksheet that contain the standard error values. Click the red arrow again to close the box. Repeat this process for **Negative Error Value**. Click **OK** to close the dialog box. Finally, click **Close** to close the **Format Error Bars** dialog box.

If your version of Excel for some unknown reason has inserted horizontal error bars, delete them by clicking any one of them and pressing **Delete**.

Data Analysis with Error Bars

The addition of error bars to the mean values changes our interpretation of the results. The larger the standard error, the less confidence we have that the mean represents the true value. Furthermore, the more the error bars overlap, it is less likely that these measurement values differ significantly from each other.

REFERENCES

Council of Science Editors, Style Manual Committee. 2006. Scientific style and format: The CSE manual for authors, editors, and publishers, 7th ed. Reston, VA: Council of Science Editors.

Appendix III

PREPARING ORAL PRESENTATIONS WITH MICROSOFT POWERPOINT 2010

INTRODUCTION

This appendix deals exclusively with Microsoft PowerPoint 2010 using the Windows 7 operating system. Although there are similarities between the PowerPoint 2010 and PowerPoint for Mac 2011 Ribbon interfaces, some of the commands are accessed differently than is shown here.

Microsoft PowerPoint allows you to create a slide show containing text, graphics, and even animations. If you have been using Office 2007, then Office 2010 will not seem that different. However, if you are a Mac user, the Ribbon is a new addition to the user interface at the top of the screen. The **Ribbon** is a single strip that displays **commands** in task-oriented **groups** on a series of **tabs** (Figure A3.1). The new **File** tab replaces the Microsoft **Office Button** that was located in the top left of the screen in PowerPoint 2007. Additional commands in some of the groups can be accessed with the **Dialog Box Launcher**, a diagonal arrow in the right corner of the group label. The **Quick Access Toolbar** comes with buttons for saving your file and undoing and redoing commands; you can also add buttons for tasks you perform frequently. The window size and screen resolution affect what you see on the Ribbon. You may see fewer command buttons or an entire group incorporated into one button on a smaller screen or if you have set your screen display to a lower resolution. If you want to see the tab names without the command buttons, click **Minimize the Ribbon**.

Clicking the **File** tab opens **Backstage View** (Figure A3.2). The left panel lists the file-related commands, the center panel provides the options for that command, and the right panel shows a preview or other options. Document commands such as **Save, Save As, Open, Close, Recent, New**, and **Print**, which were located on the **Office Button** in PowerPoint 2007 and on the **File** menu in PowerPoint 2003, are located on the **File** tab. Other commands on the **File** tab are

- **Info** provides new file management tools for setting permissions, inspecting the document for hidden properties and personal information, and recovering earlier autosaved or unsaved versions of a file.

This chapter has been adapted from *A Student Handbook for Writing in Biology*, Appendix 3. Used with permission from the publishers, Sinauer Associates, Inc. and W. H. Freeman & Co.

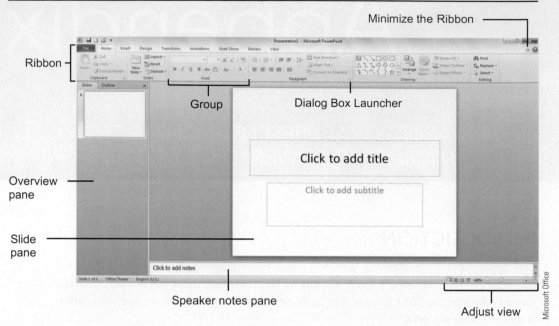

Minimize the Ribbon

Ribbon

Group   Dialog Box Launcher

Overview
pane

Slide
pane

Speaker notes pane

Adjust view

Microsoft Office

FIGURE A3.1 | Screenshot of a PowerPoint 2010 presentation in **Normal** view. The command buttons are located on Ribbon tabs at the top of the screen. View options are on the status bar in the lower right corner.

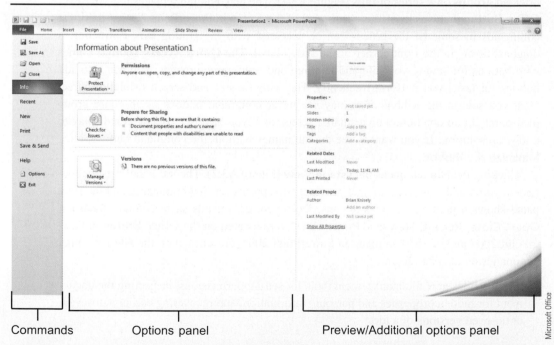

Commands   Options panel   Preview/Additional options panel

Microsoft Office

FIGURE A3.2 | A click on the the **File** tab brings up the **Backstage View**. The left panel shows file-related commands such as **Save**, **Save As**, **Open**, **Close**, and **Print**. The center and right panels show options related to the command selected on the left panel.

- **Save & Send** makes it easier to collaborate with others on files. All collaborators must use Office 2010 and possess SharePoint 2010 software or a SkyDrive account to access and save the files online. Similar to Google Docs, this feature allows two or more people to work on the same document at the same time.
- **Help**. The **Help** menu can also be accessed by pressing the **F1** function key.
- **Options** allows you to change program settings such as user name and initials, rules for checking spelling and grammar, AutoCorrect options, how often documents are saved automatically, and so on. The **Options** button is also the gateway to commands for customizing the Ribbon and the Quick Access Toolbar and for managing Microsoft Office add-ins and security settings.

Autocorrect

Unformatted AutoCorrect entries programmed in Word also work in PowerPoint, but entries containing symbols or italics do not (see the section on "AutoCorrect" in Appendix I).

Customize the Ribbon

This feature lets you create a new tab on the Ribbon, to which you can add hard-to-find commands that you use frequently. Click **File | Options | Customize Ribbon**. After clicking the **New Tab** button, add groups and commands to the new tab (see Figure A1.6 in Appendix I).

HANDLING COMPUTER FILES

The section "Housekeeping" in Appendix I applies equally to Word documents and PowerPoint presentations. Reading this section will help you develop good habits for naming, organizing, and backing up computer files. In addition, if you are making the transition from Microsoft Office 2003 (or earlier) to 2010, there are some things you should know about file compatibility, which is also covered under "Housekeeping."

COMMANDS IN POWERPOINT 2010

The best way to get up to speed with PowerPoint 2010 is to learn where the commands you used most frequently in earlier versions of PowerPoint are located. Table A3.1 is designed to help you with this task. Frequently used commands are listed in alphabetical order in the first column, and the PowerPoint 2010 command sequence is given in the second column. Many of these commands can also be accessed through right-click shortcuts. To avoid excessive complexity, no attempt has been made to include all possible ways of accessing each command. The PC keyboard shortcuts for most of the commands are shown in the third column. If you are a Mac user,

Ctrl = Command (⌘)
Alt = Option (however, some **Alt** key sequences do not have a Mac equivalent)

The nomenclature for the command sequence is as follows: **Ribbon tab** (→Group) | **Command button | Additional Commands** (if available). For example, **Home** (→Slides) | **New Slide ▼ | Title Only** means "Click the **Home** tab and in the **Slides** group, click the down arrow (▼) on the **New Slide** button to select the **Title Only** option."

TABLE A3.1 | Common commands (listed alphabetically) and how to carry them out in PowerPoint. For **Alt** commands, hold down **Alt** while pressing the first letter key, which corresponds to a tab name

File	Home	Insert	Page Layout	References	Mailings	Review	View	Add-Ins	Acrobat
F	H	N	P	S	M	R	W	X	B

. Then release **Alt** and type the next letter(s) in the sequence. **F11** is a function key.

Command	PowerPoint 2010 Command Sequence	PC Hot Key
Editing	See Table A1.3	
File	See Table A1.3	
Formatting	See Table A1.3	
Handouts		
Headers and Footers	**Insert** (→Text) | **Header & Footer** | **Notes and Handouts** tab	Alt+NH
Master	**View** (→Master Views) | **Handout Master**	Alt+WH
Print	**File** | **Print** | **Full Page Slides** | **Handouts**	Alt+PH
Print		
Handouts	**File** | **Print** | **Full Page Slides** | **Handouts**	Alt+FPH
Slides	**File** | **Print** | **Full Page Slides**	Alt+FPH
Speaker notes	**File** | **Print** | **Full Page Slides** | **Notes Pages**	Alt+FPH
Slides and notes economically	**File** | **Save & Send** | **Create Handouts**	Alt+FDH
Slide Show		
End slide show	**Right-click + End**	Esc
Go to the next slide	**Left-click**	Enter or space bar
Go to the previous slide	**Right-click + Previous**	Backspace
Mouse pointer		
As a laser pointer		Ctrl + left mouse button
As a pen or highlighter	**Right-click | Pointer Options | Pen (Highlighter)**	Right-click + OP (OH)
Start at the beginning	**Slide Show** (→Start Slide Show) | **From Beginning**	F5 or Alt+SB
Start at the current slide	**Slide Show** (→Start Slide Show) | **From Current Slide**	Shift+F5 or Alt+SC
Slides		
Animations	**Animations** tab and select effects	Alt+A
Background	**Design** (→Background) and choose background	Alt+GB (GG)
Copy	Click a slide on the **Overview** pane or in **Slide Sorter** view. **Home** (→Clipboard) | **Copy**	Ctrl+C or Alt HC

TABLE A3.1 | Continued

Command	PowerPoint 2010 Command Sequence	PC Hot Key
Delete	Click a slide on the **Overview** pane or in **Slide Sorter** view and press **Backspace** or **Delete**	
Duplicate	Click slide(s) you wish to duplicate on the **Overview** pane or in **Slide Sorter** view and press **Home** (→Slides) | **New Slide ▼** | **Duplicate Selected Slides**	Ctrl+V or Alt+HID
Headers and Footers	**Insert** (→Text) | **Header & Footer** | **Slide** tab	Alt+NH
Insert new	**Home** (→Slides) | **New Slide**	Ctrl+M or Alt+HI
Layout	**Home** (→Slides) | **Layout**	Alt+HL
Print	**File** | **Print** | **Full Page Slides**	Ctrl+P or Alt+P
Slide Master	**View** (→Master Views) | **Slide Master**	Alt+WM
Templates, design	See Themes	
Themes		
Start with new blank	**File** | **New** | **Blank presentation**	Alt+FNL
Start with built-in theme	**File** | **New** | **Themes**	Alt+FNI
Retrofit existing presentation	**Design** (→Themes) and select desired theme	Alt + GH
Transitions	**Transitions** tab and select effect	Alt+K
View		
Normal	**View** (→Presentation Views) | **Normal**	Alt+WL
Slide Sorter	**View** (→Presentation Views) | **Slide Sorter**	Alt+WI
Slide Show	**Slide Show** (→Start Slide Show) |	
Start from beginning	**From Beginning**	F5 or Alt+SB
Start from current slide	**From Current Slide**	Shift + F5 or Alt+SC
Visual elements		
Charts (graphs)	Copy and paste from Excel	
Hyperlinks, insert	First click object or text	
Action button	**Insert** (→Links) | **Action**	Alt+NK
Insert Hyperlink button	**Insert** (→Links) | **Hyperlink**	Ctrl K or Alt+NI
Line drawings	See Shapes	
Media, insert	**Insert** (→Media) | **Video (Audio)** or **Home** (→Slides) | **New Slide▼** | **Title and Content** | **Insert Media Clip** icon	Alt+NV (NO)

TABLE A3.1 | Continued

Command	PowerPoint 2010 Command Sequence	PC Hot Key
Pictures	Insert (→Images) | **Pictures**	Alt+NP
Shapes	Insert (→Illustrations) | **Shapes**	Alt+NSH
SmartArt Graphics	Insert (→Illustrations) | **SmartArt** or **Home** (→Slides) | **New Slide ▼** | **Title and Content** | **Insert SmartArt Graphic** icon	Alt+NM or Alt+HI
Tables		
Format	Click inside table. **Table Tools Design** and **Layout** tabs	Alt+JT and Alt+JL
Insert	Insert (→Tables) | **Table** or **Home** (→Slides) | **New Slide ▼** | **Title and Content** Select **Table** icon	Alt+NT
View gridlines	**Table Tools Layout** (→Table) | **View Gridlines**	Alt+JLTG
Textbox, insert	Insert (→Text) | **Text Box**	Alt + NX

Microsoft Office also has online training materials to help you transition from earlier versions of PowerPoint to PowerPoint 2010: Go to Microsoft's homepage (http://office .microsoft.com) and type "getting started with powerpoint 2010" into the search box. Choose your situation and follow the instructions on the website.

DESIGNS FOR NEW PRESENTATIONS

To start your PowerPoint presentation, click **Start | All Programs | Microsoft PowerPoint 2010** and then **File | New** to open the gallery of available templates and themes (Figure A3.3). You can start with a **Blank Presentation,** select a **Sample template** or **Theme,** use the format of an existing presentation, or download a template or theme from Microsoft Office Online. Designs with a light background make the room much brighter during the actual presentation, which has the dual advantage of keeping the audience awake and allowing you to see your listener's faces. Many professional speakers, however, prefer white text on a dark background because the text appears larger. Whatever your preference, choose a design that reflects your style, is appropriate for the topic, and complements the content of the individual slides.

When you choose one of the Themes (called "design templates" in earlier versions of PowerPoint), the slides in your presentation will have a consistent and professional appearance. The color schemes for the background, text, bullets, and illustrations (shapes and

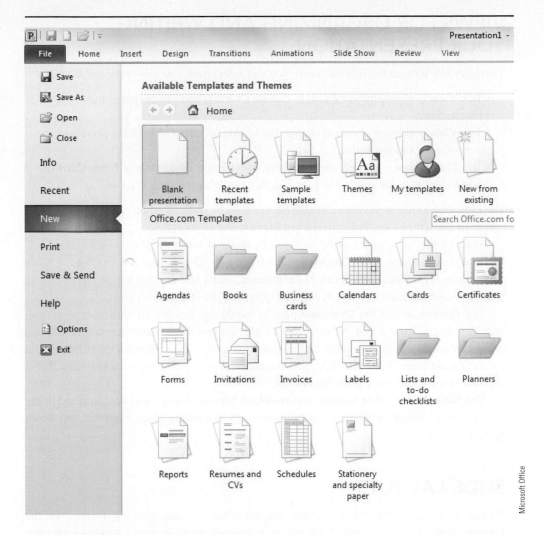

Microsoft Office

FIGURE A3.3 | Select themes and templates for new presentations by clicking File | New.

SmartArt) were designed to be aesthetically pleasing; however, some themes provide better contrast between background and content than others.

To apply a different **Theme** to a presentation-in-progress, click a button under **Design** (→Themes). Hold the mouse pointer over a **Theme** (without clicking) to activate **Live Preview**. The **Live Preview** feature allows you to see how the change affects the slide before actually applying it. Clicking a **Theme** button applies the design to all slides in the presentation.

VIEWS FOR ORGANIZING AND WRITING YOUR PRESENTATION

There are two tabs on the Ribbon where you can select how you view your presentation:

- Click **View** (→Presentation Views) to choose a view when you prepare your presentation.
- Click **View** (→Master Views) to change the format on every slide in the presentation, the handouts, and the notes.
- Click **Slide Show** (→Start Slide Show) and select **From Beginning** or **From Current Slide** to see what the slides will look like during the actual presentation.

The most frequently used views can also be accessed by clicking a button in the lower right corner of the screen (see "Adjust view" in Figure A3.1).

Most of the time, you will be working in **Normal** view. As shown in Figure A3.1, **Normal** view consists of an **Overview** pane (with two tabs, **Slides** and **Outline**), a **Slide** pane whose layout and content can be customized, and a **Notes** pane where you can write notes for your presentation, which will not appear on the slides.

The **Outline** tab on the **Overview** pane is handy for making an outline of your presentation first (Figure A3.4). As you work on the outline, think about how to present this information using as many visuals and as little text as possible. Then select a slide layout for each main point. For each slide, add the text and the visuals to the **Slide** pane. For a preview of the show, click **Slide Show** view.

The **Slide Sorter**, **Slide Master**, and **Handout Master** buttons will be described in the sections "Revising and Polishing Presentations" and "Tweaking Slide, Handout, and Notes Masters."

SLIDE LAYOUTS

Technical presentations usually follow the Introduction-Body-Closing format. In the **Introduction**, the speaker guides the audience from information that is generally familiar to them to new information, which is the focus (**Body**) of the talk. The take-home message for the audience and any acknowledgments are given in the **Closing**.

Title Slide

The default first slide in a new presentation is the **Title Slide** (see Figure A3.1). Follow the instructions in the placeholders to add text or add custom content (Figure A3.5). The same content will appear on the **Slides** tab of the **Overview** pane to the left.

The title slide is going to set the tone for your presentation. A catchy title and interesting visuals are likely to capture your audience's attention. If possible, allude to the main benefit your audience will gain from listening to your presentation.

Adding Slides

To insert a slide after the Title Slide, click **Home** (→Slides) | **New Slide ▼** and choose a slide layout (Figure A3.6). The name of the dialog box is the theme you selected for your presentation. The theme for a blank presentation is **Office Theme**.

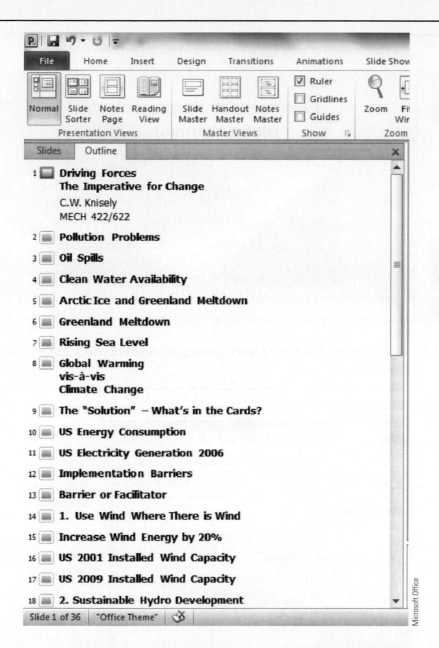

FIGURE A3.4 | The Outline tab in **Normal** view helps you structure your presentation. The slide layouts for each slide are chosen by clicking **Home** (→Slides) | **Layout**.

Basic Layouts

The basic slide layout options allow you to arrange text and content (pictures, graphs, tables, movies, and graphics) in one or two columns. The placeholders, however, can be modified to suit your needs (see the next section on "Custom Layouts"). Alternatively, if

FIGURE A3.5 | Title slide modified by repositioning placeholders and adding graphics and textboxes.

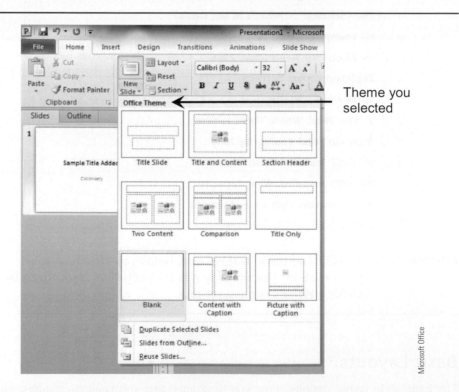

FIGURE A3.6 | Basic slide layouts are selected by clicking **Home** (→Slides) | **New Slide** ▼. There are options for selecting existing slides or outlines from other compatible files.

you decide that a different layout would work better after you have already added content to a slide, click **Home** (→Slides) | **Layout** and select a different layout.

Custom Layouts

It is easy to convert one of the basic layouts to a custom layout by **modifying placeholders** and **inserting visuals** (pictures, shapes, SmartArt Graphics, graphs, textboxes, and media clips).

Placeholders can be modified as follows:

- To **delete** one, click the dotted border of the placeholder and hit the **Delete** key.
- To **resize** one, click inside the placeholder to display selection handles. Position the mouse pointer over a handle in one of the corners to display a two-headed arrow. Hold down the left mouse button and drag the border to the desired size. Release the mouse button.
- To **move** one, position the cursor over the placeholder to display crossed arrows, hold down the left mouse button, drag the image to the desired location, and release the mouse button.
- To **format text** within one, select the text, click **Home** (→Font), and change the typeface, font size, color, and so on. See the section "Formatting Text."

Visuals can be inserted by clicking the **Insert** tab and selecting one of the more common options:

- Tables group   Table
- Images group   Picture
- Illustrations group   Shapes, SmartArt, and charts (graphs)
- Links group   Hyperlinks
- Text group   Textboxes
- Media group   Video and audio files

Graphs, tables, and images from Word, Excel, or other files can also be inserted into a PowerPoint slide using Copy and Paste. Visuals do not require a placeholder; they can be added anywhere on the slide and resized or repositioned as needed. Different kinds of visuals will be described in detail later.

Deleting Slides

To delete a slide,

- In **Normal** view, click the slide you want to delete on the **Slide** tab on the **Overview** pane and hit **Delete** or **Backspace**.
- In **Slide Sorter** view, click the slide you want to delete and hit **Delete** or **Backspace**.

Warning: PowerPoint does not ask you if you are sure that you want to delete a slide. If you delete one unintentionally, however, just click **Undo** on the **Quick Access Toolbar** to get it back.

The Last Slide

A professional slide show has a definitive ending. End the show with an acknowledgments slide or add a slide that invites the audience to ask questions. Some presenters choose to

Previous slide
Scroll box
Next slide
Previous and Next slide

Microsoft Office

FIGURE A3.7 | Ways to navigate among slides in **Normal** view.

add a black slide as the absolute last slide in their file to avoid audience eye-strain with a sudden increase in light level. To add a black slide, click **File | Options | Advanced | End with a Black Slide** (last check box at the bottom of the page).

NAVIGATING AMONG SLIDES IN NORMAL VIEW

There are several ways to move from one slide to another when preparing a presentation in **Normal** view (Figure A3.7):

- Click the **Page Up** or **Page Down** key on the keyboard.
- Use the vertical scroll bar on the right side of the **Slide** pane.
  - » Drag the scroll box inside the scroll bar up or down to move backward or forward through the presentation.
  - » Click the **Previous Slide** or **Next Slide** arrow buttons.
- Use the **Overview** pane on the left side of the screen. Click the slide you want to display. *Note*: There are two tabs on the **Overview** pane: **Outline** (text) and **Slides** (thumbnails). Choose the view you prefer.

FORMATTING TEXT

Just as in Microsoft Word, you can apply formatting such as typeface, font size, bold, italics, underline, alignment, and color to selected words or lines in a single slide by using the buttons in the **Home** (→Font) group. A good rule of thumb is to use no more than two fonts or colors, and to use them only for emphasis. For example, boldfaced, italicized, larger-sized, and colored text stands out against the default text.

A bulleted list works well if you want to give an overview of your talk first (Figure A3.8b, c, and e, for example). Each bulleted item should be a short phrase with keywords or key

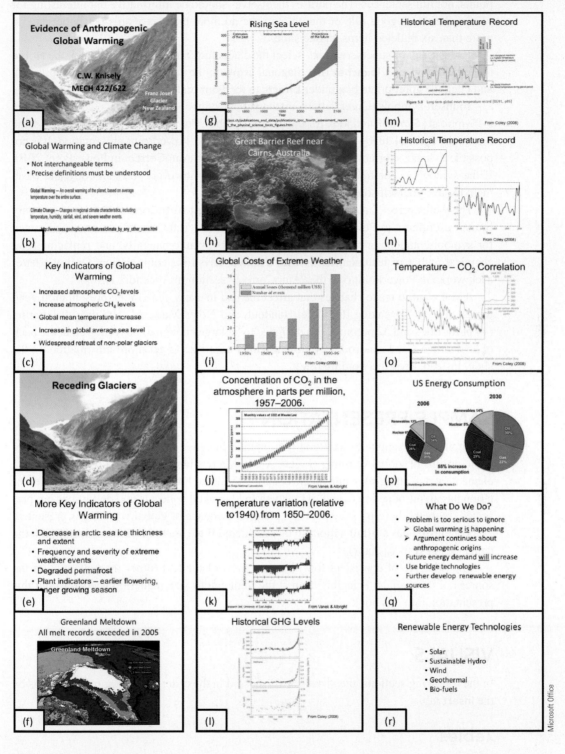

FIGURE A3.8 | A sample PowerPoint for a classroom presentation on the evidence suggesting anthropogenic global warming. Many slides consist of a title and a visual element (a, d, f, and g-p), while others consist of bulleted lists (b, c, e, q, and r). The presentation has an introduction (a and b), body (c-p), and conclusion (q and r).

points, not full sentences. The phrases themselves should be informative but interesting for your audience. A good rule of thumb is to use no more than six words per bullet and no more than six bulleted items on one slide.

To subscript or superscript text, select the character(s) to be formatted, click the **Home** (→Font) **dialog box launcher** (the diagonal arrow in the right corner of the group label), and click the appropriate box under **Effects** in the **Font** dialog box.

To insert symbols (Greek letters, mathematical symbols, Wingdings, etc.), click **Insert** (→Symbols) **| Symbol**, and look for the desired symbol. For a uniform look, select the same font in the **Symbol** dialog window as the text font in your PowerPoint presentation. It is not possible to make shortcut keys or to program symbols in AutoCorrect in PowerPoint 2010. If the symbols occur frequently in your presentation, however, you can copy and paste them to save yourself some keystrokes.

The AutoCorrect feature **(File | Options | Proofing | AutoCorrect Options)** has only limited usefulness in PowerPoint 2010. In Word 2010, AutoCorrect can be programmed for symbols, italicizing scientific names of organisms automatically, and replacing long chemical names or technical terms with a simple keystroke combination. Only the latter trick works in PowerPoint (see the "AutoCorrect" section in Appendix I).

If you want to make changes that affect the text in every slide in the presentation, read the section on "Tweaking the Slide, Handout, and Notes Masters." The default lettering (44 pt for titles and 32 pt for text) is designed to be large enough to be read even in the back of an average-sized classroom, so do not reduce font size to get more information on the slide. Instead, keep the wording brief.

SAMPLE PRESENTATION

The topic of the sample PowerPoint presentation included here is evidence for anthropogenic origins of global warming (Figure A3.8). Many slides consist of a title and a visual element (a, d, f, and g-p), while others consist of bulleted lists (b, c, e, q, and r). The presentation has an introduction (a and b) in which fundamental concepts and definitions are presented. In the body (c-p), statistical and pictorial evidence of global warming is presented. The conclusion has a list of actions to be considered if we are to heed the warnings that the evidence suggests (q and r).

The titles are of consistent font size, and most of the text (other than citations of the sources of the visuals) is legible even at the tremendous size reduction needed to put this presentation into this book.

VISUALS

In the following sections, the visuals are discussed in the order of the command buttons on the **Insert** tab.

Tables

PowerPoint allows you to insert large and small tables on a slide by selecting any of the layouts with the word *Content* or *Comparison* from the **New Slide** dropdown menu (see Figure A3.6). In the center of the *Click to add text* placeholder, click the **Insert Table**

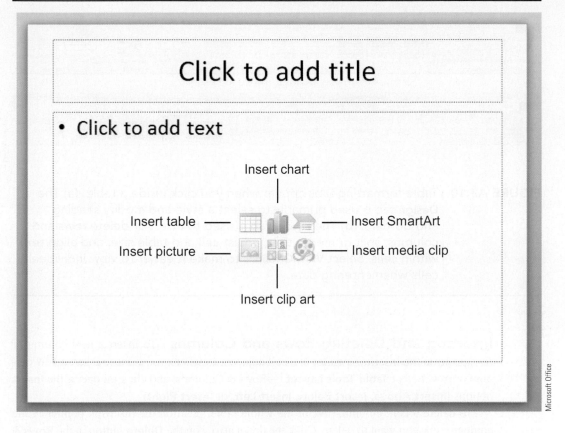

Microsoft Office

FIGURE A3.9 | To insert a large table, click **Home** (→Slides) | **New Slide ▼** | **Table and Content** and then click the **Insert Table** icon in the center of the slide.

icon (Figure A3.9) and enter the number of rows and columns in the **Insert Table** dialog box. It is also possible to add a table to a **Blank** layout with **Insert** (→Tables) | **Table**.

As in Word 2010, the table format can be modified with the **Table Tools Design** and **Layout** tabs, which only appear when the insertion pointer is inside the table (Figure A3.10). By convention, tables in technical papers do not have vertical lines to separate the columns, and horizontal lines are used only to separate the table caption from the column headings, the headings from the data, and the data from any footnotes. To format a table in this style, select all of the cells in the table and click **Design** (→Table Styles). Scroll through the styles until you reach the "Light Style 1" series. This series has the appropriate horizontal lines, and no vertical lines, as preferred in technical papers. For a presentation, the colored rows are appropriate and help the audience follow the arrangement of data in the table.

Navigating in Tables   In addition to the mouse, you can use keyboard keys to navigate within a table:

- To jump from one cell to an adjacent one, use the arrow keys.
- To move forward across the row, use the **Tab** key. *Note:* If you press **Tab** when you are in the last cell of the table, another row will be added to the table.
- To align text on a tab stop in a table cell, press **Ctrl+Tab**.

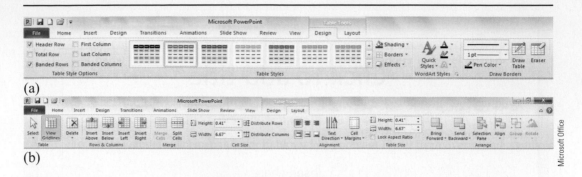

FIGURE A3.10 | Table formatting tabs appear when you click inside a table. **(a)** The **Design** tab is used primarily to select a style and modify shading and borders. **(b)** The Layout tab is used to insert or delete rows and columns, split or merge cells, adjust cell and table size, and align text within cells. Select **View Gridlines** to make it easier to view individual cells when entering data.

Inserting and Deleting Rows and Columns

To insert a new column or row, position the insertion pointer in a cell adjacent to where you want to insert a new column or row. Select **Table Tools Layout** (→Rows & Columns) and click on one of the **Insert** options (**Insert Above, Insert Below, Insert Left**, or **Insert Right**).

To delete a cell, column, row, or the whole table, position the insertion pointer in the compartment you want to delete. Click the down arrow on the **Delete** button in the **Rows & Columns** group and make your selection.

Merging or Splitting Cells

To merge two or more cells, select the cells and click **Table Tools Layout** (→Merge) | **Merge Cells**. To split a cell into two or more cells, select the cell and click **Table Tools Layout** (→Merge) | **Split Cells**. Specify in the dialog box how you want to split the cell: horizontally (enter number of rows) or vertically (enter number of columns).

Changing Column Width or Row Height

To change the width of a column, position the cursor on one of the vertical lines so that ←||→ appears. Then hold down the left mouse button and drag the line to make the column the desired width.

To change the height of a row, position the cursor on one of the horizontal lines so that ⬍ appears. Then hold down the left mouse button and drag the line to make the row the desired height.

Viewing Gridlines

It is easier to enter data in a table when the gridlines are displayed. Position the insertion pointer inside the table and click **Table Tools Layout** (→Table) | **View Gridlines**. Gridlines are not printed.

Formatting Text within Tables

You can format the text in each cell just as you would format running text. For example, to center the column headings or to put them in boldface, select the top row and click the **Center** or **Bold** button on the **Home** tab. To align numbers, select the entire column and click the **Align Text Right** button. These same options are in the **Table Tools Layout** (→Alignment) group.

To format all the cells, click the top left cell, hold down the **Shift** key, and click the bottom right cell. To format selected cells, do the same, except hold down the **Ctrl** key.

Pictures

To insert a picture, click **Insert** (→Images) | **Picture**. While that function is not new, the ability to edit the picture in PowerPoint is. When you click the picture, the **Picture Tools Format** tab appears. The buttons on this tab make it possible to remove the background, adjust brightness and contrast, apply artistic effects, and add various borders. Because high-resolution images can increase the size of a PowerPoint presentation significantly, compress the pictures if the file will be printed, posted to the Web, or shared via email (**Picture Tools Format** (→Adjust) | **Compress Pictures**).

Line Drawings

To make line drawings or add simple graphics to your document, click **Insert** (→Illustrations) | **Shapes** (Figure A3.11). To insert a line, for example, click a line to display

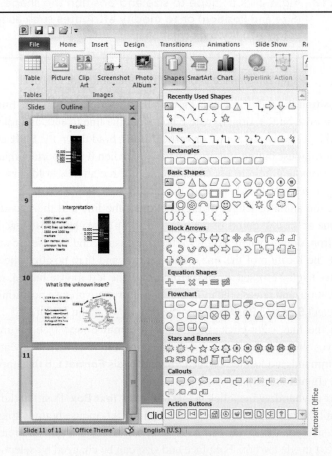

FIGURE A3.11 | Select shapes for simple graphics from the **Insert** (→Illustrations) | **Shapes** dropdown menu.

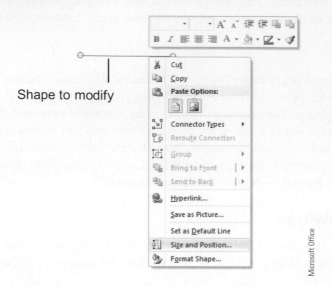

Shape to modify

Microsoft Office

FIGURE A3.12 | Right-click a shape to size and position the shape precisely (**Size and Position**) or to modify attributes such as fill, weight, color, and style (**Format Shape**).

crosshairs. Position the mouse pointer on the slide where you want the line to start, hold down the left mouse button, drag to where you want the line to end, and release the mouse button. To make perfectly horizontal or vertical lines, hold down the **Shift** key while holding down the left mouse button and drag. In general, when it is important that a shape retain perfect proportions (e.g., circle not oval; square not rectangle), hold down the **Shift** key while drawing or resizing the shape with the mouse.

To change the angle or length of the line, click the line and then position the mouse pointer over one of the end points of the line to display a two-headed arrow. Hold down the left mouse button, drag to where you want the line to end, and release the mouse button.

To change the location of the line, click the line and then position the mouse pointer over the line to display crossed arrows. Hold down the left mouse button, drag the line to the desired location, and release the mouse button. For more precise positioning, right-click the line and select **Size** and **Position** from the dropdown menu (Figure A3.12).

To format the line, right-click it and select **Format Shape** (see Figure A3.12). This dialog box gives you options for changing the appearance of the line. Many of the same formatting commands are located on the **Drawing Tools Format** tab that appears when you select the line (Figure A3.13).

To add text to your drawing, click **Insert** (→Text) | **Text Box**. Position the mouse pointer where you want the textbox to appear, hold down the left mouse button, drag to enlarge the textbox to the size you think you will need (you can resize it later), and release the mouse button. Type text inside the box. Font face and size can be changed by selecting the text and using the buttons in the **Home** (→Font) group. To format the textbox itself, click the textbox to display the selection handles on the border. Right-click the textbox to open a dropdown menu similar to the one in Figure A3.12 or use the buttons on the **Drawing Tools Format** tab.

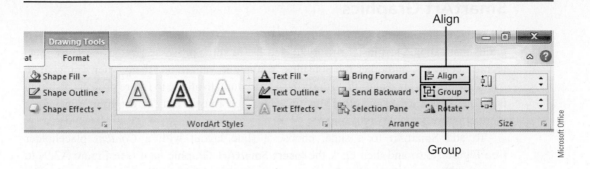

FIGURE A3.13 | To format a shape, click it to open the **Drawing Tools Format** tab. Right-clicking the shape opens a dialog box with additional options.

If several objects make up a graphic, then it may make sense to group them as a unit. Grouping allows you to copy, move, or format all the objects in the graphic at one time. To group individual objects into one unit, click each object while holding down the **Shift** key. Release the **Shift** key and then click **Group** under **Drawing Tools Format** (→Arrange). One set of selection handles surrounds the entire unit when the individual objects are grouped (Figure A3.14). To ungroup, simply select the group and click **Group | Ungroup**.

The **Align** button is handy for precisely lining up objects (shapes as well as textboxes). Click each object that you want to line up while holding down the **Shift** key. Release the **Shift** key, click **Align** under **Drawing Tools Format** (→Arrange), and select one of the alignment options. Another useful feature for arranging objects on a page is **Align | View Gridlines**. This command puts a nonprinting grid on your page. Click **Align | Grid Settings** to adjust the spacing between the gridlines or to snap objects to the grid. Check the **Snap objects to grid** box if you want to position objects on gridlines. Do not check this box if you prefer to position the objects freely. Finally, to move objects just a fraction of a millimeter from their current position, click the object, hold down the **Ctrl** key, and use the arrow keys to nudge the object exactly where you want it on the slide.

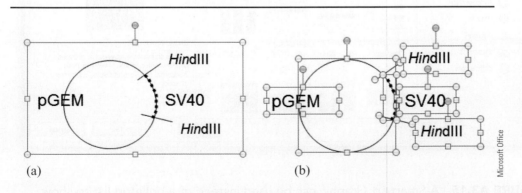

(a)   (b)

FIGURE A3.14 | Individual graphics grouped **(a)** and ungrouped **(b)**.

SmartArt Graphics

SmartArt Graphics was first introduced in Office 2007. **SmartArt** is a collection of diagrams that can be used instead of a bulleted list to show processes, stages or steps, hierarchies, concepts, and other relationships. You might want to use them when you give an overview of your presentation, to describe a procedure, to allude to future projects resulting from the current work, and other situations in which a visual aid can get your point across more effectively than words alone.

To add SmartArt to a slide, choose a slide layout with a *Content* placeholder (see Figure A3.6) and then click the **Insert SmartArt Graphic** icon (see Figure A3.9) to open the **Choose a SmartArt Graphic** gallery (Figure A3.15). Choose a category on the left or just scroll through all the graphics in the center pane. When you find a graphic that you might like to use, click it to display a detailed description in the right pane. Click **OK** to insert it on the slide. Another way to add SmartArt is by clicking **Insert** (→Illustrations) | **SmartArt**.

After you have inserted the SmartArt graphic on a slide, fill the components with text and modify the format if desired. To format the entire graphic, click inside the placeholder. To format individual components of the graphic, click the component. Either way, the **SmartArt Tools** formatting tabs appear on the Ribbon (Figure A3.16). Use the buttons to change the layout and color scheme of both the components and the text. The **Live Preview** feature allows you to see how the change affects the graphic before actually applying it. Simply hold the mouse pointer over a button on the Ribbon (without clicking) to activate **Live Preview**. When you have decided on a style or layout, click it.

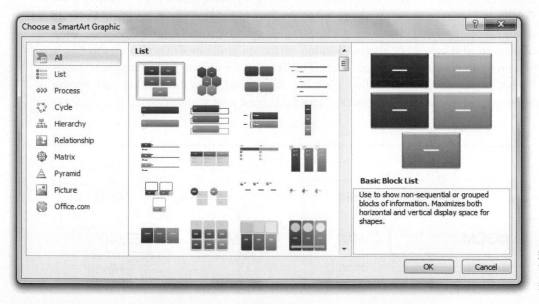

Microsoft Office

FIGURE A3.15 | A SmartArt Graphic can be used instead of a bulleted list to show relationships.

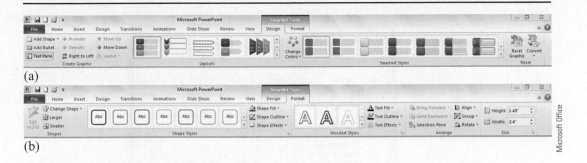

(a)

(b)

Microsoft Office

FIGURE A3.16 | SmartArt Tools formatting tabs appear when you click the graphic. **(a)** The **Design** tab is used to add components and modify the layout and color scheme. **(b)** The **Format** tab is used to change the shape, size, color, text, and arrangement of the graphic components.

To convert a bulleted list to SmartArt, right-click anywhere within the list and select **Convert to SmartArt**. Then choose an appropriate graphic.

Graphs (Charts)

If you arc alrcady familiar with making graphs in Microsoft Excel, then the easiest way to add graphs to a PowerPoint presentation is to copy and paste as follows:

- Make a graph in Excel (see Appendix II).
- Single-click the Chart Area to activate the graph and then copy it.
- In PowerPoint, click **Home** (→Slides) | **New Slide** ▼ and choose a layout suitable for the graph.
- Click **Paste**. Resize and reposition the graph as needed.

Hyperlinks

There may be times during a presentation when you would like to refer to another slide, access an Excel spreadsheet or Word document, or link to an Internet address. These connections are made by inserting hyperlinks into your presentation. Two ways to insert hyperlinks are:

- Using the **Action** button
- Using the **Insert Hyperlink** button

For both methods, you can attach the hyperlink to text (which will be underlined to identify the link) or more subtly to an illustration such as a picture, clip art, shape, or SmartArt graphic. In **Normal** view, type text or insert an object using **Insert** (→Illustrations). To attach the hyperlink to text, simply right-click the text and click **Hyperlinks** on the menu. To attach the hyperlink to a visual element (picture, clip art, etc.), right-click the element and then click **Hyperlinks**. Alternatively, click the text or visual and then **Insert** (→Links) | **Hyperlink**.

Action Button Click **Insert** (→Links) | **Action** to open the **Action Settings** dialog box (Figure A3.17). Use the **Mouse Click** tab if you would like to activate the link with a

FIGURE A3.17 | To open the **Action Settings** dialog box, click **Insert** (→Links) | **Action**. Link the active slide to other slides in the presentation, to a website, or to other Microsoft Office files (e.g., Word, Excel, or PowerPoint) by making a selection on the dropdown menu.

mouse click; if you prefer to activate the link by moving the mouse over the object, use the **Mouse Over** tab. From the **Hyperlink to** dropdown menu, select one of the options to which to link the active slide: a slide in the current presentation, a URL to connect to an Internet address, a different PowerPoint presentation, or a file with a different format (such as an Excel spreadsheet or a Word document). Use the "Insert Hyperlink Button" method (see next section) if you need to browse for the URL address or copy and paste it from another location.

Insert Hyperlink Button   Click **Insert** (→Links) | **Hyperlink** to open the **Insert Hyperlink** dialog box (Figure A3.18a). To link to another file on your computer, click the **Existing File or Web Page** button on the left side of the box. To browse for the file to link, click the **Current Folder** or the **Recent Files** button in the **Look in** area, select the desired file, and click **OK**.

To link to a webpage, click the **Existing File or Web Page** button on the left side of the box (Figure A3.18a). If you have visited the website recently, click the **Browsed Pages** button in the **Look in** area and select the desired URL. Alternatively, launch a Web browser such as Internet Explorer, Firefox, or Safari, find the desired website, and copy and paste the URL into the **Address** box. Click **OK**.

To link to another page in your current presentation, click the **Place in This Document** button on the left side of the **Insert Hyperlink** dialog box (Figure A3.18b). Select the slide to link from the **Select a place in this document** area and click **OK**.

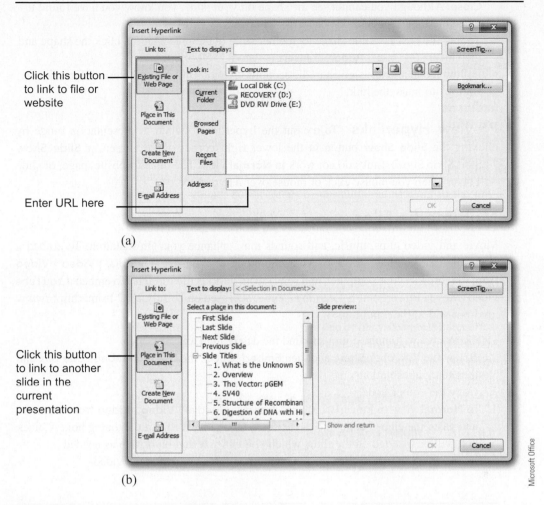

Click this button
to link to file or
website

Enter URL here

(a)

Click this button
to link to another
slide in the
current
presentation

(b)

Microsoft Office

FIGURE A3.18 | To open the **Insert Hyperlink** dialog box, click **Insert** (→Links) |
Hyperlink. **(a)** Click the **Existing File or Web Page** button to link
the active slide to a website or to another Microsoft Office file.
(b) Click the **Place in This Document** button to link to another slide
in the current presentation.

A clever way to return to the slide in the presentation where you inserted the hyperlink is to add an "invisible" return hyperlink. Assume that when you reach Slide 8 of your
presentation, you want to refer back to an earlier picture shown on Slide 5. After showing
Slide 5, you want to return to Slide 8 and continue your presentation. To make a return link
on Slide 5, position a shape with no line or color somewhere on the slide where you can
easily find it, and then link that shape to Slide 8. To add an invisible hyperlink:

- In **Normal** view, insert a shape using **Insert** (→Illustrations) | **Shapes**. Select a two-
  dimensional shape like a rectangle, oval, star, or arrow. Enlarge the shape and position
  it next to text or in a corner where you will be able to find it easily.
- To make the shape invisible, right-click it and select **Format Shape** from the dropdown menu. Under **Fill**, select **No Fill**. Under **Line Color**, select **No Line**. Then click

Close. Although you cannot see the shape on your slide, you know you have found it when the mouse pointer displays crossed arrows.

- To link the now invisible shape to another slide in the presentation, click the shape and then **Insert** (→Links) | **Action** or **Insert** (→Links) | **Hyperlink**.
- Follow the instructions given in the sections "Action Button" and "Insert Hyperlink Button" to make the link.

Preview Hyperlinks   To try out the hyperlinks, switch to presentation mode by clicking the **Slide Show** button in the lower right corner of the screen or **Slide Show** (→Start Slide Show). Links do not work in **Normal** view. The linked website, page, or slide will open when you mouse click or mouse over a hyperlink.

Multimedia Files

Movie and video clips, music, and sounds may enhance your presentation. To embed a video file that you have saved on your computer, click **Insert** (→Media) | **Video** | **Video from file**, browse your computer, and select from various file formats. To embed a YouTube video (Moyea PowerPoint E-Learning Center, accessed on 26 July 2012 from <http://www .dvd-ppt-slideshow.com/blog/>):

1. Go to <www.youtube.com> and find the desired video.
2. Below the link, click **Share** and then **Embed**. Check the **Use old embed code** box and then click a resolution.
3. Copy the embed code.
4. In **Normal** view in PowerPoint, click **Insert** (→Media) | **Video** | **Video from Website** and paste the embed code into the textbox. Click **Insert** to exit the dialog box. A black box will be inserted on the slide, which you can size and reposition as needed.
5. In presentation mode, click the play button on the icon to start the video.

Warning

1. Not all embedded videos will play in PowerPoint due to copyright restrictions. In that case, copy the URL into a slide layout with a *Content Placeholder* (see Figure A3.6). In presentation mode, click the link to the website.

2. The embed code method described above does not work with the **Insert Media Clip** button on *Content Placeholder* slides (see Figure A3.9).

3. Some videos require the installation of additional software such as QuickTime for the clip to work. If you are having problems playing multimedia files in PowerPoint, ask someone in computer services for help.

4. Embedded media files may drastically increase the size of a PowerPoint file. When possible, compress media by clicking **File | Info | Compress Media** and choosing one of the three options:
   a. **Presentation quality** results in about the same video and audio quality, but makes the file smaller
   b. **Internet quality** results in a reduction in quality comparable to streamed video
   c. **Low quality** is sufficient for attaching the PowerPoint to an e-mail

Based on Microsoft Office, http://office.microsoft.com/en-us/powerpoint-help/compress-your-media-files-HA010382163.aspx, accessed 13 March 2013

SAVING AND PRINTING PRESENTATIONS

To save a presentation for the first time, click **File | Save**. Do not wait until you are finished making the entire presentation to save the file. Save it after you have made the title slide.

1. Click **File | Save** (or **Save As**). The **Save As** dialog box appears.
2. Click a link in the breadcrumb trail in the box at the top or use the navigation pane on the left to locate the folder in which you want to save the presentation.
3. To create a new folder, click the **New Folder** button. The **New Folder** dialog box appears. Give the folder a short, descriptive name. Click **OK**.
4. Choose a descriptive name for the file.

PowerPoint 2010 automatically saves the file as a "Presentation," with the .pptx extension. If you would like to share a PowerPoint 2010 presentation with someone who has an earlier version of PowerPoint (97–2003) and who has not installed the Compatibility Pack (see the section on "Housekeeping" in Appendix I), click **File | Save As** and next to **Save as type**, select **PowerPoint 97–2003 Presentation**.

If you need to print colored slides on transparencies, because the presentation room does not have projection equipment for a PowerPoint presentation,

- Specify the desired size and orientation of the slides with **Design** (→Page setup) | **Page setup**
- On the print menu (**File | Print | Full Page Slides**), change the default **Grayscale** setting to **Color** (Figure A3.19a).

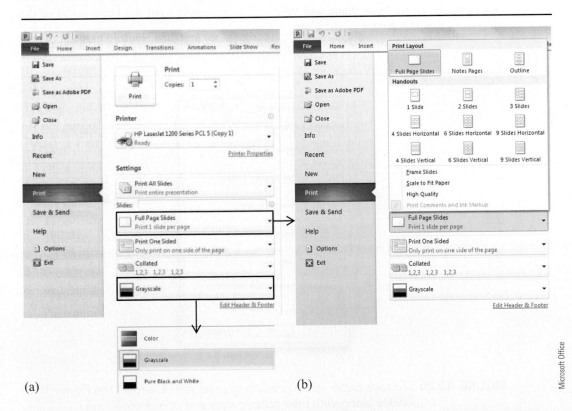

(a)   (b)

Microsoft Office

FIGURE A3.19 | File | Print provides options for printing slides, handouts, and notes pages.

As a speaker, you may wish to print out an outline of your presentation or notes pages to refer to when you deliver the presentation. Both of these options are available on the center panel upon clicking **File | Print | Full Page Slides** (Figure A3.19b). When selecting **Notes Pages**, a thumbnail of the slide is printed along with the speaker notes below it. The speaker notes are comparable to notecards in that they help you (the presenter) remember the important points to make about each slide. By sticking to the script, you are likely to stay within your time limit and avoid getting sidetracked.

There is only one option for printing notes pages from PowerPoint: Each printed page consists of one slide with its corresponding notes page below it. If you do not write notes for every slide, this format is wasteful because you end up with very little content on a pile of paper. A more economical alternative is to combine multiple slides with their speaker notes on one sheet of paper. To do so, click **File | Save & Send | File Types | Create Handouts**. Select a page layout and then click **OK** in the **Send to Microsoft Office Word** dialog box (Figure A3.20). A new Word document is created, which consists of a three-column table with slide number, thumbnail of the slide, and speaker notes in the three columns, respectively. The row height can be adjusted to fit up to five slides with corresponding speaker notes on one page. By printing your notes pages from Word instead of PowerPoint, you can significantly reduce your paper use.

Your audience might appreciate having a handout with thumbnails of the slides for note-taking purposes. To print handouts, click **File | Print | Full Page Slides | Handouts** (see Figure A3.19b). Between one and nine slide thumbnails can be printed on one page (the more thumbnails, the smaller each thumbnail). The three-slides-per-page option comes with lines next to each thumbnail to facilitate note-taking. When printing on a black-and-white printer, select **Pure Black and White** to avoid printing out the background of the slides (see Figure A3.19a).

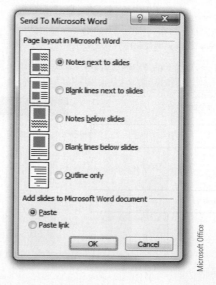

FIGURE A3.20 | To save paper when printing speaker notes, export the PowerPoint slides along with their notes pages and print from Word.

REVISING AND POLISHING PRESENTATIONS

As described previously, the most frequently used views can also be accessed by clicking a button in the lower right corner of the screen (see "Adjust view" in Figure A3.1). **Slide Sorter** view is useful for evaluating the overall appearance of a presentation because it displays thumbnails of all of the slides in the presentation, complete with slide transitions (Figure A3.21). You cannot edit the content of an individual slide in **Slide Sorter** view, but you can rearrange the slides. To select a slide to move, copy, or delete, single-click it. To edit the content of the slide, double-click it to return to **Normal** view. Slides can also be rearranged in the **Overview** pane in **Normal** view.

Before following the instructions below, click **View** (→Presentation Views) | **Slide Sorter** to change to **Slide Sorter** view.

Moving Slides

To move a slide in Slide Sorter view, position the mouse pointer on the slide you want to move, hold down the left mouse button and drag the vertical line to a new location between slides, and release the mouse button. The slide now appears at its new position. You can also move a slide using the **Cut** and **Paste** buttons on the **Home** tab, the **Ctrl+X** and **Ctrl+V** keyboard shortcuts, or the right mouse button.

To move a group of adjacent slides, click the first slide in the group, hold down **Shift** on the keyboard, and click the last slide in the group. All slides will be marked with a highlighted border. Hold down the left mouse button, drag the vertical line to a new insertion point between slides, and release the mouse button. The group of slides appears at its new position. As with a single slide, you can also use **Cut** and **Paste**.

To move a group of nonadjacent slides, follow the procedure for adjacent slides, but hold down **Ctrl** instead of **Shift** on the keyboard. The slides appear in the same consecutive order at the new location.

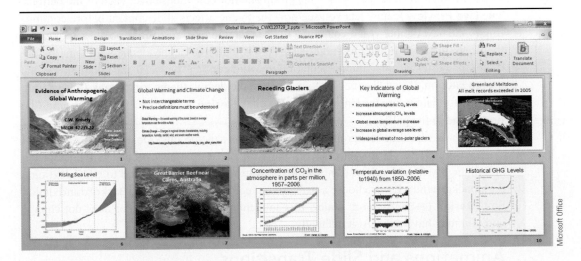

FIGURE A3.21 | Slide sorter view makes it easy to rearrange slides in a presentation.

Adding and Deleting Slides

To add a new slide in Slide Sorter view, click the insertion point between two slides. Then click **Home** (→Slides) | **New Slide ▼** and choose a slide layout. Double-click the new slide to add content.

To delete an unwanted slide, single-click the slide and press **Delete** on the keyboard or click **Home** (→Slides) | **Delete**. To correct the deletion, click **Quick Access Toolbar** | **Undo**.

Copying Slides

To copy a slide in Slide Sorter view, use the **Copy** and **Paste** buttons on the **Home** tab, the **Ctrl+C** and **Ctrl+V** keyboard shortcuts, or the right mouse button. To insert a duplicate slide immediately after the original, click the original and then **Home** (→Slides) | **New Slide ▼** | **Duplicate Selected Slides**. To change the layout on the duplicate, simply single-click the duplicate slide, click **Home** (→Slides) | **Layout**, and select a new layout for the slide.

Spell Check

Do not let an otherwise interesting and well-organized presentation be ruined by typos on the slides. As with Microsoft Word, PowerPoint gives you visual indicators to alert you to possible spelling mistakes. Words that are possibly misspelled are underlined with a wavy red line. To deal with this word, right-click it and correct the mistake, tell PowerPoint to ignore it, or, if the word is a correctly spelled technical term, add it to the dictionary. In contrast to Word, PowerPoint does not flag grammatical errors.

Applying a Theme

In an effective presentation, the slides are simple, legible, and organized logically. Although there is nothing wrong with black-and-white slides (using the default Office Theme), most people prefer color. PowerPoint offers a variety of color-coordinated design templates (called *themes*). To see how your presentation would look with a different theme, click the **More** arrow in the **Design** (→Themes) group and hold the mouse over each of the buttons. The **Live Preview** feature lets you see how the theme affects slide content before committing to the new theme. When you find a theme that you like, clicking the button applies the theme to all slides in the presentation. It may actually be better to select a theme before rather than after writing your presentation to avoid conflicts between slide content and the design later on. If you later decide to go back to the default theme for a Blank Presentation and it is too late to use the **Undo** command, click **Design** (→Themes) | **Office Theme** (the first theme in the gallery).

To change the slide background without applying a theme, click **Design** (→Background) | **Background Styles**. If you select one of the available styles, the text color will be adjusted automatically (e.g., black text on a white background will be converted to white text on a dark background). If you choose a background color from the **Format Background** dialog box (**Design** (→Background) **dialog box launcher**), however, text color does not change and you may have to adjust it manually.

Animations and Slide Transitions

Animation affects the movement of individual objects on a slide during a slide show. Without animation, each visual element or line of text is present when the slide appears.

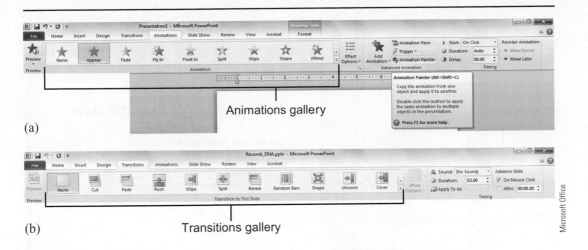

(a)

(b)   Transitions gallery

FIGURE A3.22 | (a) Animations can be applied to individual objects on a slide. (b) Slide transitions affect how slides change from one to the next.

However, when you would like to give the audience time to focus on and process one point at a time (for example, when listing objectives or summarizing conclusions), it is advantageous to use animations.

Animations are selected on the **Animations** tab (Figure A3.22a). In **Normal** view, click the **More** arrow next to the animations and hold the mouse over the buttons to preview the animation. Adjust the direction and sequence of the animation effect with the **Effect Options** button. To copy an animation to another object on the slide (or another slide), click the animated object, then select **Animations** (→Advanced Animation) | **Animation Painter**, and click the new object to animate.

Transitions affect the way slides appear in a slide show. Without slide transitions, each slide stays on the screen until you click the left mouse button, press **Enter**, or press the space bar on the keyboard to advance to the next slide. The slides in the presentation are displayed sequentially unless you press the **Esc** key to end the show, click a button on the **Slide Show** toolbar in the lower left corner of the screen, or right-click and choose an option from the menu. Transitions are selected on the **Transitions** tab (Figure A3.22b).

For a tutorial on using animations and transitions in PowerPoint 2010, see <http://office.microsoft.com/en-us/powerpoint-help/animations-and-transitions-RZ102809184.aspx?CTT=1>. If you decide to use animations and slide transitions, be careful not to overuse them. *You want your audience to be listening to you, not be distracted by the special effects!*

TWEAKING SLIDE, HANDOUT, AND NOTES MASTERS

The **Slide Master** controls the appearance of the content in every slide *including* the title slide. The Title Master in earlier versions of PowerPoint has been replaced with a Title Style layout within the **Slide Master** in PowerPoint 2010.

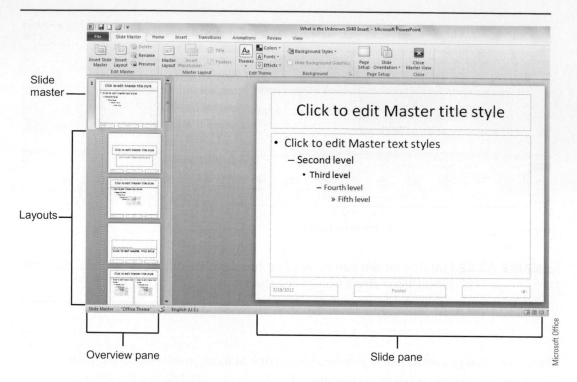

FIGURE A3.23 | Default Slide Master (Office Theme).

You can modify the Slide Master at any time—before, during, or after making the first draft of your presentation. Click **View** (→Master Views) | **Slide Master** to open an **Overview** pane, with the **Slide Master** thumbnail at the top and the layouts that belong to it beneath it (Figure A3.23).

The **Slide** pane shows the selected thumbnail in **Normal** view. If you want to make changes that affect every slide in the presentation, be sure to modify the Slide Master, not one of the layout slides.

Text, Bullets, and Illustrations

To modify the font face, style, size, or color, select the line of text you want to change. Right-click and select **Font** from the dropdown menu. Make your selection and click **OK**. To change a bullet, click anywhere on the bulleted line. Right-click and select **Bullets** from the dropdown menu. Make your selection and click **OK**. To add a logo to the Slide Master, click **Insert** (→Images) and then make the appropriate selection. **Remember that these changes affect every slide in the presentation**.

Headers and Footers

The three boxes at the bottom of the Slide Master allow you to enter the date and time, a custom footer, and the page number. While the default setting is *not* to include this

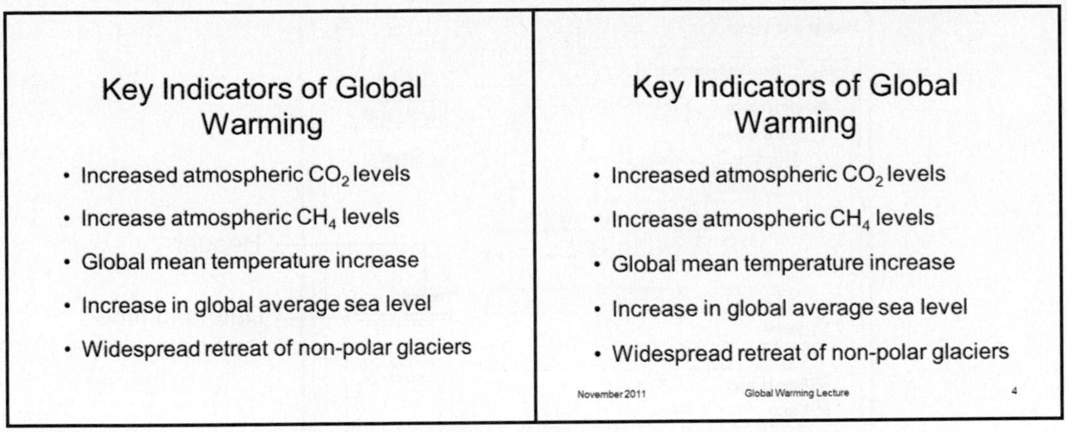

<table>
<tr><td>

Key Indicators of Global Warming

- Increased atmospheric CO_2 levels
- Increase atmospheric CH_4 levels
- Global mean temperature increase
- Increase in global average sea level
- Widespread retreat of non-polar glaciers

</td><td>

Key Indicators of Global Warming

- Increased atmospheric CO_2 levels
- Increase atmospheric CH_4 levels
- Global mean temperature increase
- Increase in global average sea level
- Widespread retreat of non-polar glaciers

November 2011   Global Warming Lecture   4

</td></tr>
</table>

(a)   (b)

© Cengage Learning 2015

FIGURE A3.24 | Footers on slides can be turned off **(a)** or on **(b)** by checking the appropriate boxes on the **Slide** tab in the **Header and Footer** dialog box.

information during a slide show, it may be handy to have it on the printed slides if they are used as handouts (Figure A3.24). To include the footer information:

- Click **View** (→Presentation Views) | **Slide Master**.
- Click **Insert** (→Text) | **Header and Footer** to open the **Header and Footer** dialog box (Figure A3.25).
- On the **Slide** tab, click the **Date and Time, Slide Number**, and/or **Footer** check boxes.
- Click **Apply to All** to print this information on every slide.
- Click **File | Print** and select **Settings | Full Page Slides** from the **Print options** panel.
- Make any other changes to the settings and then click **OK**.

Handouts and Notes Pages

Checking the footer boxes on the **Slide** tab of the **Header and Footer** dialog box causes footers to be displayed on *slides*, but not on handouts or notes pages. To print headers and/or footers on these pages:

- Click **View** (→Presentation Views) | **Handout Master**.
- Click **Insert** (→Text) | **Header and Footer** to open the **Header and Footer** dialog box.
- On the **Notes and Handouts** tab, click the relevant check boxes (see Figure A3.25). Date and time are displayed in the upper right corner and the page number in the lower right corner. A custom header and footer can be entered in the respective boxes if desired.
- Click **Apply to All** to print the selected information on every page.
- Click **File | Print** and select **Settings | Full Page Slides | Handouts** or **Notes Pages** from the **Print options** panel.
- Make any other changes to the settings and then click **OK**.

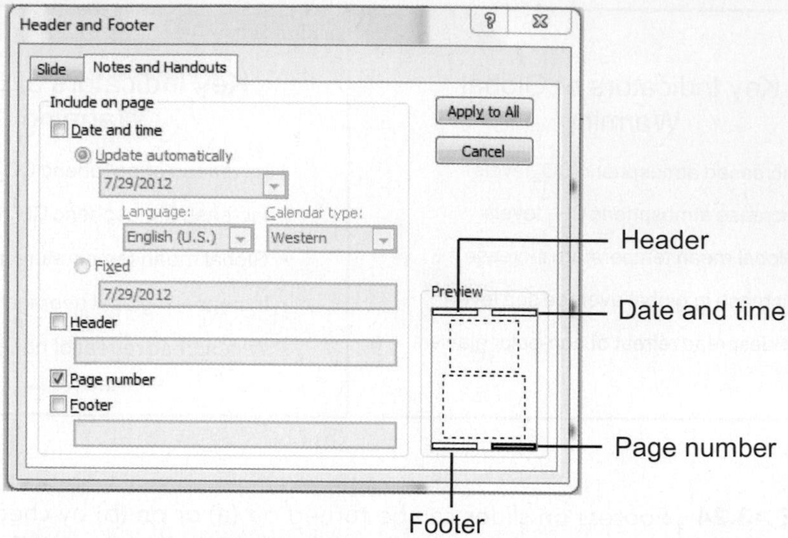

FIGURE A3.25 | The **Header and Footer** dialog box allows you to print the date and time, a page number, and a custom header and footer on Notes and Handouts pages. Similar information can be included (or not) on the Slides by making the appropriate selections on the **Slide** tab.

DELIVERING PRESENTATIONS

A PowerPoint presentation can be "delivered" electronically by attaching it to an email or posting it to a website, but most presentations are delivered in person. Table A3.2 gives some presentation tips. See Chapter 11 for detailed instructions on planning and delivering oral presentations.

TABLE A3.2 | Presentation tips

Do	Do Not
Keep the wording simple	Write every word you're going to say on the slide
Make the text large and legible	Use backgrounds that make text and images hard to read
Strive for a consistent look (use the same font and format for each slide)	Use distracting animations and slide transitions, or sound effects
Include visuals that complement and support the auditory information	Include tables when you can show the trend better with a graph
	Talk endlessly without referring to a visual
Allow enough time for each slide	Rush through the slides without mentioning the take-home message about each one

Resources in the Presentation Room

When you are invited to give a talk, find out what audiovisual equipment will be available in the room. If you only have a chalkboard, just print out the speaker notes. If you'll have access to an overhead projector and screen, print out the slides and make transparencies. These printing options are explained in the section "Saving and Printing Presentations."

To deliver a PowerPoint slide show, you will need a computer and projection equipment as well as a screen on which to display the slides. If the presentation room does not have this equipment, ask your host if he or she can provide it or, if you are affiliated with a college or university, you may be able to borrow a laptop computer and projector for your presentation.

There are several ways to run a PowerPoint presentation:

- From a Web server
- From a flash drive or CD on a computer at the presentation site
- From the hard drive of your laptop computer

To run a presentation from a Web server, first find out if you have access to the Internet at the presentation site. If so, save your file to a remote server, such as Google Drive, SkyDrive, or Dropbox. Similarly, if your college or university has networked computers, you can prepare your presentation on a networked computer in your room; for example, save the file in your Netspace, and then access the file from another networked computer in the room where you will hold your talk.

If the presentation site has a computer but it is not hooked up to the Internet, save your presentation file on a flash drive or CD and carry it with you. Ask your host if PowerPoint is installed on the presentation room's computer. If so, just plug the flash drive into the USB port, start PowerPoint, open your presentation file, and click the **Slide Show** button in the lower right corner of the screen.

If you have a laptop computer, make sure PowerPoint is installed on it, and then save your presentation file on the hard drive. At the presentation site, connect your laptop to the projector, start PowerPoint, open the file, and click the **Slide Show** button in the lower right corner of the screen to begin your presentation.

Navigating among Slides during a Slide Show

To start a slide show *from the first slide*, click the **F5** key or **Slide Show** (→Start Slide Show) | **From Beginning**. Clicking the **Slide Show** button in the lower right corner of the screen or **Shift + F5** starts your presentation *from the current slide*. It is possible to specify a time for the slides to advance automatically (**Transitions** (→Timing) | **Advance Slide After**), but it is more common to use the keyboard, the mouse, or a wireless presenter to advance to the next slide.

The following keystrokes allow you to navigate among slides when you deliver your presentation:

- To advance to the next slide, press **Enter** or the space bar.
- To go back to the previous slide, press **Backspace**.
- To end a slide show, press **Esc**.

To change slides using the mouse:

- Click a button on the **Slide Show** toolbar in the lower left corner of the screen (Figure A3.26).
- Click the left mouse button to advance to the next slide.
- Click the right mouse button to display the main menu of navigation options.

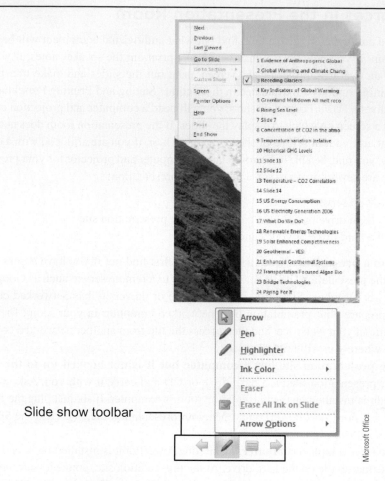

Slide show toolbar

Microsoft Office

FIGURE A3.26 | Options for navigating among slides during a slide show. The **Slide Show** toolbar has arrows to go forward and backward, a pointer button, and a slide button for navigation options.

To turn the mouse into a laser pointer:

- In **Slide Show** view, hold down the **Ctrl** key while holding down the left mouse button.
- Choose a color for the pointer by clicking **Slide Show** (→Set Up) | **Set Up Slide Show**.

A **wireless presenter** allows you to change slides from anywhere in the room. This device usually consists of a USB receiver, which you plug into a USB port on the computer, and the remote control, which has buttons for moving forward and backward, pausing, or stopping the slide show. The remote control may also have a built-in laser pointer.

To write on the slides:

- In **Slide Show** view, right-click and select **Pointer Options | Pen** (or **Highlighter**). See Figure A3.26.
- Hold down the left mouse button and write.

To erase "ink" on the slides, right-click and select **Erase All Ink on Slide**.

Index